Design of Nonplanar Microstrip Antennas and Transmission Lines

Design of Nonplanar Microstrip Antennas and Transmission Lines

KIN-LU WONG
National Sun Yat-Sen University

A WILEY-INTERSCIENCE PUBLICATION
JOHN WILEY & SONS, INC.
NEW YORK / CHICHESTER / WEINHEIM / BRISBANE / SINGAPORE / TORONTO

This book is printed on acid-free paper.

Published simultaneously in Canada.

Library of Congress Cataloging-in-Publication Data:

Wong, Kin-Lu.
Design of nonplanar microstrip antennas and transmission lines / Kin-Lu Wong.
p. cm. -- (Wiley series in microwave and optical engineering)
"A Wiley-Interscience publication."
Includes bibliographical references and index.
ISBN 0-471-18244-3 (cloth: alk. paper)
1. Strip transmission lines--Design and construction.
2. Microstrip antennas--Design and constriction. I. Title.
II. Series.
TK7876.W65 1999
621.381'331--dc21 98-35003

10 9 8 7 6 5 4 3 2 1

Contents

Preface

Due to their conformal capability, research on nonplanar microstrip antennas and transmission lines has received much attention. Many studies have been reported in the last decade in which canonical nonplanar structures such as cylindrical, spherical, and conical microstrip antennas and cylindrical microstrip transmission lines have been analyzed extensively using various theoretical techniques. These results are of great importance because from the research results of such curved microstrip structures, the characteristics of general nonplanar microstrip antennas and circuits can be deduced. The information can provide a useful reference for working engineers and scientists in the design and analysis of microstrip antennas and circuits to be installed on curved surfaces. Since the results are scattered in papers in many technical journals, it is our intention in this book to organize the research results on nonplanar microstrip antennas and circuits and provide an up-to-date overview of this area of technology.

The book is organized in eight chapters. In Chapter 1 we present an introduction and overview of recent progress in research on nonplanar microstrip antennas and transmission lines and give readers a quick guided tour of subjects treated in subsequent chapters.

In Chapters 2 and 3 we discuss, respectively, resonance problems inherent in cylindrical and spherical microstrip patches. In addition to study of single-layer microstrip patches of various shapes, structures related to microstrip patches with an air gap for bandwidth enhancement or a spaced superstrate for gain improvement are analyzed based on a full-wave formulation incorporating moment-method calculations. From the formulation, the complex resonant frequencies of a curved microstrip patches are solved whose real and imaginary parts give, respectively, information on resonant frequency and radiation loss of a curved microstrip structure. By comparison with results calculated from curve-fitting formulas for complex resonant frequencies of planar rectangular and circular microstrip patches, basic curvature effects on the characteristics of curved microstrip structures can be characterized. In addition to the resonance problems discussed, electromagnetic scattering from spherical circular microstrip patches is formulated and analyzed. Uniaxial anisotropy in the substrate of a spherical microstrip structure is included in the investigation.

Practical cylindrical microstrip patch antennas fed by coax or through a coupling slot in the ground plane of a cylindrical microstrip feed line are analyzed in Chapter 4. Various theoretical techniques, including the full-wave approach, cavity-model analysis, and generalized transmission-line model (GTLM) theory, are discussed in detail, and expressions of the input impedance and far-zone radiated fields are presented and numerical results are shown. Experiments are also conducted and measured data are shown for comparison. Circular polarization and cross-polarization characteristics of microstrip antennas due to curvature variation are also analyzed.

The results for microstrip antennas mounted on spherical or conical surfaces are discussed in Chapter 5. For spherical microstrip antennas, formulations using the different theoretical approaches of full-wave analysis, the cavity-model method, and GTLM theory are described in detail. Both input impedance and radiation characteristics due to the curvature variation are characterized. For conical microstrip antennas, available studies are based primarily on the cavity-model method. Related published results for nearly rectangular and circular wraparound patches on conical surfaces are described and summarized.

Chapter 6 is devoted to coupling problems with cylindrical and spherical microstrip array antennas. Mutual coupling coefficients between two microstrip antennas mounted on cylindrical or spherical surfaces are formulated and calculated. Bandwidth-enhancement problems of cylindrical and spherical microstrip antennas using gap-coupled parasitic patches are also discussed in this chapter.

Conformal microstrip arrays are discussed in Chapter 7. A one-dimensional or wraparound microstrip array mounted on a cylindrical body for use in omnidirectional radiation is studied first. The curvature effect on the radiation patterns of two-dimensional microstrip arrays is then formulated and investigated. A design in the feed network to compensate for curvature effects on radiation patterns is also shown. Several specific applications of spherical and conical microstrip arrays are described.

Finally, in Chapter 8, characteristics of cylindrical microstrip lines are discussed. Both quasistatic and full-wave solutions of the effective relative permittivity and characteristic impedance of inside and outside cylindrical microstrip lines are shown. Coupled coplanar cylindrical microstrip lines and slot-coupled double-sided cylindrical microstrip lines are also studied. Cylindrical microstrip open-end and gap discontinuities are formulated, and equivalent circuits describing the microstrip discontinuities are presented. The characteristics of cylindrical coplanar waveguides (CPWs) are solved using a quasistatic method based on conformal mapping and a dynamic model based on a full-wave formulation. Inside CPWs, outside CPWs, and CPWs in substrate–superstrate structures are investigated.

The information contained in this book is largely the result of many years of research at National Sun Yat-Sen University, and I would like to thank my many former graduate students who took part in the studies. This book was designed to provide information on the basic characteristics of conformal microstrip antennas and microstrip transmission lines and to serve as a useful reference for those who are interested in the analysis and design of nonplanar microstrip antennas and circuits.

KIN-LU WONG

Kaohsiung, Taiwan

CHAPTER ONE

Introduction and Overview

1.1 INTRODUCTION

The microstrip antenna concept was first proposed in the 1950s. Due to the development of printed-circuit technology, many practical applications of microstrip antennas mounted on missiles and aircraft were demonstrated in the early 1970s. Since then, the study of microstrip antennas has boomed, giving birth to a new antenna industry. Figure 1.1 shows the basic geometry of a microstrip antenna: a metallic patch printed on a grounded dielectric substrate. The metallic patch can be of any shape, but in practical applications, rectangular and circular patches are most common, although annular and triangular patches are also common. Because of its simple geometry, the microstrip antenna offers many attractive advantages, such as low profile, light weight, easy fabrication, integrability with microwave and millimeter-wave integrated circuits, and conformability to curved surfaces. Among these advantages, conformability is the most

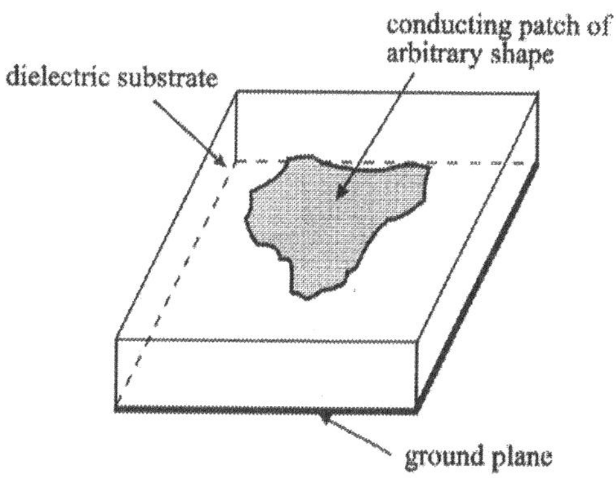

FIGURE 1.1 Geometry of a microstrip antenna with an arbitrary patch shape.

important to future applications of microstrip antennas. For example, by using conformal microstrip antennas as a substitute for conventional antennas, such as parabolic reflector antennas and wire antennas, environmental beauty can be maintained, which is a great impetus for the deployment of conformal antennas.

After over two decades of research, the development of planar microstrip antennas has now reached maturity. However, the progress of research on conformal or nonplanar microstrip antennas lags far behind that for planar microstrip antennas. This situation has been improved in the last decade, in which many theoretical works on microstrip antennas conformal to nonplanar surfaces, such as those on cylindrical, spherical, and conical bodies, have been reported. In addition to research on nonplanar microstrip antennas, printed transmission lines mounted on cylindrical surfaces have also received much attention. In this book we reorganize and review these recent publications on the theoretical modeling and experimental investigation of nonplanar microstrip antennas and transmission lines.

1.2 CYLINDRICAL MICROSTRIP ANTENNAS

The basic structures of cylindrical microstrip antennas excited through a probe feed are depicted in Figure 1.2, where various patch shapes are shown: rectangle, disk, annular ring, triangle, and wraparound. The probe-fed design provides no stray radiation from the probe current and is the simplest in geometry for theoretical analysis and practical manufacturing. Among the patch shapes, the wraparound patch can provide omnidirectional radiation in the roll plane of the cylindrical host, and the rectangular and circular patches are the most commonly used for general applications. Figure 1.3 shows the configurations of the rectangular and circular microstrip antennas excited through a coupling slot. Slot coupling is another commonly used feeding mechanism, in addition to the

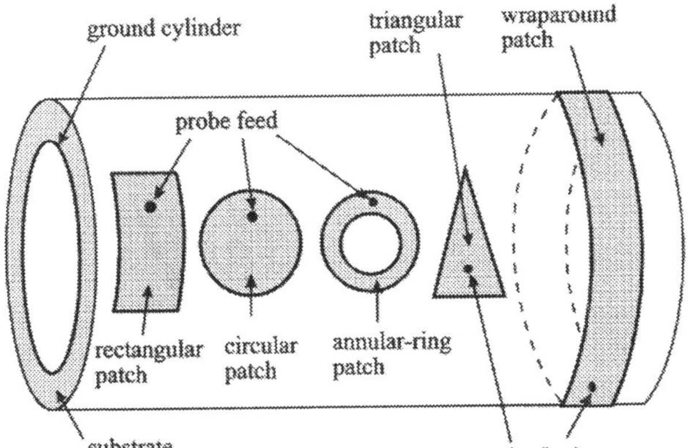

FIGURE 1.2 Basic structures of probe-fed cylindrical microstrip antennas.

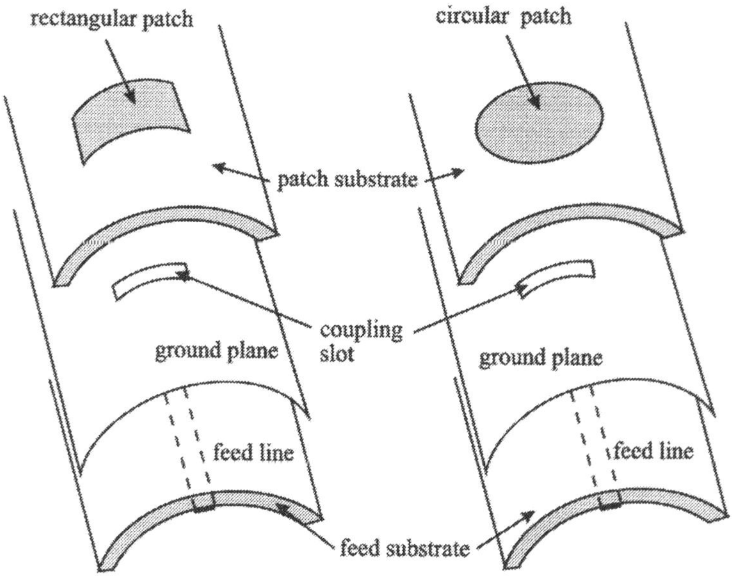

FIGURE 1.3 Configurations of slot-coupled cylindrical rectangular and circular microstrip antennas.

probe-fed case. The slot-coupling method involves two substrates separated by a ground plane; one substrate contains the radiating patch and the other contains the feeding network. Electromagnetic energy is coupled from the microstrip feed line to the microstrip patch through a coupling slot in the ground plane. A feeding mechanism using slot coupling offers the following advantages:

1. No spurious radiation from the feed network can interfere with the radiation pattern and polarization purity of the patch antenna, since a ground plane separates the feed network and the microstrip patch.
2. No direct contact with the radiating microstrip patch through the substrate is required, and the problem of a large self-reactance for a probe feed, which is critical at millimeter-wave frequencies, is avoided.
3. More degrees of freedom, such as slot size, slot position relative to the patch, feed-substrate parameters, and microstrip-feed-line parameters, are allowed in the feed design. Impedance matching can be achieved by adjusting the size of the coupling slot and the open-circuited tuning stub of the microstrip feed line. By choosing a suitable coupling slot size and adjusting its relative position to the radiating patch, a much larger antenna bandwidth can be obtained than when using a coax feed.
4. The configuration is well suited for monolithic phased-array antennas by integrating radiating patches on the low-permittivity substrate and the feed network, phase shifters, bias, and other circuitry in the high-permittivity gallium arsenide on a single monolithic chip.

Besides its advantages, slot-coupled feed is relatively costly and complex in antenna design compared to the probe-fed case. Another popular choice of feeding arrangement is feeding the microstrip patch directly through a coplanar microstrip line, which is especially suited for microstrip array design. Figure 1.4 shows two kinds of arrangements for a cylindrical rectangular microstrip antenna fed by a microstrip line. Since the input impedance at the patch edge is usually on the order of 100 to 200 Ω (for higher-permittivity substrates, the patch-edge input impedance can be greater than 200 Ω), an inset patch or quarter-wavelength impedance transformer is commonly used for impedance matching to a 50-Ω microstrip feed line.

Other feeding arrangements, using coplanar-waveguide (CPW) feed [1], buried microstrip line feed [2], and others, have also been demonstrated. CPW feed has the advantages of no via holes, easy integration with active devices, small stray radiation from the feed, and convenience in etching antenna and feed line in one step. However, CPW-feed design has less freedom in a large feed network design for antenna arrays. Buried microstrip line feed, on the other hand, provides flexibility in patch and microstrip line designs. However, this feed design has difficulty in integration with active devices.

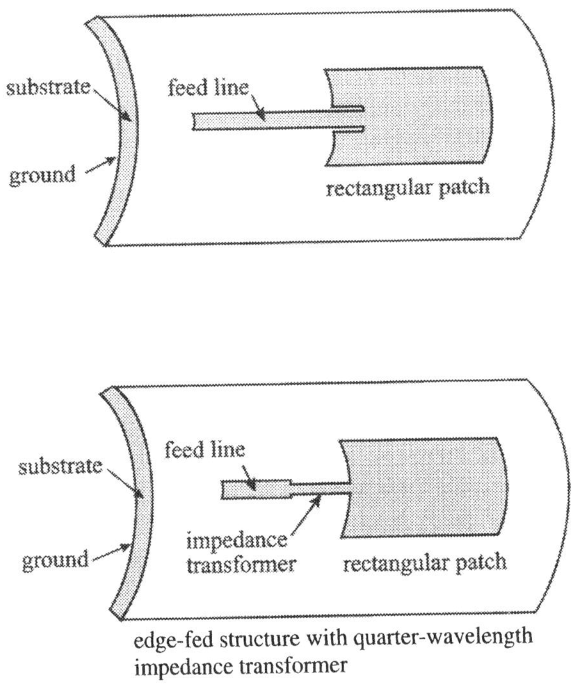

(*b*)

FIGURE 1.4 Two kinds of arrangements for the cylindrical rectangular microstrip antenna fed by a microstrip line: (*a*) inset-fed case; (*b*) edge-fed with $\lambda/4$ impedance transformer.

For the analysis and design of cylindrical microstrip antennas, a number of theoretical techniques, including the full-wave approach, cavity-model analysis, and generalized transmission-line model (GTLM) theory, have been reported. Among these models, the full-wave approach is computationally inefficient, and careful programming is usually required. Calculation of a full-wave solution may become difficult for a cylindrical microstrip antenna with a large cylinder radius (i.e., small-curvature case). However, full-wave solutions are more accurate and applicable to thick-substrate conditions. As for cavity-model analysis and GTLM theory, the theory and numerical computation are much simpler than with the full-wave approach. However, these two simple approaches are suitable only for the analysis of thin-substrate cases.

1.2.1 Full-Wave Approach

The full-wave approach for a probe-fed case is described briefly here in terms of the geometry shown in Figure 1.2. To begin with, the metallic ground cylinder and patch are assumed to be perfect conductors, and the thickness is neglected since it is much less than that of the operating wavelength. Based on the assumption, the patch can be replaced by a surface current distribution, which is unknown and needs to be solved. By noting that the radius of the feeding coax is usually a very small fraction of the operating wavelength, the probe can be treated as a line source with unit amplitude. To solve the unknown patch surface current density, the boundary condition that the total electric field tangential to the patch surface must be zero is applied; that is, on the patch,

$$\hat{\rho} \times [\mathbf{E}^D(\phi, z) + \mathbf{E}^P(\phi, z)] = 0\,, \tag{1.1}$$

where $\mathbf{E}^D(\phi, z)$ is the electric field due to the patch current and $\mathbf{E}^P(\phi, z)$ is the electric field due to the probe with the patch being absent. For deriving $\mathbf{E}^D(\phi, z)$, the theoretical formulation technique in [3] can be applied, and it gives

$$\begin{bmatrix} E^D_\phi(\phi, z) \\ E^D_z(\phi, z) \end{bmatrix} = \frac{1}{2\pi} \sum_{q=-\infty}^{\infty} e^{jq\phi} \int_{-\infty}^{\infty} dk_z\, e^{jk_z z} \bar{\bar{G}}(q, k_z) \begin{bmatrix} \tilde{J}_\phi(q, k_z) \\ \tilde{J}_z(q, k_z) \end{bmatrix}, \tag{1.2}$$

where $\bar{\bar{G}}(q, k_z)$ is the dyadic Green's function in the spectral domain for the cylindrical grounded substrate $\tilde{J}(q, k_z)$ and is the Fourier transform of the current density on the patch; the tilde denotes a Fourier transform. The subscripts ϕ and z denote, respectively, the field components in the ϕ and z directions.

For $\mathbf{E}^P(\phi, z)$, the field expression due to a point source in a layered medium needs to be derived first. Then, by imposing the boundary conditions at the cylindrical ground plane and the substrate–air interface, and after some straightforward manipulation, summing up all the field contributions from point sources along the input line-current source, an expression of $\mathbf{E}^P(\phi, z)$ can be derived which has the form of an integral equation [3]. Next, by substituting (1.2) and the derived $\mathbf{E}^P(\phi, z)$ into (1.1) and applying Galerkin's moment method to solve the resulting integral equation, a matrix equation can be obtained:

$$\begin{bmatrix} (Z_{ln}^{\phi\phi})_{N\times N} & (Z_{lm}^{\phi z})_{N\times M} \\ (Z_{kn}^{z\phi})_{M\times N} & (Z_{km}^{zz})_{M\times M} \end{bmatrix} \begin{bmatrix} (I_{\phi n})_{N\times 1} \\ (I_{zm})_{M\times 1} \end{bmatrix} = \begin{bmatrix} (V_{\phi n})_{N\times 1} \\ (V_{zm})_{M\times 1} \end{bmatrix}; \tag{1.3}$$

expressions of the matrix elements are described in subsequent chapters. By solving (1.3), the unknown patch surface currents $I_{\phi n}$ and I_{zm} are obtained, and the input impedance, radiation pattern, and other information of interest can be calculated.

To obtain full-wave solutions, numerical convergence for the moment-method calculation needs to be tested. The numerical convergence depends strongly on the basis function chosen for the expansion of the patch surface current density. A good choice of the basis functions used in moment-method calculation are the sinusoidal basis functions satisfying the edge condition that the normal component of the patch surface current must vanish at the patch edge. Details of the results are covered in Chapter 2.

1.2.2 Cavity-Model Analysis

The cavity-model analysis proposed by Lo et al. [4] offers both simplicity and physical insight into the operation of microstrip antennas. This model is valid when the substrate thickness is much smaller than the operating wavelength and is based on the following observations:

1. The close proximity between the patch and the ground plane suggests that for a cylindrical microstrip structure, the electric field has only a $\hat{\rho}$ component, and the magnetic field has only $\hat{\phi}$ and $\hat{z}$ components in the region bounded by the patch and the ground cylinder.
2. The field in the above-mentioned region is independent of the ρ coordinate for the frequency of interest.
3. The electric current on the microstrip patch must have no component normal to the edge at any point on the edge, implying a negligible component of magnetic field along the edge.

The region between the patch and the ground cylinder can therefore be treated as a cavity bounded by electric walls on the top and bottom and a magnetic wall around the perimeter of the cavity. Based on this cavity approximation, resonant frequencies of the TM_{mn} mode for cylindrical rectangular and circular microstrip antennas are given as follows: For the rectangular patch,

$$f_{mn} = \frac{c}{2\sqrt{\epsilon_1}} \left[\left(\frac{m}{2W} \right)^2 + \left(\frac{n}{2L} \right)^2 \right]^{1/2}, \tag{1.4}$$

and for the circular patch,

$$f_{mn} = \frac{c}{2\pi\sqrt{\epsilon_1}} k_{mn}, \tag{1.5}$$

where c is the speed of light and ϵ_1 is the relative permittivity of the substrate; $2L$ and $2W$ are, respectively, the length and width of the rectangular patch; k_{mn} satisfies $J'_m(k_{mn}a) = 0$, where $J_m(x)$ is a Bessel function of the first kind with order m, a is the radius of the circular patch, and the prime denotes a derivative.

The fields inside the cavity can then be expressed in terms of discrete modes individually satisfying the appropriate boundary conditions. Once the fields inside the cavity are known, the radiating field can be obtained from the effective magnetic current source flowing on the magnetic wall. After the cavity and radiated fields are determined, the radiation pattern, total radiated power, and input impedance can be calculated.

1.2.3 Generalized Transmission-Line Model Theory

Theoretical treatment based on the transmission-line model (TLM) is the first and simplest method applied for the analysis and design of microstrip antennas. Although the TLM method is relatively simple, the accuracy of TLM analysis can be made comparable to that of other more complicated methods [5]. For an analysis of mutual coupling between rectangular microstrip antennas, the TLM method can also be calculated in a fairly accurate and efficient way. However, the TLM method in its original form is applicable only for planar rectangular or square microstrip antennas. To cope with this problem, generalized transmission-line model (GTLM) theory is proposed [6], where the line parameters are the electromagnetic fields under the patch. In this case, as long as the separation of variables is possible for the wave equation expressed in that particular coordinate system, GTLM theory is applicable to microstrip antennas of any patch shape. The extension of GTLM theory to microstrip antennas with thick substrates is also possible. In the TLM method, the corresponding line parameters are the characteristic impedance and effective propagation constant. The equivalent circuits of a planar probe-fed rectangular microstrip antenna derived based on the TLM and GTLM methods are shown in Figure 1.5 for comparison. For GTLM theory, the rectangular patch is considered as a transmission line in the direction joining the radiating apertures of the patch. The effect of other apertures is considered as leakage of the transmission line. The transmission line can be further separated into two sections by the feed position, and each section of the transmission line can be replaced by an equivalent network and loaded with a wall admittance y_w, at the radiating apertures; y_m denotes the mutual admittance between two radiating apertures. When expressions are derived for these circuit elements, the input impedance of the patch antenna seen at the feed position can readily be obtained.

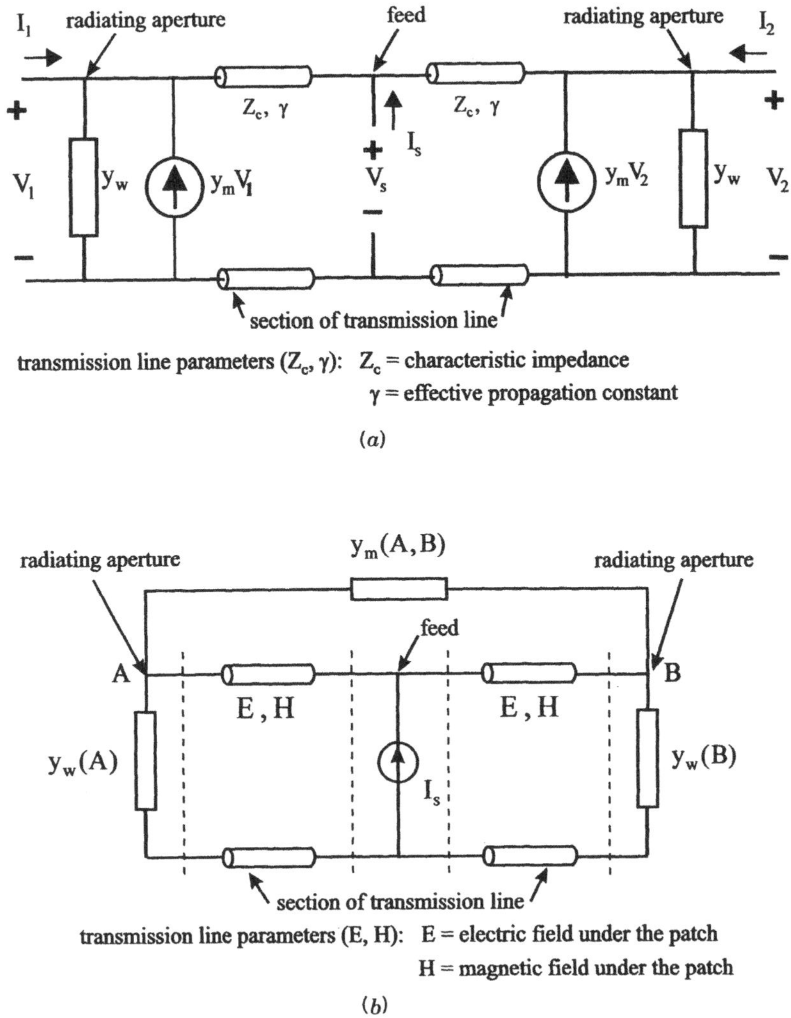

FIGURE 1.5 Equivalent circuits derived based on TLM and GTLM theory: (*a*) TLM model and (*b*) GTLM model of a planar probe-fed rectangular microstrip antenna at the TM_{01} mode.

1.3 SPHERICAL MICROSTRIP ANTENNAS

The spherical microstrip antenna is another canonical structure of conformal microstrip antennas, which can overcome the scanning problems involved with planar patch antennas at low elevations. Figure 1.6 shows the basic structures of spherical microstrip antennas with circular and annular-ring patches. The cavity-model theory has been used in the theoretical analysis of such spherical microstrip antennas [7,8], in which the curvature effects on the characteristics of microstrip patches mounted on a spherical body are analyzed. Reports on use of the full-wave

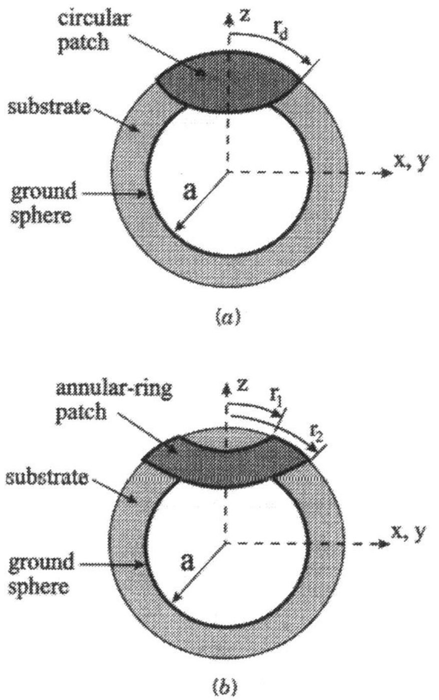

FIGURE 1.6 Basic structures of spherical microstrip antennas with (*a*) a circular patch and (*b*) an annular-ring patch.

approach and GTLM theory for analysis of spherical microstrip antennas have also been published. In full-wave analysis, the Green's function in the spectral domain for the grounded spherical substrate is formulated and Galerkin's moment method is used for the numerical calculation [9–11]. In many related reports based on the full-wave approach, the resonance problem, input impedance, radiation pattern, cross-polarization radiation, electromagnetic scattering, and surface current distribution on patches have been studied extensively [12–16]. Several modified spherical microstrip antenna structures, including a patch loaded with a superstrate layer, a microstrip structure with an air gap, and a patch with parasitic elements, have also been investigated [17–19].

GTLM theory has also been used in the analysis of spherical microstrip antennas [20,21]. According to the theory, the microstrip patches can be treated as a transmission line, taken in the direction of θ, loaded with a wall admittance evaluated at the radiation apertures. The equivalent transmission line can then be replaced by a π network. Figure 1.7 shows the corresponding equivalent circuits for circular and annular-ring patches. For a circular patch, the network of the circuit elements, Y_A, Y_B, and Y_C, represents the transmission-line section between the feed position and the radiation aperture at the patch edge. The shorted transmission-line section between the feed position and the patch center is replaced by an equivalent admittance, y_s. For an annular-ring patch, there are two sections

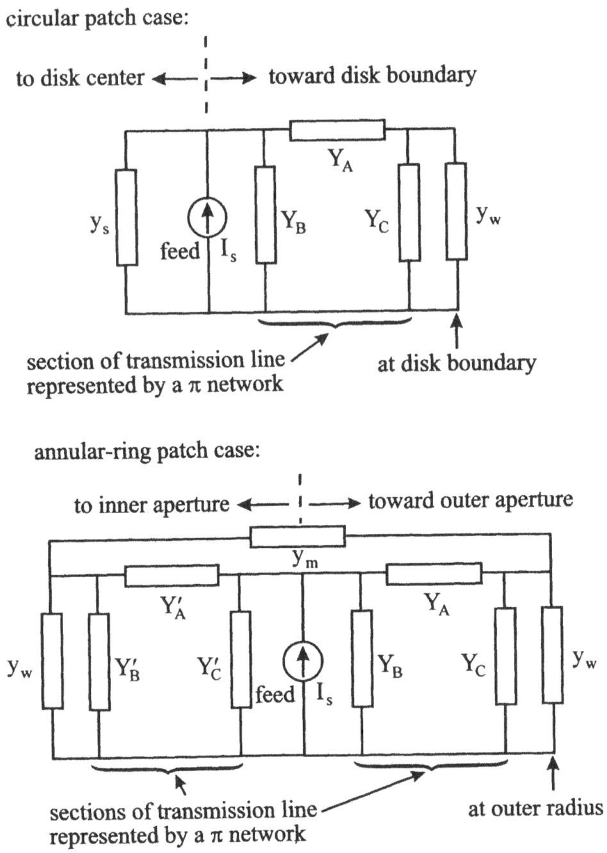

FIGURE 1.7 Equivalent circuits, derived based on GTLM theory, for the spherical circular and annular-ring microstrip antennas shown in Figure 1.6.

of transmission line and two radiation apertures, which is similar to the case of the rectangular patch (see Figure 1.5). In this case two wall admittances need to be evaluated, and the mutual admittance, y_m, between the two radiation apertures needs to be determined. Once the equivalent-circuit elements are derived, the input impedance of the antenna seen at the feed position is obtained.

1.4 CONICAL MICROSTRIP ANTENNAS

Microstrip antennas mounted on a conical surface have also been studied by several authors [22,23]. The related configurations are depicted in Figure 1.8. This kind of conical microstrip antenna can be of great interest for applications on the bodies of missiles, aircraft, and spacecraft that have conical surfaces on portions of their bodies. Due to the complexity in conical geometry compared with that of cylindrical and spherical structures, deriving the Green's function for a grounded

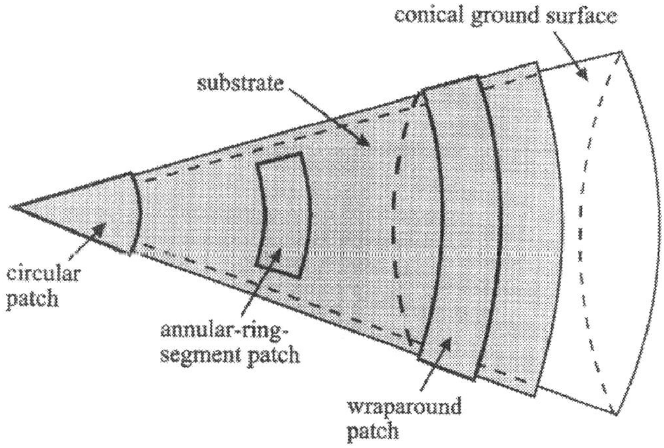

FIGURE 1.8 Basic structures of conical microstrip antennas with circular, annular-ring-segment, and wraparound patches.

conical substrate is difficult and the full-wave approach is thus not feasible. Related studies are based primarily on the cavity-model method. The main advantage of a conical microstrip antenna is its much broadened radiation pattern compared to that of a planar or cylindrical microstrip antenna.

1.5 CONFORMAL MICROSTRIP ARRAYS

Due to their conformability and light weight, microstrip patch arrays mounted on a curved surface have received much attention [24–29]. For most applications, a conformal microstrip array is usually employed on a cylindrical body such as a missile or aircraft. A typical arrangement of such a cylindrical microstrip array is shown in Figure 1.9. When the radius of the cylinder host is much larger than the operating wavelength, the effects of curvature on the characteristics of a curved microstrip array can be neglected. However, when the radius of the mounting host is comparable to that of the operating wavelength, results show that the radiation pattern of the microstrip array will be broadened, in addition to the fact that the input impedance and resonant frequency of each patch element in the array will also vary. If the radiation characteristic of a planar microstrip array is to be maintained, one needs to vary the interelement spacing and phase of each element in the cylindrical microstrip array [28]. Figure 1.10 shows another type of conformal microstrip array, with circular patches made conformal to a spherical surface. This type of spherical microstrip array can provide coverage over a wider angle of the hemisphere than that provided by a planar microstrip array and has been used in aircraft-to-satellite communication systems [25]. A microstrip array mounted on a conical surface, designed to provide a guided-weapon seeker antenna for a high-speed missile, has also been reported [30].

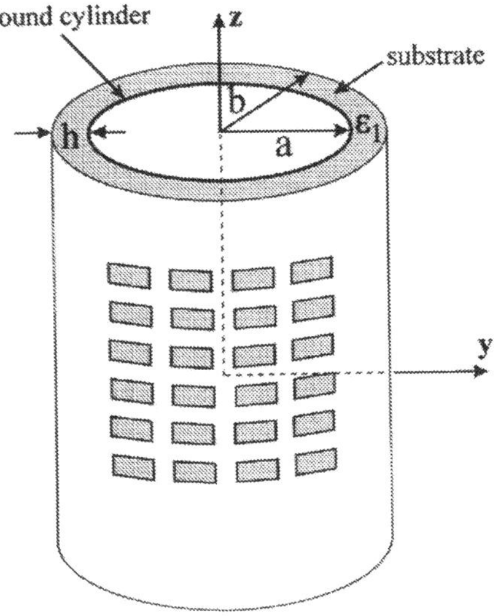

FIGURE 1.9 Microstrip array conformal to the curved surface of a cylindrical host.

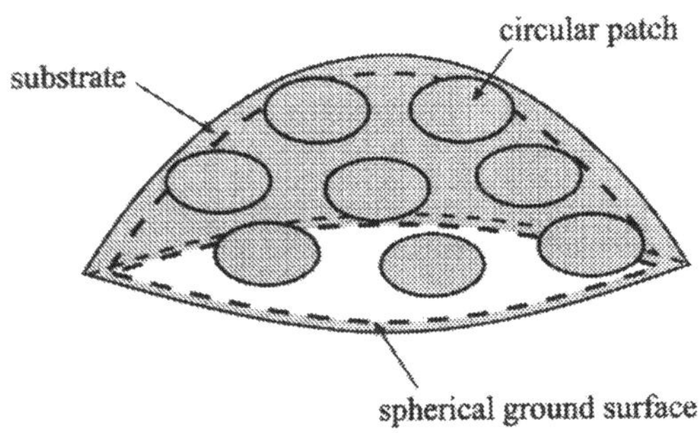

FIGURE 1.10 Circular microstrip patch array on a spherical surface.

1.6 CONFORMAL MICROSTRIP TRANSMISSION LINES

Due to the development of microstrip antennas and microstrip patch arrays mounted on conformal surfaces, an accurate design procedure becomes important not only for microstrip antennas but also for microstrip circuitry that forms the antenna or array excitation network. Many types of conformal microstrip transmission lines such as cylindrical microstrip lines [31, 32], coupled cylindrical microstrip lines [33], cylindrical microstrip discontinuities [34], slot-coupled double-sided cylindrical microstrip lines [35], and cylindrical coplanar waveguides [36] have thus been studied. These related geometries are shown in Figure 1.11. These topics are discussed in detail in Chapter 8.

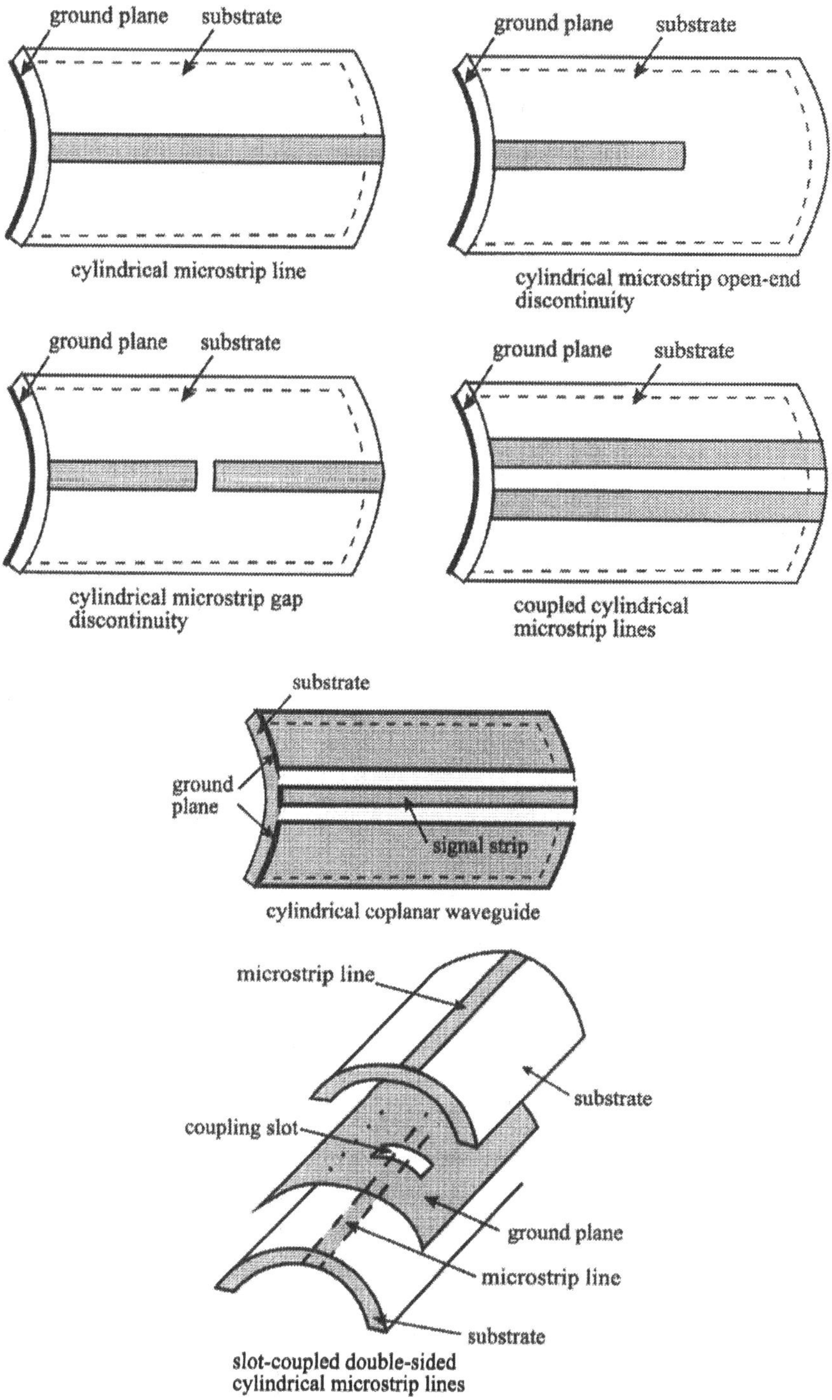

FIGURE 1.11 Geometries of typical cylindrical microstrip lines and coplanar waveguides mounted on a cylindrical body.

REFERENCES

1. W. Menzel and W. Grabherr, "A microstrip patch antenna with coplanar feed line," *IEEE Microwave Guided Wave Lett.*, vol. 1, pp. 340–342, Nov. 1991.
2. F. C. Silva, S. B. A. Fonseca, A. J. M. Soares, and A. J. Giarola, "Analysis of microstrip antennas on circular-cylindrical substrates with a dielectric overlay," *IEEE Trans. Antennas Propagat.*, vol. 39, pp. 1398–1404, Sept. 1991.
3. S. Y. Ke and K. L. Wong, "Input impedance of a probe-fed superstrate-loaded cylindrical-rectangular microstrip antenna," *Microwave Opt. Technol. Lett.*, vol. 7, pp. 232–236, Apr. 5, 1994.
4. Y. T. Lo, D. Soloman, and W. F. Richards, "Theory and experiment on microstrip antennas," *IEEE Trans. Antennas Propagat.*, vol. 27, pp. 137–145, Mar. 1979.
5. H. Pues and A. Van de Capelle, "Accurate transmission-line model for the rectangular microstrip antenna," *IEE Proc.*, pt. H, vol. 131, pp. 334–340, Dec. 1984.
6. A. K. Bhattacharyya and R. Garg, "Generalised transmission line model for microstrip patches," *IEE Proc.*, pt. H., vol. 132, pp. 93–98, Apr. 1985.
7. W. Y. Tam and K. M. Luk, "Patch antennas on a spherical body," *IEE Proc.*, pt. H, vol. 138, pp. 103–108, Feb. 1991.
8. H. T. Chen and K. L. Wong, "Analysis of probe-fed spherical-circular microstrip antennas using the cavity model theory," *Microwave Opt. Technol. Lett.*, vol. 7, pp. 205–207, May 1994.
9. W. Y. Tam and K. M. Luk, "Resonance in spherical-circular microstrip structures," *IEEE Trans. Microwave Theory Tech.*, vol. 39, pp. 700–704, Apr. 1991.
10. H. T. Chen and K. L. Wong, "Cross-polarization characteristics of a probe-fed spherical-circular microstrip patch antenna," *Microwave Opt. Technol. Lett.*, vol. 6, pp. 705–710, Sept. 20, 1993.
11. K. L. Wong and H. D. Chen, "Resonance in a spherical annular-ring microstrip structure," *Microwave Opt. Technol. Lett.*, vol. 6, pp. 852–856, Dec. 5, 1993.
12. H. D. Chen and K. L. Wong, "Analysis of a spherical annular-ring microstrip structure with an airgap," *Microwave Opt. Technol. Lett.*, vol. 7, pp. 205–207, Mar. 1994.
13. H. D. Chen and K. L. Wong, "Resonance frequency of a superstrate-loaded annular-ring microstrip structure on a spherical body," *Microwave Opt. Technol. Lett.*, vol. 7, pp. 364–367, June 5, 1994.
14. H. D. Chen and K. L. Wong, "Cross-polarization characteristics of spherical annular-ring microstrip antennas," *Microwave Opt. Technol. Lett.*, vol. 7, pp. 616–619, Sept. 1994.
15. T. J. Chang and H. T. Chen, "Full-wave analysis of scattering from a spherical-circular microstrip antenna," *Microwave Opt. Technol. Lett.*, vol. 10, pp. 49–52, Sept. 1995.
16. H. D. Chen and K. L. Wong, "Full-wave analysis of input impedance and patch current distribution of spherical annular-ring microstrip antennas excited by a probe feed," *Microwave Opt. Technol. Lett.*, vol. 7, pp. 524–528, Aug. 5, 1994.
17. K. L. Wong, S. F. Hsiao, and H. T. Chen, "Resonance and radiation of a superstrate-loaded spherical-circular microstrip patch antenna," *IEEE Trans. Antennas Propagat.*, vol. 41, pp. 686–690, May 1993.

18. K. L. Wong and H. T. Chen, "Resonance in a spherical-circular microstrip structure with an airgap," *IEEE Trans. Microwave Theory Tech.*, vol. 41, pp. 1466–1468, Aug. 1993.

19. H. T. Chen and Y. T. Cheng, "Full-wave analysis of a disk-loaded spherical annular-ring microstrip antenna," *Microwave Opt. Technol. Lett.*, vol. 12, pp. 353–358, Aug. 20, 1996.

20. B. Ke and A. A. Kishk, "Analysis of spherical circular microstrip antennas," *IEE Proc.*, pt. H., vol. 138, pp. 542–548, Dec. 1991.

21. A. A. Kishk, "Analysis of spherical annular microstrip antennas," *IEEE Trans. Antennas Propagat.*, vol. 41, pp. 338–343, Mar. 1993.

22. J. R. Descardeci and A. J. Giarola, "Microstrip antenna on a conical surface," *IEEE Trans. Antennas Propagat.*, vol. 40, pp. 460–463, Apr. 1992.

23. D. N. Meeks and P. F. Wahid, "Input impedance of a wraparound microstrip antenna on a conical surface," 1996 *IEEE AP-S International Symposium Digest*, pp. 676–679.

24. R. E. Munson, "Conformal microstrip antennas and microstrip phased arrays," *IEEE Trans. Antennas Propagat.*, vol. 22, pp. 74–78, Jan. 1974.

25. R. J. Mailloux, J. F. McIlvenna, and N. P. Kernweis, "Microstrip array technology," *IEEE Trans. Antennas Propagat.*, vol. 29, pp. 25–37, Jan. 1981.

26. J. Ashkenazy, S. Shtrikman, and D. Treves, "Conformal microstrip arrays on cylinders," *IEE Proc.*, pt. H, vol. 135, pp. 132–134, Apr. 1988.

27. C. M. Silva, F. Lumini, J. C. S. Lakava, and F. P. Richards, "Analysis of cylindrical arrays of microstrip rectangular patches," *Electron. Lett.*, vol. 27, pp. 778–780, Apr. 25, 1991.

28. R. C. Hall and D. I. Wu, "Modeling and design of circularly-polarized wraparound microstrip antennas," *1996 IEEE AP-S International Symposium Digest*, pp. 672–675.

29. E. V. Sohtell, in J. R. James and P. S. Hall, eds., *Handbook of Microstrip Antennas*, Peter Peregrinus, London, 1988, Chap. 22.

30. P. Newham and G. Morris, in J. R. James and P. S. Hall, eds., *Handbook of Microstrip Antennas*, Peter Peregrinus, London, 1988, Chap. 20.

31. N. G. Alexopoulos and A. Nakatani, "Cylindrical substrate microstrip line characterization," *IEEE Trans. Microwave Theory Tech.*, vol. 35, pp. 843–849, Sept. 1987.

32. L. R. Zeng and Y. Wang, "Accurate solutions of elliptical and cylindrical striplines and microstrip lines," *IEEE Trans. Microwave Theory Tech.*, vol. 34, pp. 259–264, Feb. 1986.

33. H. M. Chen and K. L. Wong, "Characterization of coupled cylindrical microstriplines mounted inside a ground cylinder," *Microwave Opt. Technol. Lett.*, vol. 10, pp. 330–333, Dec. 20, 1995.

34. H. M. Chen and K. L. Wong, "Characterization of cylindrical microstrip gap discontinuities," *Microwave Opt. Technol. Lett.*, vol. 9, pp. 260–263, Aug. 5, 1995.

35. J. H. Lu and K. L. Wong, "Analysis of slot-coupled double-sided cylindrical microstrip lines," *IEEE Trans. Microwave Theory Tech.*, vol. 44, pp. 1167–1170, July 1996.

36. H. C. Su and K. L. Wong, "Dispersion characteristics of cylindrical coplanar waveguides," *IEEE Trans. Microwave Theory Tech.*, vol. 44, pp. 2120–2122, Nov. 1996.

CHAPTER TWO

Resonance Problem of Cylindrical Microstrip Patches

2.1 INTRODUCTION

Since microstrip antennas are highly resonant structures, accurate determination of resonant frequency of the microstrip antenna becomes important for efficient radiation. In this chapter we describe full-wave analysis of the complex resonant frequency problem of various cylindrical microstrip structures. Numerical results are obtained using a moment-method calculation [1]. The structure of a superstrate-loaded microstrip patch is treated first. Numerical convergence for the sinusoidal basis functions with and without edge singularity is also discussed. The results obtained for the real and imaginary parts of complex resonant frequencies are analyzed as functions of the superstrate permittivity and thickness. The results are compared with those obtained for a superstrate-loaded planar microstrip structure [2] to analyze the curvature effect on the resonant frequency and quality factor of a curved microstrip structure. For fast determination of accurate resonant frequency of a planar rectangular microstrip patch antenna, curve-fitting formulas as a form of multivariable polynomial have been developed using a database generated from a full-wave approach incorporating a Galerkin's moment-method calculation, which makes possible fast, accurate determination of the complex resonant frequencies of a planar rectangular microstrip patch antenna. Details of the curve-fitting formulas are provided in Appendix A.

In addition to the superstrate layer directly loaded on a microstrip patch, a spaced superstrate has also been studied. In this case the superstrate layer is spaced away from the patch a distance of $S = n\lambda_0/2$, $n = 1, 2, 3, \ldots$ (λ_0 is the free-space wavelength), which can significantly increase the directive gain of the patch antenna. The geometry of a cylindrical microstrip structure with an air gap between the substrate layer and the ground cylinder is also discussed. By tuning the air-gap thickness, the resonant frequency of the microstrip structure is varied and the antenna bandwidth can be enhanced. The foregoing structures with a rectangular patch are considered, and a structure with a triangular patch is discussed for comparison. For the slot-coupling structure, the effects of a coupling slot in the cylindrical ground plane centered below the microstrip patch on the resonance of the cylindrical microstrip structure are described. All these microstrip structures are discussed in detail in subsequent sections.

2.2 CYLINDRICAL RECTANGULAR MICROSTRIP PATCH WITH A SUPERSTRATE

2.2.1 Theoretical Formulation

A cylindrical rectangular microstrip structure loaded with a protecting dielectric superstrate is shown in Figure 2.1, and the microstrip patch is mounted on a cylindrical ground cylinder of radius a. The cylindrical substrate (region 1) has a relative permittivity ϵ_1 and a thickness h $(= b - a)$, while the cylindrical superstrate (region 2) has a relative permittivity ϵ_2 and a thickness t $(= c - b)$. The air is in region 3 with free-space permittivity ϵ_0 and permeability μ_0. In regions 1

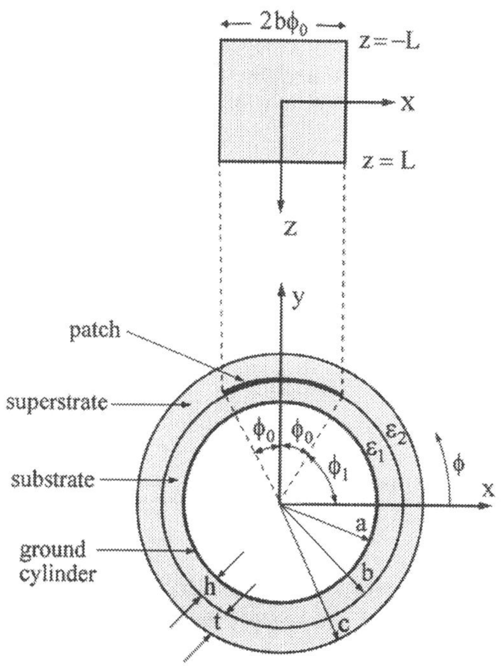

FIGURE 2.1 Geometry of a cylindrical rectangular microstrip patch with a dielectric superstrate layer.

and 2, the permeability is all assumed to be μ_0. The curved rectangular patch is at the substrate–superstrate interface of $\rho = b$ and has a straight dimension $2L$ and a curved dimension $2b\phi_0$, where $2\phi_0$ is the angle subtended by the curved patch. In this geometry with cylindrical coordinates, the z component of the electric and magnetic fields in each region can be given by [suppressing $\exp(-j\omega t)$ time dependence]

$$E_z(\rho, \phi, z) = \frac{1}{2\pi} \sum_{n=-\infty}^{\infty} e^{jn\phi} \int_{-\infty}^{\infty} dk_z\, e^{jk_z z}[A_{in} H_n^{(1)}(k_{i\rho}\rho) + B_{in} J_n(k_{i\rho}\rho)] \,, \qquad (2.1)$$

$$H_z(\rho, \phi, z) = \frac{1}{2\pi} \sum_{n=-\infty}^{\infty} e^{jn\phi} \int_{-\infty}^{\infty} dk_z\, e^{jk_z z}[C_{in} H_n^{(1)}(k_{i\rho}\rho) + D_{in} J_n(k_{i\rho}\rho)] \,, \qquad (2.2)$$

with

$$k_i^2 - k_{i\rho}^2 = k_z^2 \,, \qquad k_i = \omega\sqrt{\mu_0 \epsilon_0 \epsilon_i} \,, \qquad i = 1, 2, 3 \,,$$

$$B_{3n} = D_{3n} = 0 \,, \qquad \epsilon_3 = 1 \,,$$

where A_{in}, B_{in}, C_{in}, and D_{in} are unknown coefficients of the harmonic order n to be determined by the boundary conditions at $\rho = a$, b, and c. $H_n^{(1)}(x)$ is a Hankel function of the first kind with order n, and $J_n(x)$ is a Bessel function of the first kind with order n. Notice that the expressions above can also be replaced by linear combinations of two other linearly independent solutions of the Bessel equation. Once E_z and H_z are known, the transverse field components of E_ρ, E_ϕ, H_ρ, and H_ϕ in each region can be obtained through the following expressions:

$$E_\rho = \frac{j[k_z(\partial E_z/\partial\rho) + (\omega\mu_0/\rho)(\partial H_z/\partial\phi)]}{k_{i\rho}^2} \,, \qquad (2.3)$$

$$E_\phi = \frac{j[(k_z/\rho)(\partial E_z/\partial\phi) - \omega\mu_0(\partial H_z/\partial\rho)]}{k_{i\rho}^2} \,, \qquad (2.4)$$

$$H_\rho = \frac{j[(-\omega\epsilon_0\epsilon_i)/\rho)(\partial E_z/\partial\phi) + k_z(\partial H_z/\partial\rho)]}{k_{i\rho}^2} \,, \qquad (2.5)$$

$$H_\phi = \frac{j[\omega\epsilon_0\epsilon_i(\partial E_z/\partial\phi) + (k_z/\rho)(\partial H_z/\partial\rho)]}{k_{i\rho}^2} \,. \qquad (2.6)$$

To solve the unknown coefficients, boundary conditions at $\rho = a$, b, and c for the tangential components of the electric fields are imposed, and we have the following equations: At $\rho = a$,

$$A_{1n} H_n^{(1)}(k_{1\rho}a) + B_{1n} J_n(k_{1\rho}a) = 0 \,, \qquad (2.7)$$

$$\frac{-j\omega\mu_0}{k_{1\rho}}[C_{1n}H_n^{(1)'}(k_{1\rho}a)+D_{1n}J_n'(k_{1\rho}a)]-\frac{k_z n}{k_{1\rho}^2 a}[A_{1n}H_n^{(1)}(k_{1\rho}a)+B_{1n}J_n(k_{1\rho}a)]=0\,, \tag{2.8}$$

at $\rho = b$,

$$A_{1n}H_n^{(1)}(k_{1\rho}b)+B_{1n}J_n(k_{1\rho}b)-A_{2n}H_n^{(1)}(k_{2\rho}b)-B_{2n}J_n(k_{2\rho}b)=0\,, \tag{2.9}$$

$$\begin{aligned}&\frac{-j\omega\mu_0}{k_{1\rho}}[C_{1n}H_n^{(1)'}(k_{1\rho}b)+D_{1n}J_n'(k_{1\rho}b)]-\frac{k_z n}{k_{1\rho}^2 b}[A_{1n}H_n^{(1)}(k_{1\rho}b)+B_{1n}J_n(k_{1\rho}b)]\\&+\frac{j\omega\mu_0}{k_{2\rho}}[C_{2n}H_n^{(1)'}(k_{2\rho}b)+D_{2n}J_n'(k_{2\rho}b)]+\frac{k_z n}{k_{2\rho}^2 b}[A_{2n}H_n^{(1)}(k_{2\rho}b)+B_{2n}J_n(k_{2\rho}b)]\\&=0\,,\end{aligned} \tag{2.10}$$

and at $\rho = c$,

$$A_{3n}H_n^{(1)}(k_{3\rho}c)-A_{2n}H_n^{(1)}(k_{2\rho}c)-B_{2n}J_n(k_{2\rho}c)=0\,, \tag{2.11}$$

$$\begin{aligned}&\frac{-j\omega\mu_0}{k_{2\rho}}[C_{2n}H_n^{(1)'}(k_{2\rho}c)+D_{2n}J_n'(k_{2\rho}c)]-\frac{k_z n}{k_{2\rho}^2 c}[A_{2n}H_n^{(1)}(k_{2\rho}c)+B_{2n}J_n(k_{2\rho}c)]\\&+\frac{j\omega\mu_0}{k_{3\rho}}C_{3n}H_n^{(1)'}(k_{3\rho}c)+\frac{k_z n}{k_{3\rho}^2 c}A_{3n}H_n^{(1)}(k_{3\rho}c)=0\,.\end{aligned} \tag{2.12}$$

As for the magnetic fields at $\rho = c$, we have

$$C_{3n}H_n^{(1)}(k_{3\rho}c)-C_{2n}H_n^{(1)}(k_{2\rho}c)-D_{2n}J_n(k_{2\rho}c)=0\,, \tag{2.13}$$

$$\begin{aligned}&\frac{-j\omega\epsilon_0\epsilon_2}{k_{2\rho}}[A_{2n}H_n^{(1)'}(k_{2\rho}c)+B_{2n}J_n'(k_{2\rho}c)]+\frac{k_z n}{k_{2\rho}^2 c}[C_{2n}H_n^{(1)}(k_{2\rho}c)+D_{2n}J_n(k_{2\rho}c)]\\&+\frac{j\omega\epsilon_0\epsilon_3}{k_{3\rho}}A_{3n}H_n^{(1)'}(k_{3\rho}c)-\frac{k_z n}{k_{3\rho}^2 c}C_{3n}H_n^{(1)}(k_{3\rho}c)=0\,.\end{aligned} \tag{2.14}$$

From solving the equations above, the unknown coefficients A_{in}, B_{in}, C_{in}, and D_{in} in regions 1 and 2 can be expressed in terms of A_{3n} and C_{3n} in region 3. The expressions are as follows:

$$A_{1n}=\frac{J_n(k_{1\rho}a)[A_{2n}H_n^{(1)}(k_{2\rho}b)+B_{2n}J_n(k_{2\rho}b)]}{[J_n(k_{1\rho}a)H_n^{(1)}(k_{1\rho}b)-J_n(k_{1\rho}b)H_n^{(1)}(k_{1\rho}a)]}\,, \tag{2.15}$$

$$B_{1n} = \frac{-H_n^{(1)}(k_{1\rho}a)}{J_n(k_{1\rho}a)} A_{1n}\,, \tag{2.16}$$

$$C_{1n} = \frac{J'_n(k_{1\rho}a)[C_{2n}H_n^{(1)'}(k_{2\rho}b) + D_{2n}J'_n(k_{2\rho}b)]k_{1\rho}}{[J'_n(k_{1\rho}a)H_n^{(1)'}(k_{1\rho}b) - J'_n(k_{1\rho}b)H_n^{(1)'}(k_{1\rho}a)]k_{2\rho}}$$

$$-\frac{j\omega\epsilon_0\epsilon_2 k_z n}{k_{1\rho}k_{2\rho}^2 b}\left(\frac{\epsilon_1}{\epsilon_2} - 1\right)[A_{2n}H_n^{(1)}(k_{2\rho}b) + B_{2n}J_n(k_{2\rho}b)]\,, \tag{2.17}$$

$$D_{1n} = \frac{H_n^{(1)'}(k_{1\rho}a)}{J'_n(k_{1\rho}a)} C_{1n}\,, \tag{2.18}$$

$$A_{2n} = \alpha_1 A_{3n} + \alpha_2 C_{3n}\,, \tag{2.19}$$

$$C_{2n} = \alpha_3 A_{3n} + \alpha_4 C_{3n}\,, \tag{2.20}$$

$$B_{2n} = \frac{-\alpha_3 H_n^{(1)}(k_{2\rho}c)}{J_n(k_{2\rho}c)} A_{3n} + \frac{1 - \alpha_4 H_n^{(1)}(k_{2\rho}c)}{J_n(k_{2\rho}c)} C_{3n}\,, \tag{2.21}$$

$$D_{2n} = \frac{1 - \alpha_3 H_n^{(1)}(k_{2\rho}c)}{J_n(k_{2\rho}c)} A_{3n} - \frac{\alpha_4 H_n^{(1)}(k_{2\rho}c)}{J_n(k_{2\rho}c)} C_{3n}\,, \tag{2.22}$$

$$\alpha_0 = \frac{J_n(k_{2\rho}c)}{J_n(k_{2\rho}c)H_n^{(1)'}(k_{2\rho}c) - J'_n(k_{2\rho}c)H_n^{(1)}(k_{2\rho}c)}\,, \tag{2.23}$$

$$\alpha_1 = \alpha_0\left[\frac{\epsilon_3 k_{2\rho}H_n^{(1)'}(k_{3\rho}c)}{\epsilon_2 k_{3\rho}H_n^{(1)}(k_{3\rho}c)} - \frac{J_n^{(1)}(k_{2\rho}c)}{J_n(k_{2\rho}c)}\right], \tag{2.24}$$

$$\alpha_2 = \alpha_0 \frac{j\omega\mu_0 k_z n}{k_{3\rho}^2 k_{2\rho}c}\left(\frac{\epsilon_3}{\epsilon_2} - 1\right), \tag{2.25}$$

$$\alpha_3 = \frac{k_{2\rho}H_n^{(1)'}(k_{3\rho}c)}{k_{3\rho}H_n^{(1)}(k_{3\rho}c)} - \frac{J_n^{(1)}(k_{2\rho}c)}{J_n(k_{2\rho}c)}\,, \tag{2.26}$$

$$\alpha_4 = \frac{j\omega\epsilon_0\epsilon_3 k_z n}{k_{3\rho}^2 k_{2\rho}b}\left(1 - \frac{\epsilon_2}{\epsilon_3}\right). \tag{2.27}$$

Again, by applying the discontinuity boundary condition at $\rho = b$ for the tangential components, H_z and H_ϕ, of the magnetic field on the patch, we have

$$\tilde{J}_{\phi n}(k_z) = C_{1n}H_n^{(1)}(k_{1\rho}b) + D_{1n}J_n(k_{1\rho}b) - C_{2n}H_n^{(1)}(k_{2\rho}b) - D_{2n}J_n(k_{2\rho}b)\,, \tag{2.28}$$

$$\tilde{J}_{zn}(k_z) = \frac{-j\omega\epsilon_0\epsilon_1}{k_{1\rho}}[A_{1n}H_n^{(1)'}(k_{1\rho}b) + B_{1n}J_n'(k_{1\rho}b)]$$
$$+ \frac{k_z n}{k_{1\rho}^2 b}[C_{1n}H_n^{(1)}(k_{1\rho}b) + D_{1n}J_n(k_{1\rho}b)]$$
$$+ \frac{j\omega\epsilon_0\epsilon_2}{k_{2\rho}}[A_{2n}H_n^{(1)'}(k_{2\rho}b) + B_{2n}J_n'(k_{2\rho}b)]$$
$$- \frac{k_z n}{k_{2\rho}^2 b}[C_{2n}H_n^{(1)}(k_{2\rho}b) + D_{2n}J_n(k_{2\rho}b)], \quad (2.29)$$

where $\tilde{J}_{\phi n}(k_z)$ and $\tilde{J}_{zn}(k_z)$ are patch surface current densities in the spectral domain (the tilde denotes the spectral amplitude or a Fourier transform), and we have

$$\begin{bmatrix} \tilde{J}_{\phi n}(k_z) \\ \tilde{J}_{zn}(k_z) \end{bmatrix} = \frac{1}{2\pi}\int_{-\pi}^{\pi} e^{-jn\phi}\, d\phi \int_{-\infty}^{\infty} dz\, e^{-jk_z z} \begin{bmatrix} J_{\phi n}(\phi, z) \\ J_{zn}(\phi, z) \end{bmatrix}. \quad (2.30)$$

Then a matrix relationship between the spectral-domain patch surface current density and the field amplitudes in region 3 (air region) can be obtained and written as

$$\begin{bmatrix} \tilde{J}_{\phi n}(k_z) \\ \tilde{J}_{zn}(k_z) \end{bmatrix} = \begin{bmatrix} X_{11} & X_{12} \\ X_{21} & X_{22} \end{bmatrix} \begin{bmatrix} A_{3n}H_n^{(1)}(k_{3\rho}b) \\ C_{3n}H_n^{(1)}(k_{3\rho}b) \end{bmatrix}. \quad (2.31)$$

Furthermore, the tangential components of the electric field, $E_{\phi n}$ and E_{zn} in the spectral domain on the patch can be found to be related to the current density $J_{\phi n}$ and J_{zn} in (2.28) and (2.29), and the following equation is obtained:

$$\begin{bmatrix} \tilde{E}_{\phi n}(k_z) \\ \tilde{E}_{zn}(k_z) \end{bmatrix} = \bar{\bar{\tilde{G}}}(n, k_z) \begin{bmatrix} \tilde{J}_{\phi n}(k_z) \\ \tilde{J}_{zn}(k_z) \end{bmatrix}, \quad (2.32)$$

with

$$\bar{\bar{\tilde{G}}} = \begin{bmatrix} \tilde{G}_{\phi\phi} & \tilde{G}_{\phi z} \\ \tilde{G}_{z\phi} & \tilde{G}_{zz} \end{bmatrix} = \begin{bmatrix} Y_{11} & Y_{12} \\ Y_{21} & Y_{22} \end{bmatrix} \begin{bmatrix} X_{11} & X_{12} \\ X_{21} & X_{22} \end{bmatrix}^{-1} \quad (2.33)$$

and

$$\begin{bmatrix} \tilde{E}_{\phi n}(k_z) \\ \tilde{E}_{zn}(k_z) \end{bmatrix} = \frac{1}{2\pi}\int_{-\pi}^{\pi} e^{-jn\phi}\, d\phi \int_{-\infty}^{\infty} dz\, e^{-jk_z z} \begin{bmatrix} E_{\phi n}(\phi, z) \\ E_{zn}(\phi, z) \end{bmatrix}. \quad (2.34)$$

$\bar{\bar{\tilde{G}}}$ in (2.33) is a dyadic Green's function in the spectral domain; $\tilde{G}_{\phi z}$ denotes the

$\hat{\phi}$-directed tangential electric field at $\rho = b$ due to a unit-amplitude $\hat{z}$-directed patch surface current density. $\tilde{G}_{\phi\phi}$, $\tilde{G}_{z\phi}$, and $\tilde{G}_{zz}$ have similar meanings. The elements X_{ij} and Y_{ij} in the matrices of $\bar{\bar{X}}$ and $\bar{\bar{Y}}$, respectively, are derived as

$$X_{11} = \frac{\beta_1 X_0}{\beta_0} - y_2 \beta_4 , \tag{2.35}$$

$$X_{12} = \frac{\beta_2 X_0}{\beta_0} - \frac{J_n(k_{2\rho} b)}{J_n(k_{2\rho} c)} - y_3 \beta_4 , \tag{2.36}$$

$$X_{21} = \frac{jk_0}{120\pi} \left\{ \frac{\epsilon_{23}}{k_{2\rho}} \left[y_0 \beta_3 + \frac{J_n'(k_{2\rho} b)}{J_n(k_{2\rho} c)} \right] - \frac{\epsilon_{13} \beta_0 X_1}{k_{1\rho}} \left[y_0 \beta_4 + \frac{J_n(k_{2\rho} b)}{J_n(k_{2\rho} c)} \right] \right\}$$
$$+ \frac{nk_z}{b} \left(\frac{\beta_1}{k_{1\rho}^2 \beta_0} X_0 - \frac{y_2 \beta_4}{k_{2\rho}^2} \right), \tag{2.37}$$

$$X_{22} = \frac{jk_0}{120\pi} \left(\frac{\epsilon_{23}}{k_{2\rho}} y_1 \beta_3 + \frac{\epsilon_{13}}{k_{1\rho}} y_1 \beta_0 \beta_4 X_1 \right)$$
$$+ \frac{nk_z}{b} \left\{ \frac{\beta_5}{k_{1\rho}^2 \beta_0} X_0 - \frac{1}{k_{2\rho}^2} \left[y_3 \beta_4 + \frac{J_n(k_{2\rho} b)}{J_n(k_{2\rho} c)} \right] \right\}, \tag{2.38}$$

$$Y_{11} = \frac{-nk_z}{k_{1\rho}^2 b} \left[y_0 \beta_4 + \frac{J_n(k_{2\rho} b)}{J_n(k_{2\rho} c)} \right] - \frac{j120\pi \beta_1 k_0}{k_{1\rho}}, \tag{2.39}$$

$$Y_{12} = \frac{-nk_z y_1 \beta_4}{k_{1\rho}^2 b} - \frac{j120\pi \beta_5 k_0}{k_{1\rho}}, \tag{2.40}$$

$$Y_{21} = y_0 \beta_4 + \frac{J_n(k_{2\rho} b)}{J_n(k_{2\rho} c)}, \tag{2.41}$$

$$Y_{22} = y_1 \beta_4 , \tag{2.42}$$

where

$$X_0 = \frac{J_n'(k_{1\rho} b)[H_n^{(1)}(k_{1\rho} b) J_n'(k_{1\rho} a) - H_n^{(1)'}(k_{1\rho} a) J_n(k_{1\rho} b)]}{J_n(k_{1\rho} b)[H_n^{(1)'}(k_{1\rho} b) J_n(k_{1\rho} a) - H_n^{(1)'}(k_{1\rho} a) J_n'(k_{1\rho} b)]}, \tag{2.43}$$

$$X_1 = \frac{J_n(k_{1\rho} b)[H_n^{(1)'}(k_{1\rho} b) J_n(k_{1\rho} a) - H_n^{(1)'}(k_{1\rho} a) J_n'(k_{1\rho} b)]}{J_n'(k_{1\rho} b)[H_n^{(1)}(k_{1\rho} b) J_n'(k_{1\rho} a) - H_n^{(1)'}(k_{1\rho} a) J_n(k_{1\rho} b)]}, \tag{2.44}$$

$$\beta_0 = \frac{J_n'(k_{1\rho} b)}{J_n(k_{1\rho} b)}, \tag{2.45}$$

$$\beta_1 = \frac{nk_z}{j120\pi k_0 k_{1\rho} b}\left(\frac{k_{1\rho}^2}{k_{2\rho}^2} - 1\right)\left[y_0 \beta_4 + \frac{J_n(k_{2\rho} b)}{J_n(k_{2\rho} c)}\right] + \frac{k_{1\rho} y_2 \beta_4}{k_{2\rho}}, \tag{2.46}$$

$$\beta_2 = \frac{J_n'(k_{2\rho} c)}{J_n(k_{2\rho} c)}, \tag{2.47}$$

$$\beta_3 = H_n^{(1)'}(k_{2\rho} b) - \frac{J_n'(k_{2\rho} b)}{J_n(k_{2\rho} c)} H_n^{(1)}(k_{2\rho} c), \tag{2.48}$$

$$\beta_4 = H_n^{(1)}(k_{2\rho} b) - \frac{J_n(k_{2\rho} b)}{J_n(k_{2\rho} c)} H_n^{(1)}(k_{2\rho} c), \tag{2.49}$$

$$\beta_5 = \frac{nk_z y_1 \beta_4}{j120\pi k_0 k_{1\rho} b}\left(\frac{k_{1\rho}^2}{k_{2\rho}^2} - 1\right) + \left[y_3 \beta_3 + \frac{J_n'(k_{2\rho} b)}{J_n(k_{2\rho} c)}\right]\frac{k_{1\rho}}{k_{2\rho}}, \tag{2.50}$$

$$\beta_6 = \frac{H_n^{(1)'}(k_{2\rho} c)}{H_n^{(1)}(k_{2\rho} c)}, \tag{2.51}$$

$$\beta_7 = \frac{H_n^{(1)'}(k_{3\rho} c)}{H_n^{(1)}(k_{3\rho} c)}, \tag{2.52}$$

$$\epsilon_{13} = \frac{\epsilon_1}{\epsilon_3}, \qquad \epsilon_{23} = \frac{\epsilon_2}{\epsilon_3}, \tag{2.53}$$

$$y_0 = \frac{1}{(\beta_2 - \beta_6) H_n^{(1)}(k_{2\rho} c)}\left(\beta_2 - \frac{k_{2\rho}}{k_{3\rho}\epsilon_{23}}\right), \tag{2.54}$$

$$y_1 = \frac{1}{(\beta_2 - \beta_6) H_n^{(1)}(k_{2\rho} c)} \frac{120\pi n k_z}{jk_0 \epsilon_{23} k_{2\rho} c}\left(\frac{k_{2\rho}^2}{k_{3\rho}^2} - 1\right), \tag{2.55}$$

$$y_2 = \frac{1}{(\beta_2 - \beta_6) H_n^{(1)}(k_{2\rho} c)} \frac{nk_z}{jk_0 120\pi k_{2\rho} c}\left(1 - \frac{k_{2\rho}^2}{k_{3\rho}^2}\right), \tag{2.56}$$

$$y_3 = \frac{1}{(\beta_2 - \beta_6) H_n^{(1)}(k_{2\rho} c)}\left(\beta_2 - \frac{\beta_7 k_{2\rho}}{k_{3\rho}}\right). \tag{2.57}$$

Finally, by imposing the boundary conditions on the patch and outside the patch, the following integral equations can be obtained: On the patch,

$$\begin{bmatrix} E_\phi(\phi, z) \\ E_z(\phi, z) \end{bmatrix} = \frac{1}{2\pi} \sum_{n=-\infty}^{\infty} e^{jn\phi} \int_{-\infty}^{\infty} dk_z \, e^{jk_z z} \bar{\bar{G}}(n, k_z) \begin{bmatrix} \tilde{J}_{\phi n}(k_z) \\ \tilde{J}_{zn}(k_z) \end{bmatrix} = \begin{bmatrix} 0 \\ 0 \end{bmatrix}, \quad (2.58)$$

and outside the patch,

$$\begin{bmatrix} J_\phi(\phi, z) \\ J_z(\phi, z) \end{bmatrix} = \frac{1}{2\pi} \sum_{n=-\infty}^{\infty} e^{jn\phi} \int_{-\infty}^{\infty} dk_z \, e^{jk_z z} \begin{bmatrix} \tilde{J}_{\phi n}(k_z) \\ \tilde{J}_{zn}(k_z) \end{bmatrix} = \begin{bmatrix} 0 \\ 0 \end{bmatrix}. \quad (2.59)$$

2.2.2 Galerkin's Moment-Method Formulation

To solve the integral equations of (2.58) and (2.59), Galerkin's moment method is applied. To begin with, we first expand the unknown surface current density on the patch in terms of linear combinations of known basis or expansion functions; that is,

$$\mathbf{J}(\phi, z) = \sum_{n=1}^{N} I_{\phi n} \mathbf{J}_{\phi n}(\phi, z) + \sum_{m=1}^{M} I_{zm} \mathbf{J}_{zm}(\phi, z), \quad (2.60)$$

where $I_{\phi n}$ and I_{zm} are unknown coefficients for the basis functions $J_{\phi n}$ and J_{zm} in the ϕ and z directions, respectively. A convenient choice of the basis functions is the cavity-model function of

$$\mathbf{J}_{\phi n}(\phi, z) = \hat{\phi} \sin\left[\frac{p\pi}{2\phi_0}(\phi' + \phi_0)\right] \cos\left[\frac{q\pi}{2L}(z + L)\right], \quad (2.61)$$

$$\mathbf{J}_{zm}(\phi, z) = \hat{z} \sin\left[\frac{r\pi}{2L}(z + L)\right] \cos\left[\frac{s\pi}{2\phi_0}(\phi' + \phi_0)\right], \quad (2.62)$$

or

$$\mathbf{J}_{\phi n}(\phi, z) = \hat{\phi} \frac{1}{\sqrt{L^2 - z^2}} \sin\left[\frac{p\pi}{2\phi_0}(\phi' + \phi_0)\right] \cos\left[\frac{q\pi}{2L}(z + L)\right], \quad (2.63)$$

$$\mathbf{J}_{zm}(\phi, z) = \hat{z} \frac{1}{\sqrt{\phi_0^2 - \phi'^2}} \sin\left[\frac{r\pi}{2L}(z + L)\right] \cos\left[\frac{s\pi}{2\phi_0}(\phi' + \phi_0)\right], \quad (2.64)$$

where $2L$ is the patch length and $2\phi_0$ is the angle subtended by the curved patch; $\phi' = \phi - \pi/2$. The sinusoidal basis functions of (2.63) and (2.64) consider the edge-singularity condition for the tangential component of the surface current at the edge of the patch, while the basis functions of (2.61) and (2.62) do not consider the edge-singularity condition. The combinations of the integers p, q, r, and s depend on the mode numbers n and m. For the first three modes, $n = 1$, 2, and 3, the values of (p, q) are $(1, 0)$, $(1, 1)$, and $(1, 2)$, respectively, and the values of (r, s) are $(1, 0)$, $(1, 1)$, and $(1, 2)$ for $m = 1$, 2, and 3.

Next, by taking the spectral amplitudes of the selected basis functions and substituting into (2.58), we have

$$\sum_{u=-\infty}^{\infty} e^{ju\phi} \int_{-\infty}^{\infty} dk_z\, e^{jk_z z} \bar{\bar{G}} \begin{bmatrix} \sum_{n=1}^{N} I_{\phi n} \tilde{J}_{\phi n,u}(k_z) \\ \sum_{m=1}^{M} I_{zm} \tilde{J}_{zm,u}(k_z) \end{bmatrix} = \begin{bmatrix} 0 \\ 0 \end{bmatrix}, \tag{2.65}$$

where

$$\tilde{J}_{\phi n,u}(k_z) = \frac{1}{2\pi} \int_{-\phi_0}^{\phi_0} d\phi\, e^{-ju\phi} \int_{-L}^{L} dz\, e^{-jk_z z} J_{\phi n}(\phi, z)\,, \tag{2.66}$$

$$\tilde{J}_{zm,u}(k_z) = \frac{1}{2\pi} \int_{-\phi_0}^{\phi_0} d\phi\, e^{-ju\phi} \int_{-L}^{L} dz\, e^{-jk_z z} J_{zm}(\phi, z)\,. \tag{2.67}$$

From (2.66)–(2.67), the spectral amplitudes of the basis functions of (2.61)–(2.62) are expressed as

$$\tilde{J}_{\phi n,u}(k_z) = \frac{j^{p+q-1} p k_z}{\phi_0} \frac{\sin[(p\pi/2) - u\phi_0]}{u^2 - (p\pi/2\phi_0)^2} \frac{\sin[(q\pi/2) - k_z L]}{k_z^2 - (q\pi/2L)^2}\,, \tag{2.68}$$

$$\tilde{J}_{zm,u}(k_z) = \frac{j^{r+s-1} r u}{L} \frac{\sin[(s\pi/2) - u\phi_0]}{u^2 - (s\pi/2\phi_0)^2} \frac{\sin[(r\pi/2) - k_z L]}{k_z^2 - (r\pi/2L)^2}\,. \tag{2.69}$$

As for the basis functions of (2.63)–(2.64), the spectral amplitudes are written as

$$\tilde{J}_{\phi n,u}(k_z) = \frac{j^{p+q+1} p\pi}{4\phi_0} \frac{\sin[(p\pi/2) - u\phi_0]}{u^2 - (p\pi/2\phi_0)^2} \left[J_0\!\left(\frac{q\pi}{2} - k_z L\right) + (-1)^q J_0\!\left(\frac{q\pi}{2} + k_z L\right) \right], \tag{2.70}$$

$$\tilde{J}_{zm,u}(k_z) = \frac{j^{r+s+1} r\pi}{4L} \frac{\sin[(r\pi/2) - k_z L]}{k_z^2 - (r\pi/2L)^2} \left[J_0\!\left(\frac{s\pi}{2} - u\phi_0\right) + (-1)^s J_0\!\left(\frac{s\pi}{2} + u\phi_0\right) \right], \tag{2.71}$$

where $J_0(x)$ is a Bessel function of the first kind with order zero.

Then, using the selected basis functions as testing functions and integrating over the patch area, we can have the following homogeneous matrix equation:

$$\begin{bmatrix} (Z_{ln}^{\phi\phi})_{N\times N} & (Z_{lm}^{\phi z})_{N\times M} \\ (Z_{kn}^{z\phi})_{M\times N} & (Z_{km}^{zz})_{M\times M} \end{bmatrix} \begin{bmatrix} (I_{\phi n})_{N\times 1} \\ (I_{zm})_{M\times 1} \end{bmatrix} = \begin{bmatrix} 0 \\ 0 \end{bmatrix}, \tag{2.72}$$

where

$$Z_{ln}^{\phi\phi} = \int_{-\infty}^{\infty} dk_z \sum_{u=-\infty}^{\infty} \tilde{J}_{\phi l,-u}(k_z^*)\tilde{G}_{\phi\phi}\tilde{J}_{\phi n,u}(k_z), \tag{2.73}$$

$$Z_{lm}^{\phi z} = \int_{-\infty}^{\infty} dk_z \sum_{u=-\infty}^{\infty} \tilde{J}_{\phi l,-u}(k_z^*)\tilde{G}_{\phi z}\tilde{J}_{zm,u}(k_z), \tag{2.74}$$

$$Z_{kn}^{z\phi} = \int_{-\infty}^{\infty} dk_z \sum_{u=-\infty}^{\infty} \tilde{J}_{zk,-u}(k_z^*)\tilde{G}_{z\phi}\tilde{J}_{\phi n,u}(k_z), \tag{2.75}$$

$$Z_{km}^{zz} = \int_{-\infty}^{\infty} dk_z \sum_{u=-\infty}^{\infty} \tilde{J}_{zk,-u}(k_z^*)\tilde{G}_{zz}\tilde{J}_{zm,u}(k_z), \tag{2.76}$$

$$k, m = 1, 2, \ldots, M, \qquad l, n = 1, 2, \ldots, N.$$

There exist nontrivial solutions for the unknown amplitudes $I_{\phi n}$ and I_{zm} if the determinant of the [Z] matrix in (2.72) vanishes; that is,

$$\det\begin{bmatrix} (Z_{ln}^{\phi\phi})_{N\times N} & (Z_{lm}^{\phi z})_{N\times M} \\ (Z_{kn}^{z\phi})_{M\times N} & (Z_{km}^{zz})_{M\times M} \end{bmatrix} = 0. \tag{2.77}$$

The solutions to (2.77) are found to be satisfied by complex frequencies for a particular mode. This complex frequency, $f = f' + jf''$, gives the resonant frequency f' and the quality factor $f'/2f''$ for the microstrip patch. The imaginary part of the complex resonant frequency also represents the radiation loss (the loss includes the surface-wave loss) of the microstrip structure. Since the microstrip patch is a resonant structure, the inverse of the quality factor also represents the half-power operating bandwidth for the microstrip patch as a radiator.

2.2.3 Complex Resonant Frequency Results

To obtain full-wave solutions of complex resonant frequencies of the superstrate-loaded cylindrical rectangular microstrip patch, numerical convergence of the moment-method calculation incorporating the sinusoidal basis functions with and without edge singularity is first studied. The fundamental mode TM_{01} (to the ρ direction) is considered; that is, the patch is excited in the direction along with the cylinder axis. It should also be noted that in a cylindrical structure, TM and TE waves are coupled together [see (2.17) in which E_z and H_z are coupled together with the exception of the $n = 0$ case]; that is, the TM_{01} mode considered here is not pure TM polarized wave. Calculated results of the real and imaginary parts of the complex resonant frequencies for different numbers of the sinusoidal basis functions are presented in Figure 2.2. The obtained frequencies are normalized with respect to the cavity-model resonant frequency given by

$$f_{01} = \frac{c}{4L\sqrt{\epsilon_1}}, \tag{2.78}$$

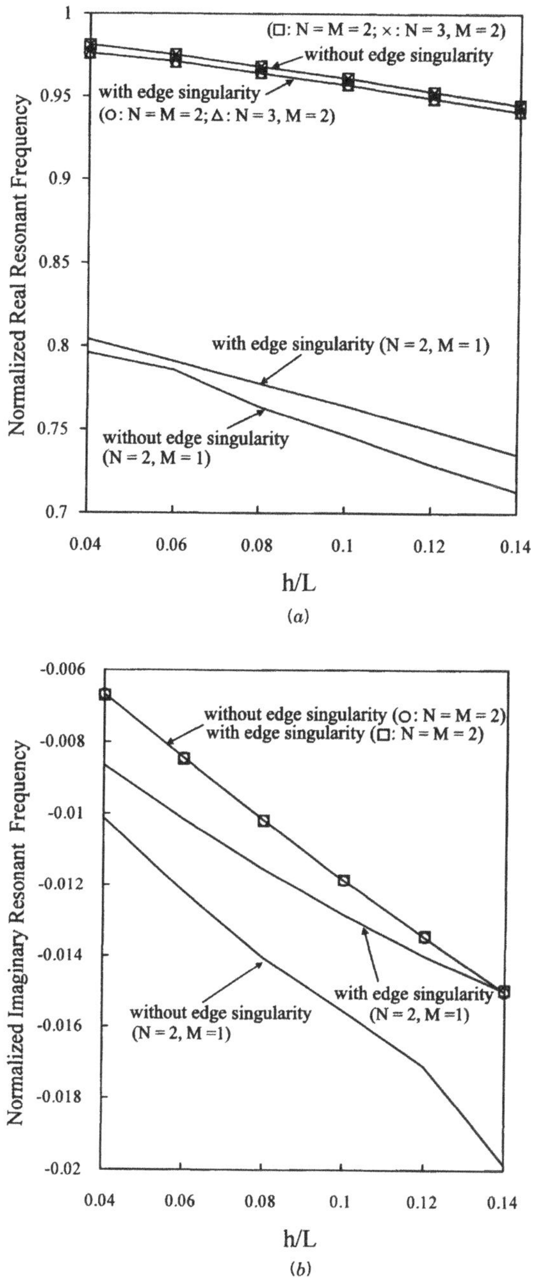

FIGURE 2.2 Normalized complex resonant frequency for different numbers of the sinusoidal basis functions with and without edge-singularity condition versus normalized substrate thickness; $a = 20$ cm, $2L = 8$ cm, $2b\phi_0 = 16.8$ cm, $\epsilon_1 = 2.3$. (*a*) Real part of complex resonant frequency; (*b*) imaginary part of complex resonant frequency. (From Ref. [3], © 1994 IEEE, reprinted with permission.)

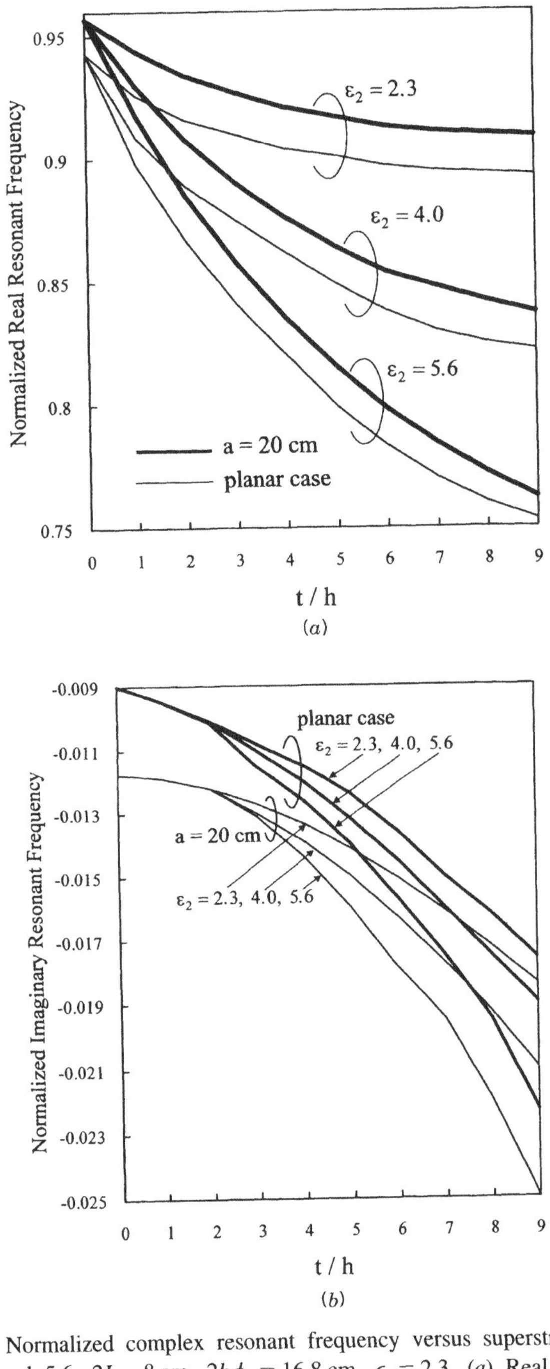

FIGURE 2.3 Normalized complex resonant frequency versus superstrate thickness for $\epsilon_2 = 2.3$, 4.0, and 5.6; $2L = 8$ cm, $2b\phi_0 = 16.8$ cm, $\epsilon_1 = 2.3$. (*a*) Real part of complex resonant frequency; (*b*) imaginary part of complex resonant frequency. The planar results are obtained from [2]. (From Ref. [3], © 1994 IEEE, reprinted with permission.)

where c is the speed of light. It is observed that both the real and imaginary resonant frequencies can reach convergent solutions for both the sinusoidal basis functions with $N \geq 2$ and $M \geq 2$. Real parts of the convergent solutions using basis functions with edge singularity are also found to differ from those without edge singularity with only about 0.5%, while the imaginary parts of the convergent solutions for both kinds of basis functions are almost exactly the same. This suggests that both the sinusoidal basis functions with and without considering the edge-singularity condition are suitable to be used for expansion of the unknown patch surface current on the cylindrical rectangular microstrip patch.

Figure 2.3 shows the normalized real and imaginary parts of the complex resonant frequencies versus superstrate thickness. The results for the planar patch case are obtained using the curve-fitting formulas described in Appendix A. The basis functions considering the edge singularity are used to obtain the results. It should be noted that the results obtained, shown in Figure 2.3, are almost the same as the convergent solutions obtained using the basis functions without considering the edge singularity. It can be seen that the resonant frequency, the real part of the complex resonant frequency, decreases as the superstrate permittivity increases and the curved patch has a higher resonant frequency than the planar patch. The imaginary resonant frequency results imply that the radiation loss of the curved patch is higher than that for the planar patch and can be increased further when a

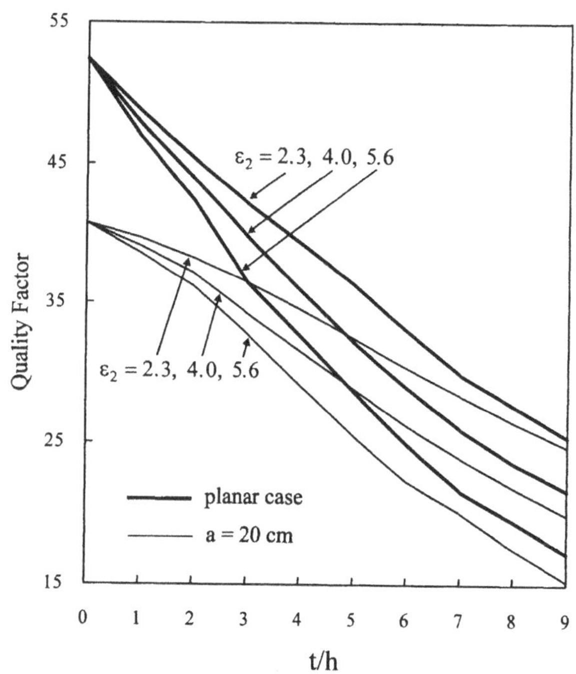

FIGURE 2.4 Variations of the quality factor with superstrate thickness for the case shown in Figure 2.3. (From Ref. [3], © 1994 IEEE, reprinted with permission.)

larger superstrate permittivity is used. Variations of the quality factor with superstrate thickness are shown in Figure 2.4. The quality factor is seen to decrease when the superstrate permittivity increases, and the curved patch has a lower quality factor than the planar patch.

2.3 CYLINDRICAL RECTANGULAR MICROSTRIP PATCH WITH A SPACED SUPERSTRATE

In this section we present a related cylindrical microstrip structure: a cylindrical rectangular microstrip patch covered with a spaced superstrate layer. To keep the superstrate layer spaced away from the microstrip patch in a practical design, foam material is usually used between the superstrate layer and the microstrip patch. Since the foam has a relative permittivity close to 1.0, spacing with such a material can be treated approximately as an air layer. The thickness of the spacing is usually selected to be a half-wavelength or its multiples. With this spacing the directivity of the microstrip patch as a radiator can be enhanced significantly. Both resonance and radiation problems are solved and discussed in this section.

2.3.1 Theoretical Formulation

Figure 2.5 shows the geometry of the microstrip patch covered with a spaced dielectric superstrate. The cylindrical superstrate is spaced away from the patch a distance of a half-wavelength or its multiples ($S = n\lambda_0/2$, $n = 1, 2, 3, \ldots$); that is, the region between the superstrate and the microstrip patch is an air region with free-space permittivity ϵ_0 and free-space permeability μ_0. The reason for setting S to be multiples of a half-wavelength is that the optimal directive gain enhancement of the patch antenna using a spaced superstrate can be obtained only at such spacing distances [5]. The curved patch is again assumed to have a straight dimension of $2L$ and a curved dimension $2b\phi_0$. Other parameters are illustrated in Figure 2.5.

We first follow the theoretical approach in Section 2.2.1. The field expressions in each region are first solved by imposing the continuity boundary conditions at $\rho = a$, b, c, and d for the tangential electric and magnetic fields. Next, by applying a discontinuity boundary condition at $\rho = b$ for the tangential magnetic field, we can have a similar matrix relationship, as shown by (2.31), between the surface patch current density and the field amplitudes in the outer region ($\rho > d$). The tangential components of the electric field on the patch can further be found to be related to the surface current density as shown by (2.32). Finally, a set of vector integral equations can be obtained by imposing the boundary conditions on the patch and outside the patch at $\rho = b$; that is, on the patch we have (2.58), and outside the patch we have (2.59). The integral equations are then solved using a Galerkin's moment-method calculation. Basis functions of either (2.61)–(2.62) or (2.63)–(2.64) can be used in the calculation. Then, by taking the spectral amplitudes of the basis functions selected and substituting them into (2.58), using

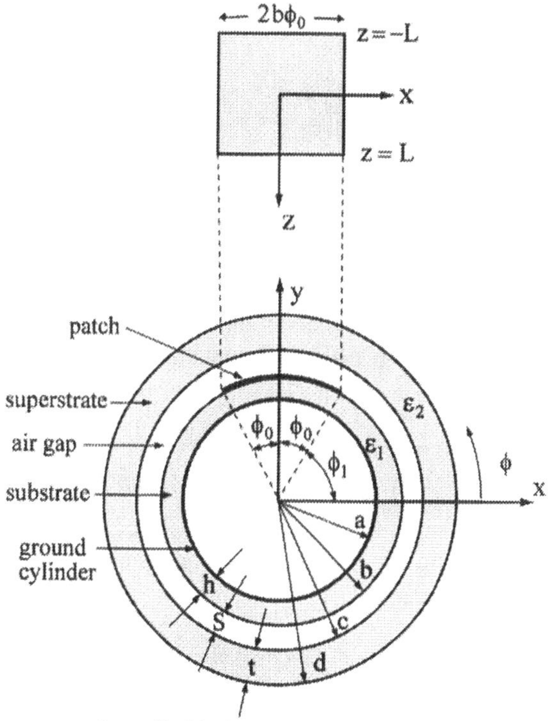

FIGURE 2.5 Geometry of a cylindrical rectangular microstrip structure covered with a spaced superstrate.

the selected basis functions as testing functions and integrating over the patch area, the homogeneous matrix equation (2.72) is derived. The complex resonant frequencies can then be obtained by solving

$$\det[Z(f)] = 0\,, \tag{2.79}$$

which has the form of (2.77).

Radiation patterns at the resonant frequency can also be calculated. After some straightforward manipulation, the E_z and H_z fields in the outer region ($\rho > d$) can be written as

$$\begin{bmatrix} E_z \\ H_z \end{bmatrix} = \frac{1}{2\pi} \sum_{n=-\infty}^{\infty} e^{jn\phi} \int_{-\infty}^{\infty} dk_z\, e^{jk_z z} \frac{H_n^{(1)}(k_\rho \rho)}{H_n^{(1)}(k_\rho d)} \bar{\bar{X}}^{-1} \begin{bmatrix} \tilde{J}_{\phi n}(k_z) \\ \tilde{J}_{zn}(k_z) \end{bmatrix}, \tag{2.80}$$

where $k_\rho = \sqrt{\omega^2 \mu_0 \epsilon_0 - k_z^2}$, and the elements in the $\bar{\bar{X}}$ matrix have similar forms as shown by (2.35)–(2.38), which have been derived in [4]. $J_{\phi n}$ and J_{zn} are, respectively, the nth basis functions in the ϕ and z directions. By further applying the stationary-phase evaluation, the far-zone radiated fields in spherical coordinates are given approximately as

$$\begin{bmatrix} E_\theta \\ E_\phi \end{bmatrix} = \frac{1}{\sin\theta} \begin{bmatrix} -1 & 0 \\ 0 & \eta_0 \end{bmatrix} \begin{bmatrix} E_z \\ H_z \end{bmatrix}$$

$$\simeq \frac{1}{\sin\theta} \begin{bmatrix} -1 & 0 \\ 0 & \eta_0 \end{bmatrix} \frac{e^{jk_0 r}}{\pi r} \sum_{n=-\infty}^{\infty} \frac{(-j)^{n+1} e^{jn\phi}}{H_n^{(1)}(k_0 d \sin\theta)} \bar{\bar{X}}^{-1} \begin{bmatrix} J_\phi \\ J_z \end{bmatrix}, \qquad (2.81)$$

where J_ϕ and J_z are the patch surface current densities obtained in the ϕ and z directions, respectively; η_0 is free-space intrinsic impedance. From (2.81), the directive gain of the microstrip patch as a radiator can be calculated as

$$D(\theta, \phi) = \frac{4\pi(|E_\theta|^2 + |E_\phi|^2)}{\int_0^{2\pi} \int_0^{\pi} (|E_\theta|^2 + |E_\phi|^2) \sin\theta \, d\theta \, d\phi}. \qquad (2.82)$$

The directivity, defined to be the maximum value of (2.82), is analyzed and discussed in the next section.

2.3.2 Resonance and Radiation Characteristics

Typical numerical results of the resonant frequency, half-power bandwidth, radiation pattern, and directivity for the TM_{01} mode are presented. Figure 2.6 shows the dependence of the normalized resonant frequency (the resonant frequency is set to be unity when the substrate thickness approaches zero and the superstrate layer is absent) and the half-power bandwidth on the spacing S. The case of a superstrate layer directly loaded on the patch ($S = 0$) is also shown. It is seen that superstrate loading strongly reduces the resonant frequency of the microstrip patch for the case of $S = 0$. However, for $S = 0.5\lambda_0$ and $1.0\lambda_0$, variations of the resonant frequency with superstrate thickness are much smaller than that for $S = 0$. It can also be seen that for $S = 1.0\lambda_0$, the resonant frequency of the microstrip patch is almost not affected when the superstrate thickness is less than $0.08\lambda_2$ ($\lambda_2 = \lambda_0/\sqrt{\epsilon_2}$); that is, the effect of superstrate loading on the resonant frequency is reduced significantly if the superstrate layer is spaced away from the microstrip patch. The results of bandwidth variations due to the superstrate loading are shown in Figure 2.6*b*. The spaced superstrate is seen to decrease the operating bandwidth of the microstrip patch as a radiator, and the decrease is smaller for larger spacing distances. This phenomenon is valid only when S is about $n\lambda_0/2$. (It should also be noted that the bandwidth variations for the spacing distances of $S \neq n\lambda_0/2$, not shown in the figure, vary with no general rules.) On the other hand, for $S = 0$, the bandwidth varies slightly for thin superstrate layers and increases significantly when the superstrate layer becomes thicker. This is probably due to the surface-wave loss introduced by the adding of a superstrate layer directly on

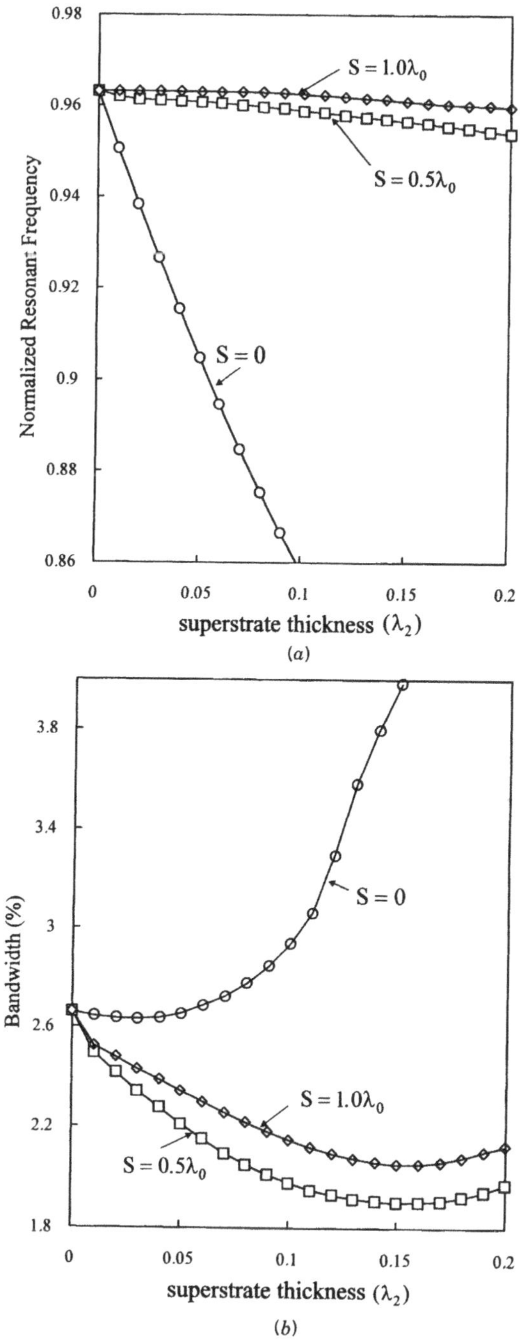

FIGURE 2.6 Dependence of (*a*) normalized resonant frequency and (*b*) half-power bandwidth on superstrate thickness; $\epsilon_1 = 2.3$, $\epsilon_2 = 5.6$, $h = 0.24$ mm, $2L = 8$ cm, $2b\phi_0 =$ 16.8 cm, $a = 20$ cm.

the microstrip patch. Higher superstrate permittivity will also increase the surface wave loss, which results in an increased antenna bandwidth.

Figure 2.7 shows some typical radiation patterns at resonance in the $\phi = 90°$ plane (y–z plane). The half-power beamwidths (HPBWs) of the patterns are reduced from 76° ($t = 0$) to 45° for $S = 0.5\lambda_0$ and 29° for $S = 1.0\lambda_0$. The HPBW is narrowed with the adding of a spaced superstrate, which in this case serves as a directive parasitic antenna. The dependence of the directivity, calculated from (2.82), on superstrate thickness is shown in Figure 2.8. With a superstrate layer, the directivity of the microstrip patch as a radiator increases with increasing superstrate thickness and reaches a maximum value around 0.14 to $0.18\lambda_2$. This maximum directivity also increases with increasing superstrate permittivity and spacing thickness. For $S = 1.0\lambda_0$ and $\epsilon_2 = 13.2$, the directivity increases to be almost 11.9 dB. Figure 2.9 shows the curvature effect on the directivity. The three cases of $a = 10$ cm, 20 cm, and ∞ (planar microstrip patch) with $S = 0.5\lambda_0$ and $1.0\lambda_0$ are shown. Maximum directivity also occurred at around 0.14 to $0.18\lambda_2$ and is higher for larger curvature radius. For the planar microstrip patch, the directivity can reach about 18.2 dB for $S = 1.0\lambda_0$, which is 10 dB higher than for the case of

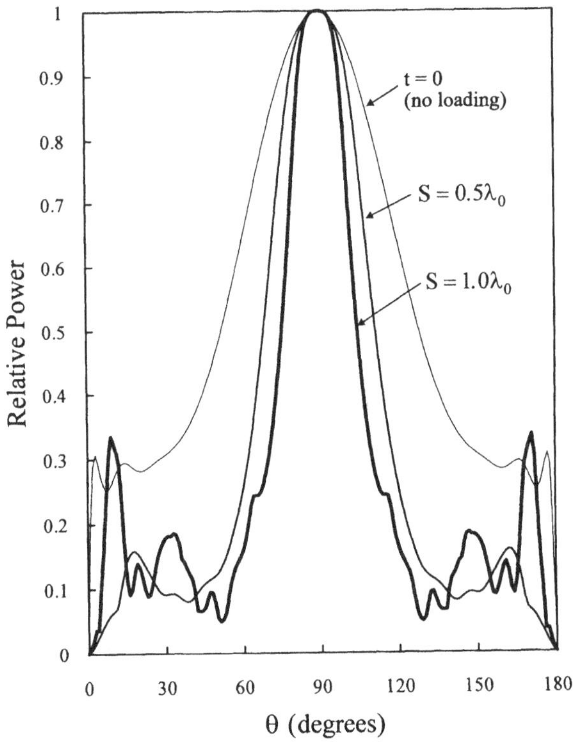

FIGURE 2.7 Radiation patterns against θ in the $\phi = 90°$ plane; $\epsilon_1 = 2.3$, $\epsilon_2 = 5.6$, $h = 0.24$ mm, $t = 0.15\lambda_2$, $2L = 8$ cm, $2b\phi_0 = 16.8$ cm, $a = 20$ cm.

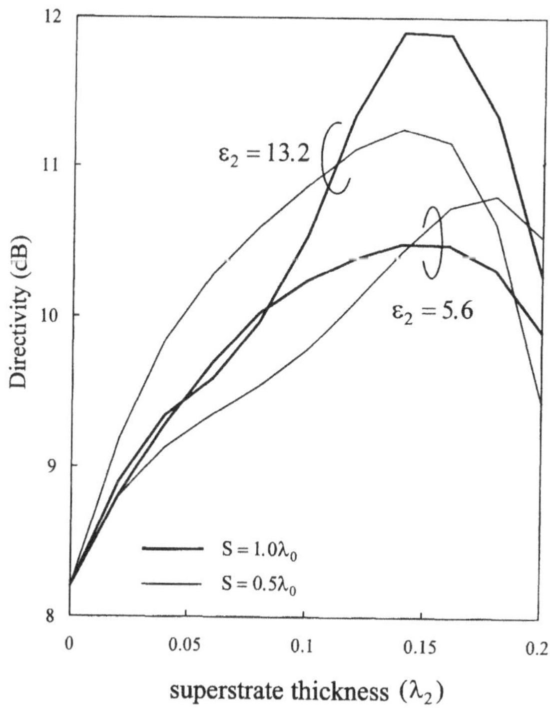

FIGURE 2.8 Dependence of directivity for the microstrip patch as a radiator on superstrate thickness; $\epsilon_1 = 2.3$, $h = 0.24$ mm, $2L = 8$ cm, $2b\phi_0 = 16.8$ cm, $a = 20$ cm.

no superstrate presence ($t = 0$). This suggests that the directive gain enhancement of the microstrip patch using a spaced superstrate is more effective for the planar microstrip patch than for the cylindrical microstrip patch.

2.4 CYLINDRICAL RECTANGULAR MICROSTRIP PATCH WITH AN AIR GAP

In Section 2.3 we presented the results of a microstrip patch covered with a spaced superstrate; that is, an air gap is present between the microstrip patch and the superstrate layer. This air region can also be placed between the microstrip patch and the substrate layer. The resulting microstrip structure (see Figure. 2.10) has a decreased quality factor, and the operating bandwidth for the microstrip patch as a radiator can thus be enhanced. Following the derivation procedure described in Section 2.2, the similar homogeneous matrix equation (2.77) can be obtained, whose solutions are also satisfied by complex frequencies. From the complex frequency calculated, information on the resonant frequency and half-power bandwidth of the structure is obtained.

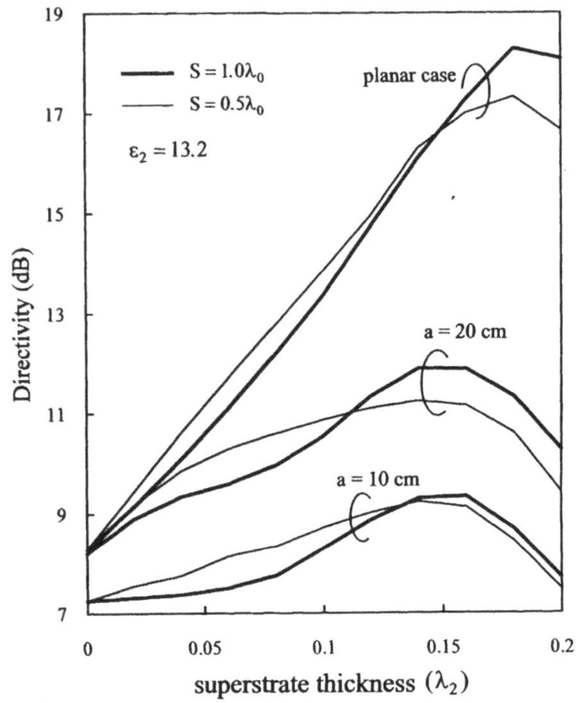

FIGURE 2.9 Dependence of directivity for the microstrip patch as a radiator on superstrate thickness; $\epsilon_1 = 2.3$, $h = 0.24$ mm, $2L = 8$ cm, $2b\phi_0 = 16.8$ cm.

2.4.1 Complex Resonant Frequency Results

The TM_{01} mode is studied next. The effect of introducing an air gap between the substrate and the ground plane is exemplified in Figure 2.11, where the real and imaginary parts of complex resonant frequencies are shown. The results for the planar case ($a = \infty$) are presented for comparison, which are calculated using a full-wave approach in Cartesian coordinates. For both cylindrical and planar structures, the two cases $\epsilon_1 = 2.3$, $\epsilon_s = 1.0$ (an air-gap layer) and $\epsilon_1 = 2.3$, $\epsilon_s = 2.3$ (a dielectric layer with the same parameters as the substrate layer; that is, the substrate thickness is now $S + h$) are presented. For the case of $\epsilon_1 = 2.3$, $\epsilon_s = 1.0$, it is seen that the resonant frequency increases quickly as S increases. This behavior is quite different from the case of $\epsilon_1 = 2.3$, $\epsilon_s = 2.3$. This is probably due to the fact that the effective permittivity of the region under the patch is lowered with the air-gap presence. However, when the air-gap thickness is large enough (about 5.2 mm for the typical case discussed here), the resonant frequency has a decreasing trend. This is probably because in this case the effective permittivity of the region under the patch varies slightly and the thickness of the region under the patch starts to dominate the effect, which reduces the resonant frequency. As for the imaginary part of complex resonant frequencies (Figure 2.11*b*), it is seen that

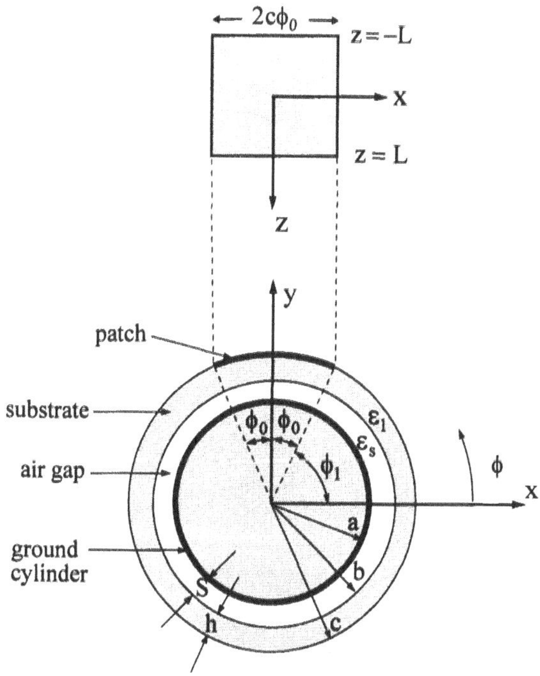

FIGURE 2.10 Geometry of a cylindrical rectangular microstrip structure with an air gap.

the radiating loss of the microstrip structure increases with increasing S. Also, the cylindrical structure is seen to be a more efficient radiator than the planar structure. The half-power bandwidth of the microstrip structure is presented in Figure 2.12. The bandwidth is found to be increased considerably due to the existence of an air gap, and the cylindrical structure also has a larger bandwidth than the planar structure.

2.5 CYLINDRICAL RECTANGULAR MICROSTRIP PATCH WITH A COUPLING SLOT

In this section we present a study of a slot-coupled rectangular microstrip structure as shown in Figure 2.13. In this structure a coupling slot is cut in the cylindrical ground plane and placed under the patch. This slot-coupling feed mechanism thus involves two substrates separated by a ground plane, one for the microstrip patch and another for the feed line. One can choose a low-permittivity substrate for the microstrip patch to increase its radiation efficiency, and on the other hand, select a high-permittivity substrate for the microstrip feed line to reduce its feed-energy loss. The electromagnetic energy is then coupled from the feed line to the microstrip patch through the coupling slot. Also, due to the presence of the

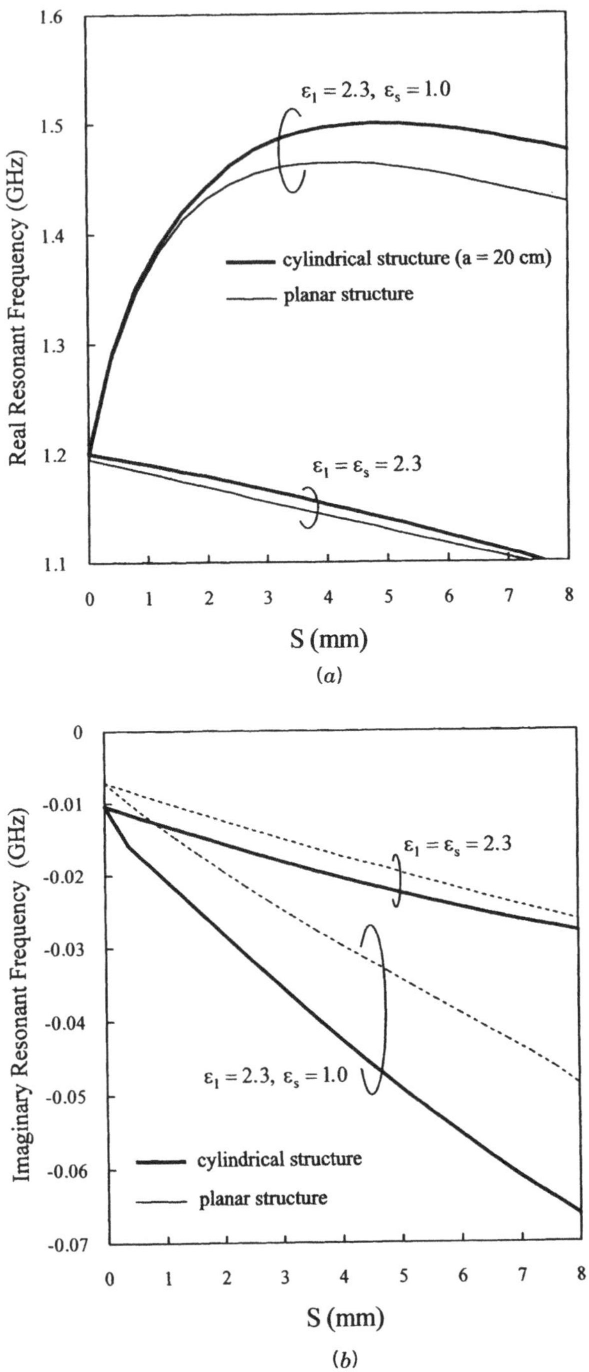

FIGURE 2.11 (*a*) Real and (*b*) imaginary parts of complex resonant frequencies as a function of spacing thickness; $\epsilon_1 = 2.3$, $h = 0.24$ mm, $\epsilon_s = 1.0$ (air gap) or 2.3, $2L = 8$ cm, $2b\phi_0 = 16.8$ cm. (From Ref. [6], © 1994 IEEE, reprinted with permission.)

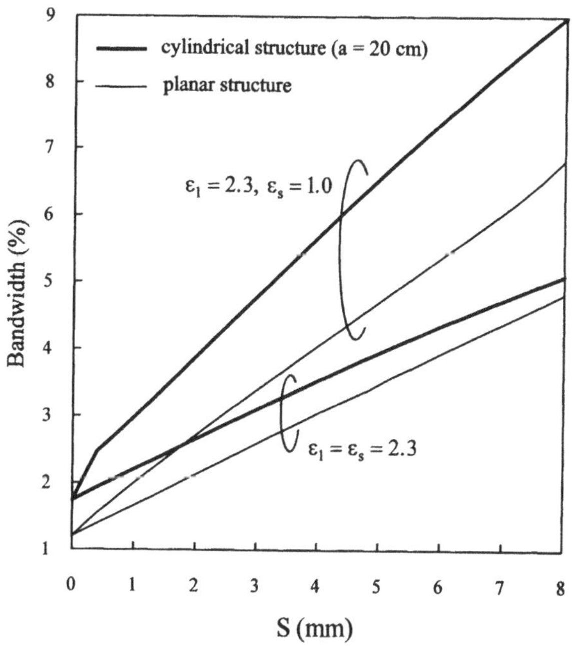

FIGURE 2.12 Half-power bandwidth for the case shown in Figure 2.11. (From Ref. [6], © 1994 IEEE, reprinted with permission.)

coupling slot, the resonant frequency of the microstrip structure is strongly affected. Accurate determination of the resonant frequency of the microstrip structure with a coupling slot is thus important. This problem can be solved using a full-wave approach. The theoretical formulation and results are given below.

2.5.1 Theoretical Formulation

The slot-coupled microstrip structure is depicted in Figure 2.13. The coupling slot of length L_s and width W_s is assumed to be narrow ($L_s \gg W_s$) and centered below the patch. This structure can be excited through the coupling slot by using a microstrip feed line printed on a substrate of thickness h_f and relative permittivity ϵ_f. On a substrate of thickness h and relative permittivity ϵ_1 is a rectangular patch of length $2L$ and width $2W$ ($=2b\phi_0$). Other regions ($\rho > b$ and $\rho < b_f$) are assumed to be air.

To begin, we assume that the coupling slot is electrically small ($L_s \ll \lambda_0$). In this case the electric field in the slot, $\mathbf{e}_z^a(\phi, z)$, can be given approximately by a single piecewise-sinusoidal (PWS) expansion mode with an unit amplitude; that is, we have

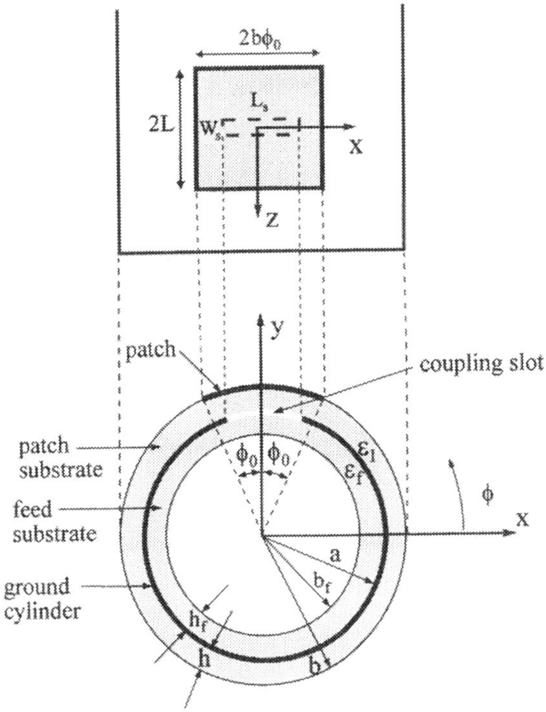

FIGURE 2.13 Geometry of a slot-coupled cylindrical rectangular microstrip structure.

$$\mathbf{e}_z^a(\phi, z) = \hat{z}\,\frac{\sin[k_e b(\phi_s - |\phi'|)]}{W_s \sin(k_e b\phi_s)}, \qquad |\phi'| < \phi_s, \quad |z| < \frac{W_s}{2}, \quad \phi' = \phi - \frac{\pi}{2}, \tag{2.83}$$

with

$$k_e = k_0\sqrt{\frac{\epsilon_1 + \epsilon_f}{2}}, \qquad k_0 = \omega\sqrt{\mu_0\epsilon_0}, \tag{2.84}$$

where ϕ_s is half the angle subtended by the coupling slot and k_e is the equivalent propagation constant in the slot considering the two dielectric substrates adjacent to the slot. Then, by invoking the equivalence principle, the slot can be closed off and replaced by two oppositely directed equivalent magnetic surface currents [$\mathbf{M}_\phi$ $(= \mathbf{e}_z^a \times \hat{\rho})$ and $-\mathbf{M}_\phi$ $(= -\mathbf{e}_z^a \times \hat{\rho})$] just above and below the ground plane. In this structure there are thus four current elements that need to be considered for derivation of the required Green's functions. These four current elements are:

1. J_z, the $\hat{z}$-directed component of the electric current on the patch.
2. M_ϕ, the $\hat{\phi}$-directed magnetic current just above the ground plane ($\rho = a^+$).

3. J_ϕ, the $\hat{\phi}$-directed component of the electric current on the patch.
4. $-M_\phi$, the $\hat{\phi}$-directed magnetic current just below the ground plane ($\rho = a^-$).

The required Green's functions in this study have been derived and given in [7], which are interpreted as follows:

1. $G^{EJ}_{\phi\phi}$: E_ϕ at $\rho = b$ due to a unit $\hat{\phi}$-directed electric current element at $\rho = b$.
2. $G^{EJ}_{\phi z}$: E_ϕ at $\rho = b$ due to a unit $\hat{z}$-directed electric current element at $\rho = b$.
3. $G^{EJ}_{z\phi}$: E_z at $\rho = b$ due to a unit $\hat{\phi}$-directed electric current element at $\rho = b$.
4. G^{EJ}_{zz}: E_z at $\rho = b$ due to a unit $\hat{z}$-directed electric current element at $\rho = b$.
5. $G^{EM}_{\phi\phi}$: E_ϕ at $\rho = b$ due to a unit $\hat{\phi}$-directed magnetic current element at $\rho = a^+$.
6. $G^{EM}_{z\phi}$: E_z at $\rho = b$ due to a unit $\hat{\phi}$-directed magnetic current element at $\rho = a^+$.
7. $G^{HM(p)}_{\phi\phi}$: H_ϕ at $\rho = a^+$ due to a unit $\hat{\phi}$-directed magnetic current element at $\rho = a^+$.
8. $G^{HM(f)}_{\phi\phi}$: H_ϕ at $\rho = a^-$ due to a unit $\hat{\phi}$-directed magnetic current element at $\rho = a^-$.
9. $G^{HJ}_{\phi\phi}$: H_ϕ at $\rho = a^+$ due to a unit $\hat{\phi}$-directed electric current element at $\rho = b$.
10. $G^{HJ}_{\phi z}$: H_ϕ at $\rho = a^+$ due to a unit $\hat{z}$-directed electric current element at $\rho = b$.

Next, by expanding the unknown surface current on the patch with a set of sinusoidal basis functions of (2.61)–(2.62) and imposing the boundary condition that the total E_ϕ and E_z fields, due to the excitation of the equivalent magnetic current on the slot and the electric current element on the patch, must vanish on the patch surface, we can obtain an electric-field integral equation. By further applying the Galerkin's moment-method formulation as described in Section 2.2.2, the following matrix equation is obtained:

$$\begin{bmatrix} (Z^{\phi\phi}_{ln})_{N\times N} & (Z^{\phi z}_{lm})_{N\times M} \\ (Z^{z\phi}_{kn})_{M\times N} & (Z^{zz}_{km})_{M\times M} \end{bmatrix} \begin{bmatrix} (I_{\phi n})_{N\times 1} \\ (I_{zm})_{M\times 1} \end{bmatrix} = \begin{bmatrix} (V^s_{\phi n})_{N\times 1} \\ (V^s_{zm})_{M\times 1} \end{bmatrix}, \tag{2.85}$$

where

$$Z^{\phi\phi}_{ln} = -2\pi b^2 \int_{-\infty}^{\infty} dk_z \sum_{\nu=-\infty}^{\infty} \tilde{J}_{\phi l}(-\nu, -k_z)\tilde{G}^{EJ}_{\phi\phi}(\nu, k_z)\tilde{J}_{\phi n}(\nu, k_z), \tag{2.86}$$

$$Z^{\phi z}_{lm} = -2\pi b^2 \int_{-\infty}^{\infty} dk_z \sum_{\nu=-\infty}^{\infty} \tilde{J}_{\phi l}(-\nu, -k_z)\tilde{G}^{EJ}_{\phi z}(\nu, k_z)\tilde{J}_{zm}(\nu, k_z), \tag{2.87}$$

$$Z_{kn}^{z\phi} = -2\pi b^2 \int_{-\infty}^{\infty} dk_z \sum_{\nu=-\infty}^{\infty} \tilde{J}_{zk}(-\nu, -k_z)\tilde{G}_{z\phi}^{EJ}(\nu, k_z)\tilde{J}_{\phi n}(\nu, k_z), \tag{2.88}$$

$$Z_{km}^{zz} = -2\pi b^2 \int_{-\infty}^{\infty} dk_z \sum_{\nu=-\infty}^{\infty} \tilde{J}_{zk}(-\nu, -k_z)\tilde{G}_{zz}^{EJ}(\nu, k_z)\tilde{J}_{zm}(\nu, k_z), \tag{2.89}$$

$$V_{\phi n}^{s} = 2\pi ab \int_{-\infty}^{\infty} dk_z \sum_{\nu=-\infty}^{\infty} \tilde{J}_{\phi n}(-\nu, -k_z)\tilde{G}_{\phi\phi}^{EM}(\nu, k_z)\tilde{e}_z^{a}(\nu, k_z), \tag{2.90}$$

$$V_{zm}^{s} = 2\pi ab \int_{-\infty}^{\infty} dk_z \sum_{\nu=-\infty}^{\infty} \tilde{J}_{zm}(-\nu, -k_z)\tilde{G}_{z\phi}^{EM}(\nu, k_z)\tilde{e}_z^{a}(\nu, k_z), \tag{2.91}$$

where $V_{\phi n}^{s}$ and V_{zm}^{s} are the voltages due to the excitation of the equivalent magnetic current on the slot, and the tilde in the equation denotes a Fourier transform. By solving (2.85), the unknown coefficients, $I_{\phi n}$ and I_{zm}, of the expansion basis functions can be obtained, and the patch surface current is determined.

Next, the input impedance Z_{in} $[= (Y^s + Y^p)^{-1}]$ seen looking into the microstrip structure at the slot is calculated to determine the resonant frequency of the slot-coupled microstrip structrure, where Y^s is the slot self-admittance and Y^p is the admittance due to the patch current contribution. The slot admittance can be calculated from the reaction of the slot electric field and the magnetic field on the patch side $[H_\phi^{s(p)}(\phi, z)]$ and the feed side $[H_\phi^{s(f)}(\phi, z)]$ of the cylindrical ground plane due to the equivalent magnetic current on the slot; that is,

$$\begin{aligned} Y^s &= Y^{s(p)} + Y^{s(f)} \\ &= \int_{S_a} e_z^a(\phi, z)[H_\phi^{s(p)}(\phi, z) + H_\phi^{s(f)}(\phi, z)]\, dS_a \\ &= 2\pi a^2 \int_{-\infty}^{\infty} \sum_{\nu=-\infty}^{\infty} \tilde{e}_z^a(-\nu, -k_z)\tilde{e}_z^a(\nu, k_z)[\tilde{G}_{\phi\phi}^{HM(p)}(\nu, k_z) + \tilde{G}_{\phi\phi}^{HM(f)}(\nu, k_z)]\, dk_z\,, \end{aligned} \tag{2.92}$$

where S_a is the slot area. For Y^p we have

$$\begin{aligned} Y^p &= \int_{S_a} e_z^a(\phi, z) H_\phi^p(\phi, z)\, dS_a \\ &= \begin{bmatrix} (V_{\phi n}^p)_{N\times 1} \\ (V_{zm}^p)_{M\times 1} \end{bmatrix}^{\mathrm{T}} \begin{bmatrix} (I_{\phi n})_{N\times 1} \\ (I_{zm})_{M\times 1} \end{bmatrix}, \end{aligned} \tag{2.93}$$

where

$$V_{\phi n}^{p} = 2\pi a^2 \int_{-\infty}^{\infty} \sum_{\nu=-\infty}^{\infty} \tilde{e}_z^a(\nu, k_z)\tilde{J}_{\phi n}(-\nu, -k_z)\tilde{G}_{\phi\phi}^{HJ}(\nu, k_z)\, dk_z\,, \tag{2.94}$$

$$V^{p}_{zm} = 2\pi a^2 \int_{-\infty}^{\infty} \sum_{\nu=-\infty}^{\infty} \tilde{e}^{a}_{z}(\nu, k_z)\tilde{J}_{zm}(-\nu, -k_z)\tilde{G}^{HJ}_{\phi z}(\nu, k_z)\, dk_z\,. \tag{2.95}$$

$H^{p}_{\phi}(\phi, z)$ is the magnetic field at the slot due to the patch surface current. It is also noted that since $\tilde{G}^{HJ}_{\phi\phi} = a\tilde{G}^{EM}_{\phi\phi}/b$ and $\tilde{G}^{HJ}_{\phi z} = a\tilde{G}^{EM}_{z\phi}/b$ [7], (2.93) can also be rewritten as

$$Y^{p} = \frac{a}{b}\begin{bmatrix} (V^{s}_{\phi n})_{N\times 1} \\ (V^{s}_{zm})_{M\times 1} \end{bmatrix}^{\mathrm{T}} \begin{bmatrix} (I_{\phi n})_{N\times 1} \\ (I_{zm})_{M\times 1} \end{bmatrix}. \tag{2.96}$$

2.5.2 Resonance Characteristics

Figure 2.14 presents typical results of numerical convergence of the moment-method calculation for the input impedance $Z_{\rm in}$ seen at the slot for $a = 7$ cm. The microstrip structure is excited in the TM_{01} mode, and the resonant frequency is determined to be the frequency where the input impedance $Z_{\rm in}$ $(= Y_{\rm in}^{-1})$ is real. Three cases of using $N = 1$, $M = 3$ (four basis functions), $N = 1$, $M = 5$ (six basis functions), and $N = 2$, $M = 6$ (eight basis functions) are shown, where N and M are, respectively, the numbers of basis functions selected for the $\hat{\phi}$- and $\hat{z}$-directed patch surface currents. Since for the TM_{01} mode, the excitation of the patch is mainly in the z direction, more basis functions in the z direction need to be selected and only one basis function is usually enough for modeling the excited patch surface current in the ϕ direction. From the results it is seen that when more than six basis functions are used, the calculated results have reached convergent solutions. The resonant frequency in the case is determined to be 2.194 GHz.

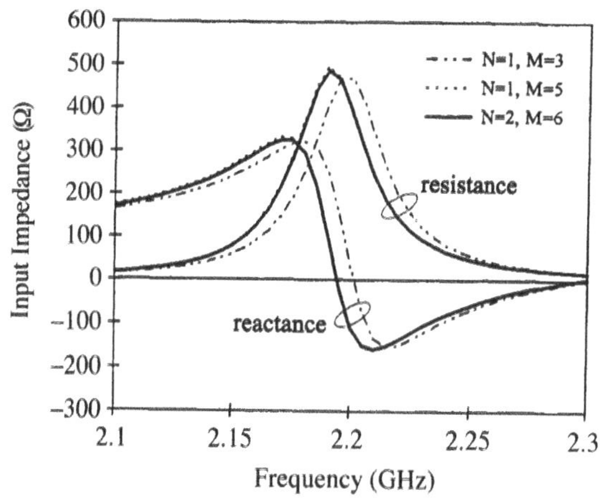

FIGURE 2.14 Input impedance of the microstrip structure, seen at the slot for the TM_{01} mode, calculated for different numbers of the expansion sinusoidal basis functions; $a = 7$ cm, $\epsilon_1 = \epsilon_f = 2.54$, $h = h_f = 1.6$ mm, $2L = 8$ cm, $2W = 6$ cm, $L_s = 14$ mm, $W_s = 1.1$ mm. (From Ref. [7], © 1994 John Wiley & Sons, Inc.)

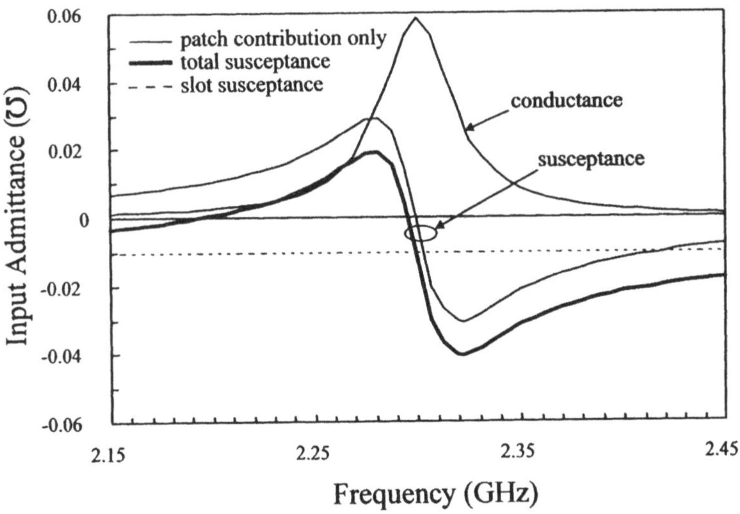

FIGURE 2.15 Behavior before and after adding the slot susceptance: Values of input conductance and susceptance seen at the slot (with and without slot susceptance) versus frequency. The parameters are as given in Figure 2.14.

Figure 2.15 presents the results of input conductance and susceptance seen at the slot (with and without slot contribution) versus frequency; the parameters are the same as given in Figure 2.14. It is clear that the effect of the coupling slot needs to be considered for accurate determination of the resonant frequency of a slot-coupled microstrip structure. Figure 2.16 shows the resonant frequency results versus slot length for different cylinder radii of $a = 7$, 10, and 15 cm. From the results it is seen that the resonant frequency decreases with increasing slot length. The resonant frequency is also increased with decreasing cylinder radius.

2.6 CYLINDRICAL TRIANGULAR MICROSTRIP PATCH

In this section the theoretical analysis of the cylindrical triangular microstrip structure using a full-wave formulation is presented. A set of basis functions similar to the true current distribution on the patch surface is selected for efficient and accurate numerical calculation. Numerical results for the complex resonant frequencies of the microstrip structure can then be calculated based on the theoretical approach described in Section 2.2. The formulation is described briefly below.

2.6.1 Theoretical Formulation

Figure 2.17 shows the geometry of a cylindrical triangular microstrip structure. The side length of the triangular patch in the ϕ direction is d_2 $(= 2b\phi_0)$, and the

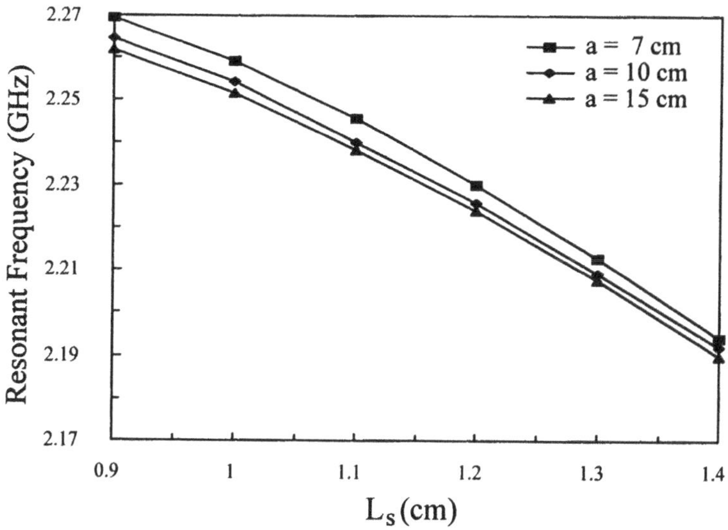

FIGURE 2.16 Variations of the resonant frequency with slot length; $a = 7$, 10, 15 cm, $\epsilon_1 = \epsilon_f = 2.54$, $h = h_f = 1.6$ mm, $2L = 8$ cm, $2W = 6$ cm, $W_s = 1.1$ mm. (From Ref. [7], © 1994 John Wiley & Sons, Inc.)

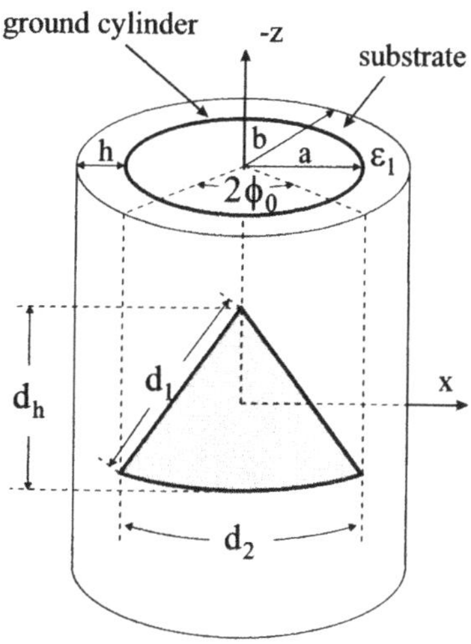

FIGURE 2.17 Geometry of a cylindrical triangular microstrip structure.

other two sides are assumed to be of the same length, d_1. The relationship of d_1 and d_2 is given by

$$d_1 = \sqrt{\left(\frac{d_2}{2}\right)^2 + d_h^2}, \tag{2.97}$$

where d_h is the distance from the tip of the triangle to the bottom side of the triangle. To begin with, we follow the full-wave formulation in Section 2.2 and the following equation is obtained:

$$\begin{bmatrix} \tilde{E}_\phi(n, k_z) \\ \tilde{E}_z(n, k_z) \end{bmatrix} = \tilde{\bar{\bar{G}}}(n, k_z) \begin{bmatrix} \tilde{J}_\phi(n, k_z) \\ \tilde{J}_z(n, k_z) \end{bmatrix}, \tag{2.98}$$

where $\tilde{\bar{\bar{G}}} = \hat{\phi}\tilde{G}_{\phi\phi}\hat{\phi} + \hat{\phi}\tilde{G}_{\phi z}\hat{z} + \hat{z}\tilde{G}_{z\phi}\hat{\phi} + \hat{z}\tilde{G}_{zz}\hat{z}$ is a dyadic Green's function in the spectral domain whose expression has been derived and interpreted in Section 2.2. E_ϕ and E_z are the tangential components of the electric field at $\rho = b$, and J_ϕ and J_z are the tangential components of the patch surface current density. To solve the unknown patch surface current, we select a set of basis functions satisfying the criterion that the normal component of the electric surface current density should vanish at the patch edges. The basis functions selected for the fundamental mode (TM_{10} mode [8]) excitation of the triangular microstrip patch are expressed as

$$\mathbf{J}_\phi(\phi, z) = \hat{\phi}\, \frac{b\phi'}{z' - d_h/2} \sin \frac{(z' + d_h/2)\pi}{d_h}, \tag{2.99}$$

$$\mathbf{J}_z(\phi, z) = \sin \frac{(z' + d_h/2)\pi}{d_h}, \tag{2.100}$$

where $\phi' = \phi - \pi/2$, $z' = -z$. In this case the unknown patch surface current density can be written as

$$\mathbf{J}(\phi, z) = I_\phi \mathbf{J}_\phi(\phi, z) + I_z \mathbf{J}_z(\phi, z)\,, \tag{2.101}$$

where I_ϕ and I_z are unknown coefficients to be determined. It should be noted that since the next-higher-order mode (TM_{11} mode) has a resonant frequency about 1.73 times that of the fundamental TM_{10} mode [8], the effects of the TM_{11} mode and other higher-order modes on the characteristics of the TM_{10} mode can be ignored. Thus it is appropriate to use only one basis function in the ϕ and z directions, respectively, to model the excited patch surface current density for the TM_{10} mode. Then, by taking the spectral amplitude of the basis function selected, imposing the boundary condition that the patch surface current and the electric field are complementary to each other at $\rho = b$ and using the basis function selected as a testing function and integrating over the patch area, we have

$$\begin{bmatrix} Z_{\phi\phi} & Z_{\phi z} \\ Z_{z\phi} & Z_{zz} \end{bmatrix} \begin{bmatrix} I_\phi \\ I_z \end{bmatrix} = \begin{bmatrix} 0 \\ 0 \end{bmatrix}, \tag{2.102}$$

where

$$Z_{\phi\phi} = \sum_{n=-\infty}^{\infty} \int_{-\infty}^{\infty} dk_z\, \tilde{J}_\phi(-n, -k_z) \tilde{G}_{\phi\phi}(n, k_z) \tilde{J}_\phi(n, k_z)\,, \tag{2.103}$$

$$Z_{\phi z} = \sum_{n=-\infty}^{\infty} \int_{-\infty}^{\infty} dk_z \, \tilde{J}_\phi(-n, -k_z)\tilde{G}_{\phi z}(n, k_z)\tilde{J}_z(n, k_z) \,, \tag{2.104}$$

$$Z_{z\phi} = \sum_{n=-\infty}^{\infty} \int_{-\infty}^{\infty} dk_z \, \tilde{J}_z(-n, -k_z)\tilde{G}_{z\phi}(n, k_z)\tilde{J}_\phi(n, k_z) \,, \tag{2.105}$$

$$Z_{zz} = \sum_{n=-\infty}^{\infty} \int_{-\infty}^{\infty} dk_z \, \tilde{J}_z(-n, -k_z)\tilde{G}_{zz}(n, k_z)\tilde{J}_z(n, k_z) \,, \tag{2.106}$$

with

$$\tilde{J}_\phi(n, k_z) = \frac{j}{2\pi} \int_{-d_h/2}^{d_h/2} dz' \, e^{-jk_z z'} \left\{ \frac{2\phi_0}{nd_h} \sin\left[\frac{\pi}{d_h}\left(z' + \frac{d_h}{2}\right)\right] \cos\left[\frac{n\phi_0}{d_h}\left(z' - \frac{d_h}{2}\right)\right] - \frac{2}{n^2(z' - d_h/2)} \sin\left[\frac{n\phi_0}{d_h}\left(z' - \frac{d_h}{2}\right)\right]\right\}. \tag{2.107}$$

As for $\tilde{J}_z(n, k_z)$ in (2.104) and (2.106), we have, for $n \neq 0$,

$$\tilde{J}_z(n, k_z) = \frac{-1}{nd_h} \left\{ \sin\frac{n\phi_0}{2} \left[\frac{\cos(k_z d_h/2 + n\phi_0/2)}{(k_z + n\phi_0/d_h)^2 - (\pi/d_h)^2} + \frac{\cos(k_z d_h/2 - n\phi_0/2)}{(k_z - n\phi_0/d_h)^2 - (\pi/d_h)^2} \right] - j\cos\frac{n\phi_0}{2} \left[\frac{\cos(k_z d_h/2 + n\phi_0/2)}{(k_z + n\phi_0/d_h)^2 - (\pi/d_h)^2} - \frac{\cos(k_z d_h/2 - n\phi_0/2)}{(k_z - n\phi_0/d_h)^2 - (\pi/d_h)^2} \right] \right\}, \tag{2.108}$$

and for $n = 0$,

$$\tilde{J}_z(n, k_z) = \frac{-\phi_0}{\pi d_h} \left\{ \frac{j}{(\pi/d_h)^2 - k_z^2} \left[\frac{4k_z \pi \cos(k_z d_h/2)}{d_h[(\pi/d_h)^2 - k_z^2]} - \pi \sin\frac{k_z d_h}{2} \right] - \frac{\pi \cos(k_z d_h/2)}{(\pi/d_h)^2 - k_z^2} \right\}. \tag{2.109}$$

To have nontrivial solutions for I_ϕ and I_z in (2.102), it is required that

$$\det\begin{bmatrix} Z_{\phi\phi} & Z_{\phi z} \\ Z_{z\phi} & Z_{zz} \end{bmatrix} = 0 \,. \tag{2.110}$$

By solving (2.110), the complex resonant frequencies ($f = f' + jf''$) of the cylindrical triangular microstrip structure are obtained, where f' and f'' represent, respectively, the resonant frequency and radiation loss of the microstrip structure, and $f'/2f''$ is the quality factor of the microstrip structure.

2.6.2 Complex Resonant Frequency Results

The calculated complex resonant frequencies of the cylindrical triangular microstrip structure at the fundamental mode are presented. Figure 2.18 shows the results for an equilateral triangular microstrip patch with a side length of 5.78 cm mounted on ground cylinders having $a = 7.8$, 15, and 20 cm. Measured data for the resonant frequencies of the triangular patch printed on a substrate of thickness $h = 0.762$ mm and relative permittivity $\epsilon_1 = 3.0$ are shown for comparison. Good agreement between the measured and calculated results is obtained. From the results it is seen that the resonant frequency (the real part of the complex resonant frequency) increases with decreasing cylinder radius. This behavior is due primarily to the decrease in effective patch size when the microstrip patch is mounted on a ground cylinder of smaller radius. As for the imaginary part of the complex resonant frequency, it is seen that it also increases with decreasing curvature. [It should be noted that the imaginary resonant frequencies shown in Figure 2.18 are their absolute value; the results calculated have a negative sign with $\exp(-j\omega t)$ assumption.] This implies that the radiation loss of the microstrip structure increases when the cylindrical triangular patch has a smaller cylinder radius. This behavior is similar to that observed for cylindrical rectangular

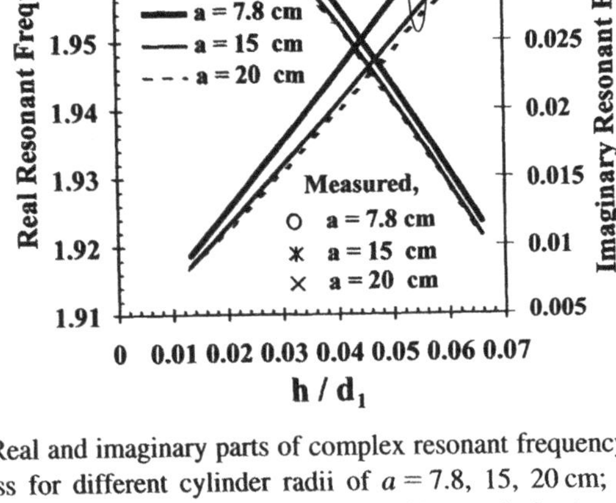

FIGURE 2.18 Real and imaginary parts of complex resonant frequency versus normalized substrate thickness for different cylinder radii of $a = 7.8$, 15, 20 cm; $\epsilon_1 = 3.0$, $d_1 = d_2 =$ 5.78 cm. (From Ref. [9], © 1997 IEEE, reprinted with permission.)

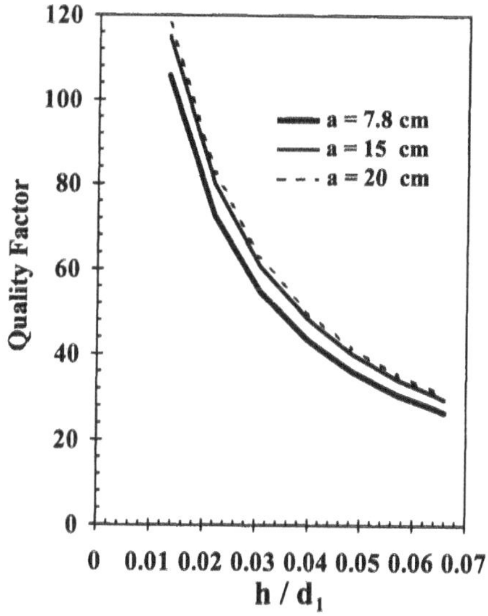

FIGURE 2.19 Quality factor versus normalized substrate thickness for the case shown in Figure. 2.18. (From Ref. [9], © 1997 IEEE, reprinted with permission.)

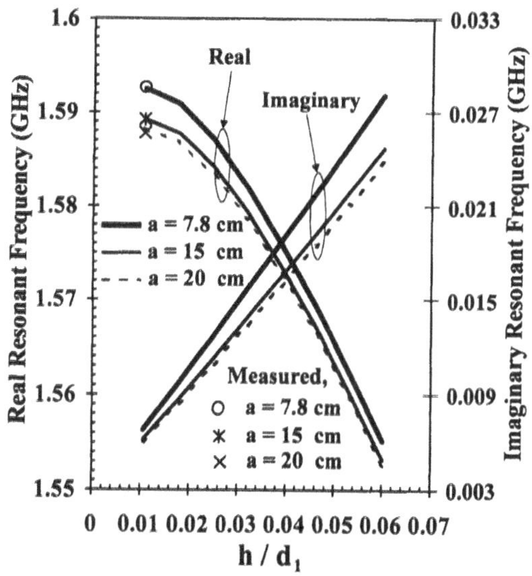

FIGURE 2.20 Real and imaginary parts of complex resonant frequency versus normalized substrate thickness for different cylinder radii of $a = 7.8$, 15, 20 cm; $\epsilon_1 = 3.0$, $d_1 = d_2 = 7.18$ cm. (From Ref. [9], © 1997 IEEE, reprinted with permission.)

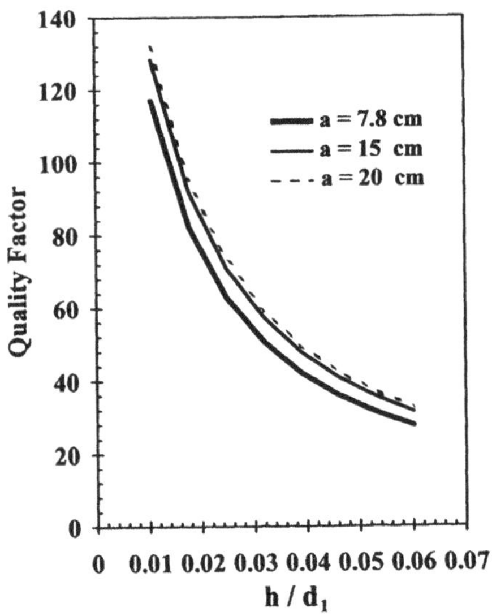

FIGURE 2.21 Quality factor versus normalized substrate thickness for the case shown in Figure. 2.20. (From Ref. [9], © 1997 IEEE, reprinted with permission.)

microstrip structures. From the complex resonant frequency results, the quality factor of the microstrip structure can readily be calculated and is presented in Figure 2.19. It is seen that the quality factor decreases with decreasing cylinder radius. This suggests that a cylindrical triangular microstrip structure is more suitable than a planar triangular microstrip structure for use as a radiator. Another case of an equilateral triangular patch with a side length of 7.18 cm has also been studied, with the results presented in Figures 2.20 and 2.21. Behavior similar to that shown in Figures 2.18 and 2.19 is also observed. The measured resonant frequencies are in good agreement with the results calculated.

2.7 CYLINDRICAL WRAPAROUND MICROSTRIP PATCH

The cylindrical wraparound microstrip structure is also a commonly used cylindrical microstrip structure. The geometry is depicted in Figure 2.22, where a wraparound patch in a substrate–superstrate geometry is shown. Resonance problems of the microstrip structure as a radiator in the TM_{01} mode and a resonator in the TM_{10} mode are solved in this section. For TM_{01}-mode operation, the wraparound patch antenna has an omnidirectional radiation characteristic, which is very suitable for use on missiles, rockets, and satellites to provide omnidirectional coverage.

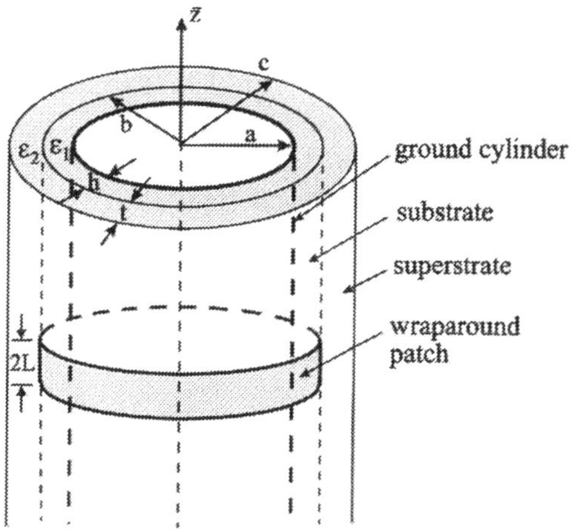

FIGURE 2.22 Geometry of a cylindrical wraparound microstrip structure with a superstrate.

2.7.1 Theoretical Formulation

As shown by the geometry given in Figure 2.22, the wraparound patch has a dimension of $2L \times 2\pi b$. By applying the formulation in Section 2.2, the following vector integral equation can be obtained: On the wraparound patch,

$$\begin{bmatrix} E_\phi(\phi, z) \\ E_z(\phi, z) \end{bmatrix} = \frac{1}{2\pi} \sum_{n=-\infty}^{\infty} e^{jn\phi} \int_{-\infty}^{\infty} dk_z \, e^{jk_z z} \begin{bmatrix} \tilde{G}_{\phi\phi} & \tilde{G}_{\phi z} \\ \tilde{G}_{z\phi} & \tilde{G}_{zz} \end{bmatrix} \begin{bmatrix} \tilde{J}_{\phi n}(k_z) \\ \tilde{J}_{zn}(k_z) \end{bmatrix} = \begin{bmatrix} 0 \\ 0 \end{bmatrix}, \tag{2.111}$$

and outside the wraparound patch,

$$\begin{bmatrix} J_\phi(\phi, z) \\ J_z(\phi, z) \end{bmatrix} = \frac{1}{2\pi} \sum_{n=-\infty}^{\infty} e^{jn\phi} \int_{-\infty}^{\infty} dk_z \, e^{jk_z z} \begin{bmatrix} \tilde{J}_{\phi n}(k_z) \\ \tilde{J}_{zn}(k_z) \end{bmatrix} = \begin{bmatrix} 0 \\ 0 \end{bmatrix}, \tag{2.112}$$

where $\tilde{G}_{\phi\phi}$, $\tilde{G}_{\phi z}$, $\tilde{G}_{z\phi}$, and $\tilde{G}_{zz}$ have been interpreted and expressed in Section 2.2.1, and other variables have meanings similar to those discussed in previous sections. To apply Galerkin's moment method, we expand the unknown surface current density on the wraparound patch in terms of known sinusoidal basis functions satisfying the edge-singularity condition; that is,

$$\mathbf{J}(\phi, z) = \sum_{n=1}^{N} I_{\phi n} \mathbf{J}_{\phi n}(\phi, z) + \sum_{m=1}^{M} I_{zm} \mathbf{J}_{zm}(\phi, z), \tag{2.113}$$

with

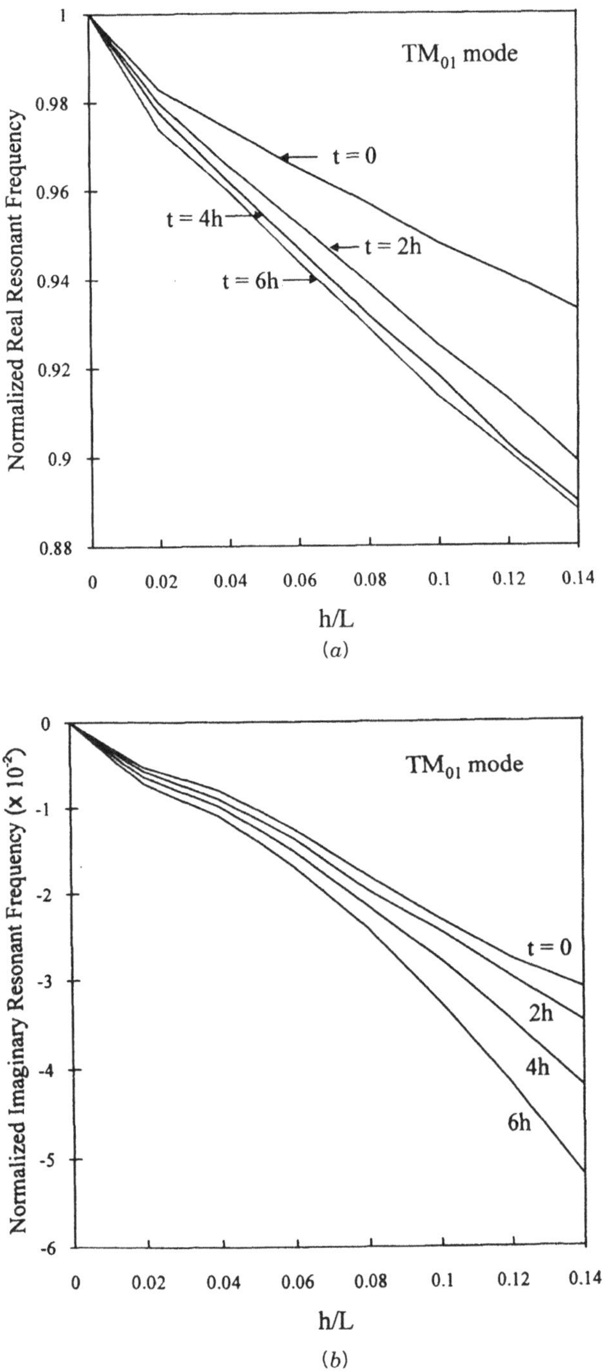

FIGURE 2.23 (*a*) Real and (*b*) imaginary parts of complex resonant frequencies as a function of superstrate thickness for the TM_{01} mode; $a = 20$ cm, $\epsilon_1 = 2.3$, $\epsilon_2 = 2.3$, $2L = 8$ cm. (From Ref. [10], © 1994 IEEE, reprinted with permission.)

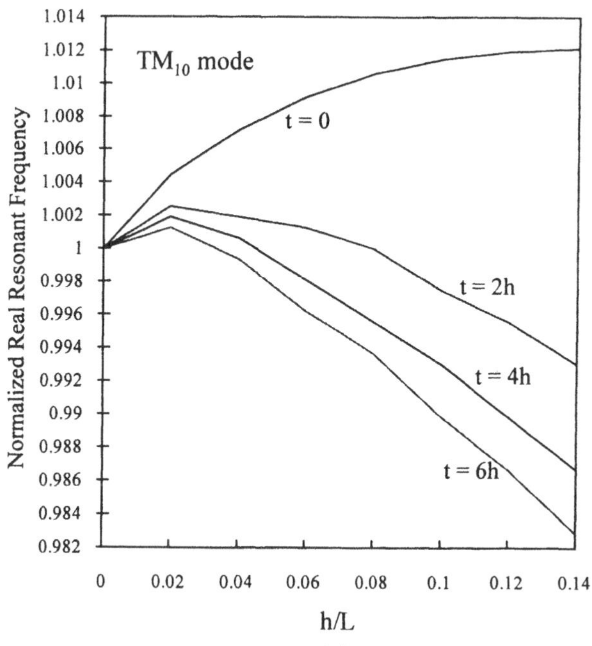

(*a*)

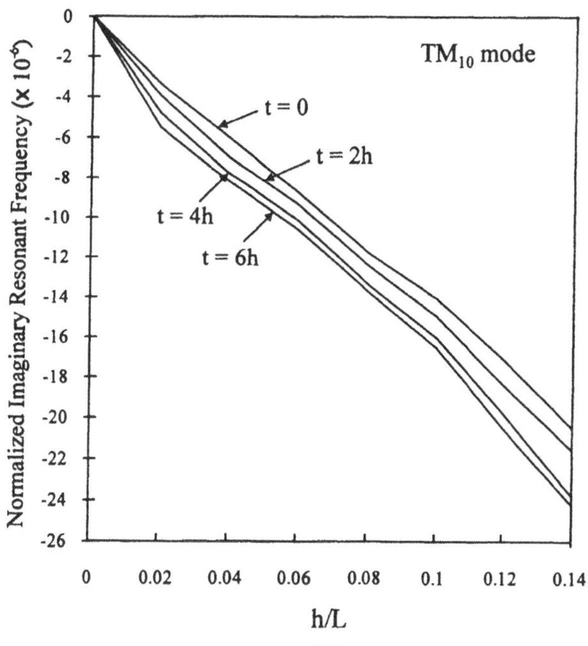

(*b*)

FIGURE 2.24 (*a*) Real and (*b*) imaginary parts of complex resonant frequencies as a function of superstrate thickness for the TM_{10} mode; $a = 20$ cm, $\epsilon_1 = 2.3$, $\epsilon_2 = 2.3$, $2L = 8$ cm. (From Ref. [10], © 1994 IEEE, reprinted with permission.)

$$\mathbf{J}_{\phi n}(\phi, z) = \hat{\phi} \frac{e^{jp\phi}}{\sqrt{L^2 - z^2}} \cos\left[\frac{q\pi}{2L}(z + L)\right], \tag{2.114}$$

$$\mathbf{J}_{zm}(\phi, z) = \hat{z} e^{jr\phi} \sin\left[\frac{s\pi}{2L}(z + L)\right], \tag{2.115}$$

where $I_{\phi n}$ and I_{zm} are coefficients for the basis functions selected in the ϕ and z directions, respectively. Next, by taking the spectral amplitude of the basis functions selected and substituting into (2.111), using these basis functions as testing functions, and integrating over the wraparound patch area, a homogeneous matrix equation with the same form as (2.72) can be derived. From this matrix equation, the resonant frequency and radiation loss of the microstrip structure are solved.

2.7.2 Complex Resonant Frequency Results

Figure 2.23 presents the results of the TM_{01} mode for various superstrate thicknesses. The basis functions of (2.114)–(2.115) with $N = 1$, $M = 2$ are used in the moment-method calculation, which shows good convergent solutions. From the results of Figure 2.23*a*, it is seen that when a superstrate layer is loaded, a significant decrease in the resonant frequency is observed. However, when the superstrate thickness is greater than $4h$, the decrease in the resonant frequency due to superstrate loading becomes small. As for the imaginary part of complex resonant frequency shown in Figure 2.23*b*, the results obtained also reveal that the radiation loss in the microstrip structure is increased with the loading of a superstrate layer.

Results for the TM_{10} mode are presented in Figure 2.24. It is seen that when the superstrate layer is absent, the resonant frequency increases with increasing substrate thickness. However, when a superstrate layer is added, the resonant frequency increases slightly for thin substrates and then decreases for thick substrates. It is also observed that due to the superstrate loading, the maximum variation of the resonant frequency shift for the TM_{10} mode shown in Figure 2.24*a* is only about 2%, which is much smaller than the corresponding resonant frequency shift (about 11%, see Figure 2.23*a*) of the TM_{01} mode. This suggests that superstrate loading has a relatively smaller effect for the cylindrical wraparound microstrip structure as a resonator in the TM_{10} mode. As for the imaginary part of resonant frequency, the result again increases with increasing superstrate loading, which implies that superstrate loading increases the radiation loss of the microstrip structure, similar to that observed for the TM_{10} mode.

REFERENCES

1. R. F. Harrington, *Field Computation by Moment Method*, Macmillan, New York, 1968.
2. H. J. Lin, Curve-Fitting Formulas for Fast Determination of Accurate Resonant

Frequency of a Rectangular Microstrip Patch Antenna, M.S. thesis, National Sun Yat-Sen University, Kaohsiung, Taiwan, 1993.

3. K. L. Wong, Y. T. Cheng, and J. S. Row, ''Resonance in a superstrate-loaded cylindrical-rectangular microstrip structure,'' *IEEE Trans. Microwave Theory Tech.*, vol. 41, pp. 814–819, May 1993.
4. K. L. Wong, Y. T. Cheng, and J. S. Row, ''Resonance and radiation of a superstrate-loaded cylindrical-rectangular microstrip patch antenna with an airgap,'' *Proc. Natl. Sci. Counc. ROC(A)*, vol. 17, pp. 365–371, Sept. 1993.
5. R. Afzalzadeh and R. N. Karekar, ''X-band directive microstrip patch antenna using dielectric parasite,'' *Electron. Lett.*, vol. 28, pp. 17–19, Jan. 2, 1992.
6. K. L. Wong, Y. T. Cheng, and J. S. Row, ''Analysis of a cylindrical-rectangular microstrip structure with an airgap,'' *IEEE Trans. Microwave Theory Tech.*, vol. 42, pp. 1032–1037, June 1994.
7. K. L. Wong and Y. C. Chen, ''Resonant frequency of a slot-coupled cylindrical-rectangular microstrip structure,'' *Microwave Opt. Technol. Lett.*, vol. 7, pp. 566–570, Aug. 20, 1994.
8. J. Helszajn and D. S. James, ''Planar triangular resonators with magnetic walls,'' *IEEE Trans. Microwave Theory Tech.*, vol. 26, pp. 95–100, Feb. 1978.
9. K. L. Wong and S. C. Pan, ''Resonance in a cylindrical-triangular microstrip structure,'' *IEEE Trans. Microwave Theory Tech.*, vol. 45, pp. 1270–1272, Aug. 1997.
10. K. L. Wong, R. B. Tsai, and J. S. Row, ''Resonance in a cylindrical wraparound microstrip structure with superstrate,'' *IEEE Trans. Microwave Theory Tech.*, vol. 42, pp. 1097–1100, June 1994.

CHAPTER THREE

Resonance Problem of Spherical Microstrip Patches

3.1 INTRODUCTION

In this chapter we present another canonical conformal microstrip structure: microstrip patches mounted on a spherical ground surface. Such a spherical microstrip structure provides more freedom of curvature variation in the θ and ϕ directions than does the cylindrical microstrip structure, which has only one degree of curvature freedom in the ϕ direction. Among microstrip patches of various shapes, circular and annular-ring patches are most suitable for employment on a spherical host; thus spherical circular and annular-ring microstrip structures are analyzed here. Rigorous theoretical formulation for the analysis of spherical microstrip structures is described, and the complex resonant frequency of the microstrip structure is then calculated and discussed. Curvature effects on radiation patterns and scattering characteristics of spherical microstrip antennas are analyzed. Also, cases with superstrate dielectric loading and an air gap between the substrate layer and the ground sphere are included in the study.

In addition, since some substrates used for the fabrication of printed antennas exhibit a considerable amount of uniaxial anisotropy, a study of uniaxially anisotropic substrates is included. The uniaxial anisotropy of the substrate is either natural or due to the shear introduced in the surface plane during processing [1]. It is well known that uniaxial anisotropy can affect the performance of printed circuits and antennas, and thus accurate characterization and design must account for this effect, which is discussed in this chapter.

3.2 SPHERICAL CIRCULAR MICROSTRIP PATCH ON A UNIAXIAL SUBSTRATE

In this section we present a full-wave analysis of a spherical circular microstrip structure on a uniaxial substrate. Wave propagation inside a uniaxially anisotropic medium is first derived in spherical coordinates. We start with Maxwell's equations to obtain wave equations inside a uniaxial substrate in spherical coordinates. The wave equations obtained are then solved for both the source-free and ideal point-source cases. Based on the obtained results, a formulation for the microstrip structure with a uniaxial substrate can easily be derived. Numerical calculations are then performed using Galerkin's moment method, and complex resonant frequency, radiation characteristics, and scattering behavior are investigated.

3.2.1 Fundamental Wave Equations in a Uniaxial Medium

In this section, wave propagation inside a uniaxial medium is discussed in spherical coordinates. The uniaxial medium is characterized by a free-space permeability μ_0 and a permittivity tensor of the form

$$\bar{\bar{\epsilon}} = \epsilon_0 \begin{bmatrix} \epsilon_r & 0 & 0 \\ 0 & \epsilon_\theta & 0 \\ 0 & 0 & \epsilon_\theta \end{bmatrix}, \tag{3.1}$$

where ϵ_r and ϵ_θ are the relative permittivities perpendicular and parallel to a spherical surface of the uniaxial medium, and ϵ_0 is the free-space permittivity. In this case we have the following relationship for the electric flux $\mathbf{D}$ and magnetic flux $\mathbf{B}$:

$$\mathbf{D} = \bar{\bar{\epsilon}} \cdot \mathbf{E}, \qquad \mathbf{B} = \mu_0 \mathbf{H}. \tag{3.2}$$

The electric ($\mathbf{J}$) and magnetic ($\mathbf{M}$) current sources are also assumed to be within the medium, and we have the two curl Maxwell's equations written as [assuming $\exp(j\omega t)$ time dependence]

$$\nabla \times \mathbf{E} = -j\omega \mathbf{B} - \mathbf{M}, \tag{3.3}$$

$$\nabla \times \mathbf{H} = j\omega \mathbf{D} + \mathbf{J}, \tag{3.4}$$

where $\mathbf{B}$ and $\mathbf{D}$ are, respectively, the magnetic flux and electric flux. By substituting (3.2) into (3.3)–(3.4) and dividing the total field into two parts, one produced by $\mathbf{J}$ and the other by $\mathbf{M}$; that is,

$$\mathbf{D} = \mathbf{D}_J + \mathbf{D}_M, \qquad \mathbf{B} = \mathbf{B}_J + \mathbf{B}_M, \tag{3.5}$$

we have

$$\nabla \times (\bar{\bar{\epsilon}}^{-1} \cdot \mathbf{D}_J) = -j\omega \mathbf{B}_J, \tag{3.6}$$

$$\nabla \times \mathbf{B}_J = j\omega\mu_0 \mathbf{D}_J + \mu_0 \mathbf{J}, \tag{3.7}$$

and

$$\nabla \times (\bar{\bar{\epsilon}}^{-1} \cdot \mathbf{D}_M) = -j\omega \mathbf{B}_M - \mathbf{M}, \tag{3.8}$$

$$\nabla \times \mathbf{B}_M = j\omega\mu_0 \mathbf{D}_M. \tag{3.9}$$

By taking the divergence operations on (3.6) and (3.9), it can be seen that the divergences of both $\mathbf{B}_J$ and $\mathbf{D}_M$ are identically zero from the vector identities, and we have

$$\mathbf{B}_J = \nabla \times \boldsymbol{\phi}_J, \tag{3.10}$$

$$\mathbf{D}_J = \frac{1}{j\omega\mu_0}(\nabla \times \nabla \times \boldsymbol{\phi}_J - \mu_0 \mathbf{J}), \tag{3.11}$$

for the waves generated by $\mathbf{J}$, and

$$\mathbf{D}_M = -\nabla \times \boldsymbol{\phi}_M, \tag{3.12}$$

$$\mathbf{B}_M = \frac{1}{j\omega}[\nabla \times (\bar{\bar{\epsilon}}^{-1} \cdot (\nabla \times \boldsymbol{\phi}_M)) - \mathbf{M}], \tag{3.13}$$

for the waves generated by $\mathbf{M}$; $\boldsymbol{\phi}_J$ and $\boldsymbol{\phi}_M$ are electric and magnetic vector potentials, respectively. The complete field equations are, therefore, the sum of (3.10)–(3.13); that is,

$$\mathbf{D} = -\nabla \times \boldsymbol{\phi}_M + \frac{1}{j\omega\mu_0}(\nabla \times \nabla \times \boldsymbol{\phi}_J - \mu_0 \mathbf{J}), \tag{3.14}$$

$$\mathbf{B} = \nabla \times \boldsymbol{\phi}_J + \frac{1}{j\omega}[\nabla \times (\bar{\bar{\epsilon}}^{-1} \cdot (\nabla \times \boldsymbol{\phi}_M)) - \mathbf{M}]. \tag{3.15}$$

By further substituting (3.10)–(3.13) into (3.6) and (3.9) and introducing two arbitrary scalar functions, f_J and f_M, which satisfy

$$\nabla \times \nabla \times \boldsymbol{\phi}_J - \omega^2 \mu_0 \bar{\bar{\epsilon}} \cdot \boldsymbol{\phi}_J = \bar{\bar{\epsilon}} \cdot \nabla f_J + \mu_0 \mathbf{J} \tag{3.16}$$

and

$$\nabla \times [\bar{\bar{\epsilon}}^{-1} \cdot (\nabla \times \boldsymbol{\phi}_M)] - \omega^2 \mu_0 \boldsymbol{\phi}_M = \nabla f_M + \mathbf{M}, \tag{3.17}$$

the expressions of (3.14) and (3.15) can, respectively, be rewritten as

$$\mathbf{D} = -\nabla \times \boldsymbol{\phi}_M + \frac{1}{j\omega\mu_0}(\bar{\bar{\epsilon}} \cdot \nabla f_J + \omega^2 \mu_0 \bar{\bar{\epsilon}} \cdot \boldsymbol{\phi}_J), \tag{3.18}$$

$$\mathbf{B} = \nabla \times \boldsymbol{\phi}_J + \frac{1}{j\omega}[\nabla f_M + \omega^2 \mu_0 \boldsymbol{\phi}_M]. \tag{3.19}$$

Also note that (3.16) and (3.17) can be considered as the vector wave equations for $\boldsymbol{\phi}_J$ and $\boldsymbol{\phi}_M$. To solve these two wave equations, we still need to specify f_J and f_M in the equations, which will be given in the next section.

3.2.2 Spherical Wave Functions in a Uniaxial Medium

In spherical coordinates, an arbitrary field is usually expressed as the sum of a TM-to-$\hat{r}$ wave and a TE-to-$\hat{r}$ wave. To form the TM- and TE-to-$\hat{r}$ waves, some particular choices of the vector potentials and sources are considered here. To formulate the TM-to-$\hat{r}$ wave, we assume that $\mathbf{J} = \hat{r}J_r$ and $\boldsymbol{\phi}_J = \hat{r}\phi_{\mathrm{TM}}$ and substitute them into the vector wave equation of (3.16). By expanding (3.16) in spherical coordinates and taking account of the permittivity tensor $\bar{\bar{\epsilon}}$ given in (3.1), it can be seen that the vector equation of (3.16) can be written as three scalar equations for the r, θ, and ϕ components; that is,

$$-\nabla_t^2 \phi_{\mathrm{TM}} - k_0^2 \epsilon_r \phi_{\mathrm{TM}} = \epsilon_0 \epsilon_r \frac{\partial f_J}{\partial r} + \mu_0 J_r\,, \tag{3.20}$$

$$\frac{1}{r}\frac{\partial^2 \phi_{\mathrm{TM}}}{\partial r\, \partial\theta} = \epsilon_0 \epsilon_\theta \frac{1}{r}\frac{\partial f_J}{\partial \theta}\,, \tag{3.21}$$

$$\frac{1}{r\sin\theta}\frac{\partial^2 \phi_{\mathrm{TM}}}{\partial r\, \partial\phi} = \epsilon_0 \epsilon_\theta \frac{1}{r\sin\theta}\frac{\partial f_J}{\partial \phi}\,, \tag{3.22}$$

where k_0 $(= \omega\sqrt{\mu_0\epsilon_0})$ is the wavenumber in air, and

$$\nabla_t^2 = \frac{1}{r^2\sin\theta}\frac{\partial}{\partial\theta}\left(\sin\theta\frac{\partial}{\partial\theta}\right) + \frac{1}{r^2\sin^2\theta}\frac{\partial^2}{\partial\phi^2}\,. \tag{3.23}$$

By comparing (3.20) and (3.21), we may take

$$f_J = \frac{1}{\epsilon_0\epsilon_\theta}\frac{\partial \phi_{\mathrm{TM}}}{\partial r}\,, \tag{3.24}$$

which is the condition that needs to be specified, as mentioned in Section 3.2.1. By substituting (3.24) into (3.20), a scalar wave equation for ϕ_{TM} can be obtained:

$$\frac{\epsilon_r}{\epsilon_\theta}\frac{\partial^2 \phi_{\mathrm{TM}}}{\partial r^2} + \nabla_t^2 \phi_{\mathrm{TM}} + k_0^2\epsilon_r\phi_{\mathrm{TM}} = -\mu_0 J_r\,. \tag{3.25}$$

Similarly, by taking the same procedure for (3.17)—substituting $\mathbf{M} = \hat{r}M_r$ and $\boldsymbol{\phi}_M = \hat{r}\phi_{\mathrm{TE}}$ into (3.17), expanding the resulting equation in spherical coordinates, and separating the r, θ, and ϕ components into three scalar equations—we have

$$-\nabla_t^2 \phi_{\mathrm{TE}} - k_0^2 \epsilon_\theta \phi_{\mathrm{TE}} = \epsilon_0 \epsilon_\theta \frac{\partial f_M}{\partial r} + \epsilon_0\epsilon_\theta M_r\,, \tag{3.26}$$

$$\frac{1}{r}\frac{\partial^2 \phi_{\mathrm{TE}}}{\partial r\,\partial\theta} = \epsilon_0\epsilon_\theta \frac{1}{r}\frac{\partial f_M}{\partial\theta}, \tag{3.27}$$

$$\frac{1}{r\sin\theta}\frac{\partial^2 \phi_{\mathrm{TE}}}{\partial r\,\partial\phi} = \epsilon_0\epsilon_\theta \frac{1}{r\sin\theta}\frac{\partial f_M}{\partial\phi}. \tag{3.28}$$

Solving (3.26)–(3.28), the scalar wave equation for ϕ_{TE} can be obtained:

$$\frac{\partial^2 \phi_{\mathrm{TE}}}{\partial r^2} + \nabla_t^2 \phi_{\mathrm{TE}} + k_0^2\epsilon_\theta\phi_{\mathrm{TE}} = -\epsilon_0\epsilon_\theta M_r, \tag{3.29}$$

with the condition

$$f_M = \frac{1}{\epsilon_0\epsilon_\theta}\frac{\partial\phi_{\mathrm{TE}}}{\partial r}. \tag{3.30}$$

From (3.25) and (3.29), ϕ_{TM} and ϕ_{TE} can be solved, and the components of **E** and **H** can then be obtained from

$$E_r = \frac{1}{Q}\left(\frac{\partial^2}{\partial r^2} + k_0^2\epsilon_\theta\right)\phi_{\mathrm{TM}}, \tag{3.31}$$

$$E_\theta = \frac{-1}{\epsilon_0\epsilon_\theta r\sin\theta}\frac{\partial\phi_{\mathrm{TE}}}{\partial\phi} + \frac{1}{Qr}\frac{\partial^2\phi_{\mathrm{TM}}}{\partial r\,\partial\theta}, \tag{3.32}$$

$$E_\phi = \frac{1}{\epsilon_0\epsilon_\theta r}\frac{\partial\phi_{\mathrm{TE}}}{\partial\theta} + \frac{1}{Qr\sin\theta}\frac{\partial^2\phi_{\mathrm{TM}}}{\partial r\,\partial\phi}, \tag{3.33}$$

$$H_r = \frac{1}{Q}\left(\frac{\partial^2}{\partial r^2} + k_0^2\epsilon_\theta\right)\phi_{\mathrm{TE}}, \tag{3.34}$$

$$H_\theta = \frac{1}{Qr}\frac{\partial^2\phi_{\mathrm{TE}}}{\partial r\,\partial\theta} + \frac{1}{\mu_0 r\sin\theta}\frac{\partial\phi_{\mathrm{TM}}}{\partial\phi}, \tag{3.35}$$

$$H_\phi = \frac{1}{Qr\sin\theta}\frac{\partial^2\phi_{\mathrm{TE}}}{\partial r\,\partial\phi} + \frac{-1}{\mu_0 r}\frac{\partial\phi_{\mathrm{TM}}}{\partial\theta}, \tag{3.36}$$

$$Q = j\omega\mu_0\epsilon_0\epsilon_\theta.$$

Now let's go on with solving the two scalar wave equations of (3.25) and (3.29). Examining the wave equations, it is seen that both ϵ_r and ϵ_θ are included in the equation for ϕ_{TM}, while only ϵ_θ appears in the equation for ϕ_{TE}. This implies that the fields of TM to $\hat{r}$ are affected by the uniaxial anisotropy, and for the TE-to-$\hat{r}$ case, the uniaxial material can be treated as an isotropic medium characterized by ϵ_θ. Therefore, we have only to work on the solutions to ϕ_{TM} and the solutions to $\phi_{\rm TE}$ can be obtained by replacing $\mu_0 J_r$ with $\epsilon_0 \epsilon_\theta M_r$ and taking the special case of $\epsilon_r = \epsilon_\theta$.

In the following derivation, (3.25) is solved indirectly. The equation of $\phi_{\rm TM}$ is converted into a Helmholtz-like partial differential equation by introducing a $\pi_{\rm TM}$ that satisfies

$$\phi_{\rm TM} = r\pi_{\rm TM}\,. \tag{3.37}$$

By substituting (3.37) into (3.25), it becomes

$$(\nabla^2_{\rm AR} + k_0^2 \epsilon_r)\pi_{\rm TM} = \frac{-\mu_0 J_r}{r}\,, \tag{3.38}$$

where the operator $\nabla^2_{\rm AR}$ is defined to be

$$\nabla^2_{\rm AR} = \frac{1}{\epsilon_{\rm AR}}\frac{\partial}{r^2\,\partial r}\left(r^2\frac{\partial}{\partial r}\right) + \nabla_t^2\,. \tag{3.39}$$

In (3.39), $\epsilon_{\rm AR}$ ($= \epsilon_\theta/\epsilon_r$) is the anisotropic ratio of the uniaxial medium. For the isotropic case ($\epsilon_{\rm AR} = 1$), the operator $\nabla^2_{\rm AR}$ is reduced to the Laplacian operator and (3.38) becomes a Helmholtz equation. Since solutions to the Helmholtz equations are well known, once the solutions to $\pi_{\rm TM}$ are obtained, it is easy to check the correctness by setting $\epsilon_{\rm AR} = 1$. In the following, both the source-free and ideal point-source cases are considered.

A. Source-Free Case By setting the source term, the right-hand side of (3.38), to zero and expanding the resulting homogeneous equation using the method of separation of variables, the following equations are obtained:

$$\frac{d}{dr}\left(r^2\frac{dR}{dr}\right) + [(kr)^2 - n(n+1)\epsilon_{\rm AR}]R = 0\,, \tag{3.40}$$

$$\frac{1}{\sin\theta}\frac{d}{d\theta}\left(\sin\theta\frac{d\Theta}{d\theta}\right) + \left[n(n+1) - \frac{m^2}{\sin^2\theta}\right]\Theta = 0\,, \tag{3.41}$$

$$\frac{d^2\Phi}{d\phi^2} + m^2\Phi = 0\,, \tag{3.42}$$

where $k = k_0\sqrt{\epsilon_\theta}$ and

$$\pi_{\rm TM}(r, \theta, \phi) = R(r)\Theta(\theta)\Phi(\phi)\,. \tag{3.43}$$

In the equations above, (3.40) is closely related to Bessel's equation and its solutions are derived in Appendix B; (3.41) is the associated Legendre equation and its solutions are called associated Legendre functions; (3.42) is the familiar harmonic equation, giving rise to solutions called harmonic functions. By composing Bessel functions, associated Legendre functions, and harmonic functions, the solutions to π_{TM} are

$$\pi_{TM} = \sum_{m,n} C_{mn} \frac{1}{\sqrt{r}} B_\nu(kr) P_n^m(\cos\theta) \begin{Bmatrix} \cos m\phi \\ \sin m\phi \end{Bmatrix}, \tag{3.44}$$

where C_{mn} are the expansion coefficients; $B_\nu(kr)$, depending on the boundary condition, can be one of the following functions or a linear combination of any two of them: $J_\nu(kr)$, a Bessel function of the first kind with order ν; $Y_\nu(kr)$ [or $N_\nu(kr)$], a Bessel function of the second kind with order ν; $H_\nu^{(1)}(kr)$, a Hankel function of the first kind with order ν; $H_\nu^{(2)}(kr)$, a Hankel function of the second kind with order ν. The quantity ν is given in Appendix B and is written as

$$\nu = \sqrt{n(n+1)\epsilon_{AR} + \tfrac{1}{4}}\,. \tag{3.45}$$

Hence the ϕ_{TM} can be obtained from (3.37); that is,

$$\phi_{TM} = r\pi_{TM} = \sum_{m,n} C'_{mn} \tilde{\tilde{B}}_n(kr) P_n^m(\cos\theta) \begin{Bmatrix} \cos m\phi \\ \sin m\phi \end{Bmatrix}, \tag{3.46}$$

where the C'_{mn} are again expansion coefficients and $\tilde{\tilde{B}}_n(kr)$, for convenience, is defined to be

$$\tilde{\tilde{B}}_n(kr) = \sqrt{\frac{\pi kr}{2}}\, B_\nu(kr)\,, \tag{3.47}$$

and we refer it as a modified spherical Bessel function [2]. For the case of $\epsilon_{AR} = 1$ (i.e., $\nu = n + \frac{1}{2}$), the modified spherical Bessel function, $\tilde{\tilde{B}}_n(kr)$, becomes a spherical Bessel function of Schelkunoff type, $\hat{B}_n(kr)$, described in [3].

B. Ideal Point-Source Case To facilitate the Green's function of π_{TM}, the right-hand side of (3.38) is replaced by a point source; that is,

$$(\nabla_{AR}^2 + k_0^2\epsilon_r) g_{TM} = -\delta(\mathbf{r} - \mathbf{r}')\,. \tag{3.48}$$

g_{TM} in (3.48) represents the Green's function of π_{TM}, and we have

$$\pi_{TM} = \iiint_{V'} g_{TM} \frac{\mu_0 J_r(\mathbf{r}')}{r'}\, dV'\,, \tag{3.49}$$

where V' is the source region and the prime used here denotes the source coordinates. The delta function $\delta(\mathbf{r} - \mathbf{r}')$ on the right-hand side of (3.48) is given by

$$\delta(\mathbf{r} - \mathbf{r}') = \frac{\delta(r - r')\delta(\theta - \theta')\delta(\phi - \phi')}{r^2 \sin\theta}\,. \tag{3.50}$$

It is well known that Green's function can be expressed in terms of the solutions of a homogeneous equation, so we expand g_{TM} in Fourier series for the ϕ dependence and in a series of Legendre functions for the θ dependence. We then have

$$g_{\mathrm{TM}} = \sum_{n=0}^{\infty} \sum_{m=-n}^{n} G_{nm}(r, r')T_{nm}(\theta, \phi)T^*_{nm}(\theta', \phi'), \tag{3.51}$$

where $T_{nm}(\theta, \phi)$ is the spherical harmonics and defined to be

$$T_{nm}(\theta, \phi) = \left[\frac{2n+1}{4\pi}\frac{(n-m)!}{(n+m)!}\right]^{1/2} P_n^m(\cos\theta)e^{jm\phi}. \tag{3.52}$$

We now expand the delta function of (3.50) as a function of $T_{nm}(\theta,\phi)$; that is,

$$\frac{\delta(\theta-\theta')\delta(\phi-\phi')}{\sin\theta} = \sum_{n=0}^{\infty}\sum_{m=-n}^{n} T_{nm}(\theta, \phi)T^*_{nm}(\theta', \phi'). \tag{3.53}$$

Substituting (3.51) and (3.53) into (3.48), we have

$$\left[\frac{d}{dr}\left(r^2\frac{d}{dr}\right) + (kr)^2 - n(n+1)\epsilon_{\mathrm{AR}}\right]G_n(r, r') = -\delta(r-r'). \tag{3.54}$$

Note that since (3.54) contains only n, we use G_n instead of G_{nm}. Appendix B shows the derivation procedure on solving (3.54), and the solution is

$$G_n(r, r') = \begin{cases} \dfrac{1}{jkrr'}\tilde{\tilde{J}}_n(kr)\tilde{\tilde{H}}_n^{(2)}(kr'), & r<r', \\ \dfrac{1}{jkrr'}\tilde{\tilde{J}}_n(kr')\tilde{\tilde{H}}_n^{(2)}(kr), & r>r', \end{cases} \tag{3.55}$$

where $\tilde{\tilde{J}}_n(x)$ and $\tilde{\tilde{H}}_n^{(2)}(x)$ are defined in (3.47).

By using the condition

$$P_n^{-m}(x) = (-1)^m\frac{(n-m)!}{(n+m)!}P_n^m(x), \tag{3.56}$$

and changing the order of summation, (3.51) can be rewritten as

$$g_{\mathrm{TM}} = \sum_{m=0}^{\infty}\cos m(\phi-\phi')\sum_{n=m}^{\infty} S_g(n, m)G_n(r, r')P_n^m(\cos\theta)P_n^m(\cos\theta'), \tag{3.57}$$

where

$$S_g(n, m) = \frac{2n+1}{4\pi}\frac{(n-m)!}{(n+m)!}e_m, \qquad e_m = \begin{cases}1, & m=0, \\ 2, & m>0.\end{cases} \tag{3.58}$$

For the case of $\epsilon_{AR} = 1$ (i.e., $\epsilon_r = \epsilon_\theta$), $\tilde{\tilde{H}}_n^{(2)}(x)$ and $\tilde{\tilde{J}}_n(x)$ in (3.55) become $\hat{H}_n^{(2)}(x)$ and $\hat{J}_n(x)$, respectively. In this case the solution to g_{TM} can be reduced to the isotropic results described in [3]. Following the derivation procedure above, the corresponding results for TE-to-$\hat{r}$ waves can readily be obtained by taking the special case of $\epsilon_r = \epsilon_\theta$ from the results of TM-to-$\hat{r}$ waves.

Now that the solutions to the wave equations of (3.25) and (3.29) have been constructed completely, we may take some brief conclusions for the results obtained. As we have discussed before, for an arbitrary wave propagating inside a uniaxial medium expressed as the sum of TM-to-$\hat{r}$ and TE-to-$\hat{r}$ waves, only the TM-to-$\hat{r}$ wave is affected by the uniaxial anisotropy. On the other hand, for the TE-to-$\hat{r}$ wave, the uniaxial material can be treated as an isotropic medium characterized by ϵ_θ. Also, for the TM-to-$\hat{r}$ wave, considering the solution described by (3.46) for the source-free case and the Green's function described by (3.57) for the point-source case, only the r dependence is affected by the uniaxial anisotropy. From the results obtained, the effect of the uniaxial anisotropy on the TM-to-$\hat{r}$ component can be represented simply by introducing a modified spherical Bessel function, defined by (3.47), into the r dependence of the field expression. That is, a modified spherical Bessel function is used for the solution to the uniaxial case, and a spherical Bessel function of Schelkunoff type for the solution to the isotropic case. By setting $\epsilon_{AR} = 1$, a modified Bessel function can also be reduced to a spherical Bessel function of Schelkunoff type.

3.2.3 Full-Wave Formulation for a Spherical Circular Microstrip Structure

Given a spherical circular microstrip antenna as shown in Figure 3.1, a metallic sphere of radius a is coated with a dielectric substrate of thickness h $(=b-a)$. A circular patch of radius r_d is mounted on the substrate and subtended by an angle θ_0. The substrate is a layer of uniaxial material characterized by free-space permeability μ_0 and permittivity tensor $\bar{\bar{\epsilon}}$ of the form written in (3.1). The outer region $(r > b)$ is free space with permeability μ_0 and permittivity ϵ_0.

The metallic ground sphere and the patch are assumed to be perfect conductors. The thickness of the patch is neglected since it is usually much smaller than that of the operating wavelength. Based on these assumptions, the patch can be replaced by a surface current distribution. This patch surface current distribution needs to be solved to analyze the characteristics of a microstrip structure as a radiator. Since this patch surface current distribution is usually not of a simple form, a set of orthogonal functions is taken to be the expansion functions for the current distribution. To solve the unknown current distribution on the patch, the spectral-domain approach (SDA) [4] and Galerkin's moment method are applied to simplify the problem.

To apply SDA technique, a transform pair must be introduced first. In spherical coordinates, the vector Legendre transform is commonly used; that is, all the transverse fields are expressed in the form of vector Legendre series. The vector Legendre transform pair is defined in matrix form as [5]

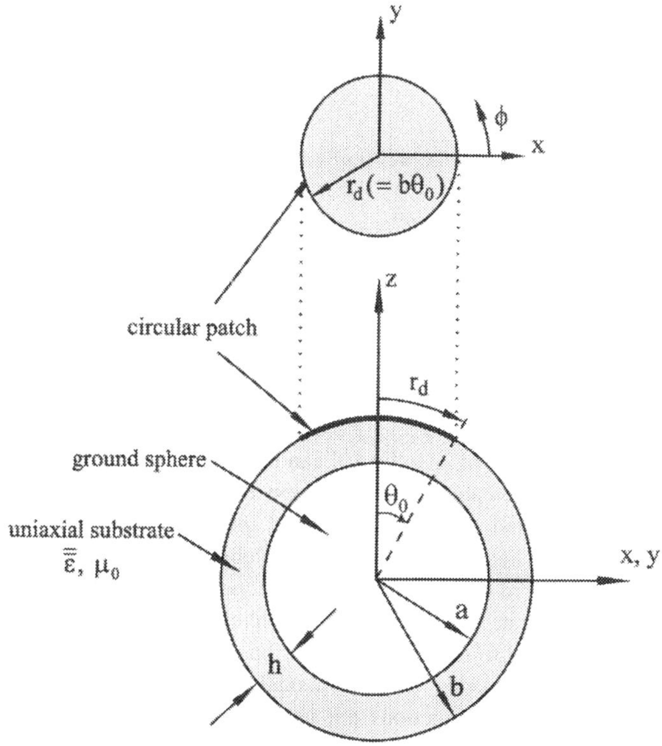

FIGURE 3.1 Geometry of a spherical circular microstrip structure with a uniaxial substrate.

$$\bar{\bar{F}}(\theta) = \sum_{n=m}^{\infty} \bar{\bar{L}}(n, m, \theta)\tilde{F}(n) , \tag{3.59}$$

$$\tilde{F}(n) = \frac{1}{S(n, m)} \int_0^{\pi} \bar{\bar{L}}(n, m, \theta)\bar{\bar{F}}(\theta) \sin\theta \, d\theta , \tag{3.60}$$

with

$$S(n, m) = \frac{2n(n+1)(n+m)!}{(2n+1)(n-m)!} . \tag{3.61}$$

$\bar{\bar{F}}(\theta)$ and $\tilde{F}(n)$ in (3.59)–(3.60) are column matrices given by

$$\bar{\bar{F}}(\theta) = \begin{bmatrix} F_\theta(\theta) \\ F_\phi(\theta) \end{bmatrix} , \tag{3.62}$$

$$\tilde{F}(n) = \begin{bmatrix} \tilde{F}_U(n) \\ \tilde{F}_L(n) \end{bmatrix}, \tag{3.63}$$

and $\bar{\bar{L}}(n, m, \theta)$ is a 2×2 matrix written as

$$\bar{\bar{L}}(n, m, \theta) = \begin{bmatrix} \dfrac{\partial P_n^m(\cos\theta)}{\partial\theta} & \dfrac{-jmP_n^m(\cos\theta)}{\sin\theta} \\ \dfrac{jmP_n^m(\cos\theta)}{\sin\theta} & \dfrac{\partial P_n^m(\cos\theta)}{\partial\theta} \end{bmatrix}. \tag{3.64}$$

As for the Galerkin's moment-method procedure, details of the formulation are described in Section 3.2.4. By applying the SDA technique and Galerkin's moment method, a matrix equation of $[Z][I] = [0]$ can be derived where the elements in $[I]$ are the unknown coefficients of the expansion functions. Once $[I]$ is solved, the patch surface current and other information of interest can be calculated.

To begin with, the field components in the substrate region and the outer (air) region can be calculated from ϕ_{TM} and ϕ_{TE}, according to the results in Section 3.2.2. Inside the substrate region, spherical Hankel functions of both the first and second kinds are needed to represent the solution since there are two boundaries to be matched. In addition, due to the uniaxial anisotropy of the substrate, the modified spherical Hankel functions are used in the TM-to-$\hat{r}$ wave; that is, we have

$$\phi_{\text{TM}}^s = e^{jm\phi} \sum_{n=m}^{\infty} [A_n \tilde{\tilde{H}}_n^{(1)}(kr) + B_n \tilde{\tilde{H}}_n^{(2)}(kr)] P_n^m(\cos\theta), \tag{3.65}$$

$$\phi_{\text{TE}}^s = e^{jm\phi} \sum_{n=m}^{\infty} [C_n \hat{H}_n^{(1)}(kr) + D_n \hat{H}_n^{(2)}(kr)] P_n^m(\cos\theta). \tag{3.66}$$

In the air region, only a spherical Hankel function of the second kind is used, as it represents an outward-traveling wave; that is, we have

$$\phi_{\text{TM}}^a = e^{jm\phi} \sum_{n=m}^{\infty} E_n \hat{H}_n^{(2)}(k_0 r) P_n^m(\cos\theta), \tag{3.67}$$

$$\phi_{\text{TE}}^a = e^{jm\phi} \sum_{n=m}^{\infty} F_n \hat{H}_n^{(2)}(k_0 r) P_n^m(\cos\theta). \tag{3.68}$$

By applying the foregoing potential functions to (3.31)–(3.36), the field components in each region can be obtained. Also, in (3.65)–(3.68), A_n, B_n, C_n, D_n, E_n, and F_n are unknown coefficients to be determined by the following boundary conditions at $r = a$ and b:

$$\hat{r} \times \mathbf{E}^s = 0, \text{ on the surface } r = a\,, \tag{3.69}$$

$$\hat{r} \times (\mathbf{E}^a - \mathbf{E}^s) = 0, \text{ on the surface } r = b\,, \tag{3.70}$$

$$\hat{r} \times (\mathbf{H}^a - \mathbf{H}^s) = \mathbf{J}, \text{ on the surface } r = b\,, \tag{3.71}$$

where $\mathbf{J}$ is the patch surface current density, and the superscripts s and a denote the substrate and air regions, respectively. Next, by using the vector Legendre transform, the spectral coefficients of the transverse field components in the substrate and air regions are, respectively, given by

$$\tilde{E}^s(n) = \begin{bmatrix} \dfrac{k}{j\omega\mu_0\epsilon_0\epsilon_\theta r}[A_n\tilde{\hat{H}}_n^{(1)\prime}(kr) + B_n\tilde{\hat{H}}_n^{(2)\prime}(kr)] \\ \dfrac{1}{\epsilon_0\epsilon_\theta r}[C_n\hat{H}_n^{(1)}(kr) + D_n\hat{H}_n^{(2)}(kr)] \end{bmatrix}, \tag{3.72}$$

$$\tilde{H}^s(n) = \begin{bmatrix} \dfrac{k}{j\omega\mu_0\epsilon_0\epsilon_\theta r}[C_n\hat{H}_n^{(1)\prime}(kr) + D_n\hat{H}_n^{(2)\prime}(kr)] \\ \dfrac{-1}{\mu_0 r}[A_n\tilde{\hat{H}}_n^{(1)}(kr) + B_n\tilde{\hat{H}}_n^{(2)}(kr)] \end{bmatrix}, \tag{3.73}$$

and

$$\tilde{E}^a(n) = \begin{bmatrix} \dfrac{k_0}{j\omega\mu_0\epsilon_0 r}E_n\hat{H}_n^{(2)\prime}(k_0 r) \\ \dfrac{-1}{\epsilon_0 r}F_n\hat{H}_n^{(2)}(k_0 r) \end{bmatrix}, \tag{3.74}$$

$$\tilde{H}^a(n) = \begin{bmatrix} \dfrac{k_0}{j\omega\mu_0\epsilon_0 r}F_n\hat{H}_n^{(2)\prime}(k_0 r) \\ \dfrac{-1}{\mu_0 r}E_n\hat{H}_n^{(2)}(k_0 r) \end{bmatrix}. \tag{3.75}$$

In the spectral domain, the corresponding boundary conditions are

$$\tilde{E}^s(n) = \begin{cases} [0]\,, & \text{at } r = a\,, \quad (3.76) \\ \tilde{E}^a(n)\,, & \text{at } r = b\,, \quad (3.77) \end{cases}$$

$$\tilde{J}(n) = \begin{bmatrix} 0 & -1 \\ 1 & 0 \end{bmatrix}[\tilde{H}^a(n) - \tilde{H}^s(n)]\,, \quad \text{at } r = b\,, \tag{3.78}$$

where $\tilde{J}(n)$ is the spectral amplitude of the surface current density on the circular patch. By substituting (3.72)–(3.75) to (3.76)–(3.78), the relationship of the surface current density and the tangential electric field on the patch can be obtained and given by

$$\tilde{J}(n) = \bar{\bar{Y}}(n)\tilde{E}(n)\,, \tag{3.79}$$

with

$$\bar{\bar{Y}}(n) = \begin{bmatrix} Y_{11}(n) & 0 \\ 0 & Y_{22}(n) \end{bmatrix}, \tag{3.80}$$

$$Y_{11}(n) = j\sqrt{\frac{\epsilon_0}{\mu_0}}\frac{\hat{H}_n^{(2)}(k_0b)}{\hat{H}_n^{(2)\prime}(k_0b)} - j\sqrt{\frac{\epsilon_0\epsilon_\theta}{\mu_0}}\frac{\tilde{\hat{H}}_n^{(1)}(kb) - \alpha_{TM}\tilde{\hat{H}}_n^{(2)}(kb)}{\tilde{\hat{H}}_n^{(1)\prime}(kb) - \alpha_{TM}\tilde{\hat{H}}_n^{(2)\prime}(kb)}, \tag{3.81}$$

$$Y_{22}(n) = -j\sqrt{\frac{\epsilon_0}{\mu_0}}\frac{\hat{H}_n^{(2)\prime}(k_0b)}{\hat{H}_n^{(2)}(k_0b)} + j\sqrt{\frac{\epsilon_0\epsilon_\theta}{\mu_0}}\frac{\hat{H}_n^{(1)\prime}(kb) - \alpha_{TE}\hat{H}_n^{(2)\prime}(kb)}{\hat{H}_n^{(1)}(kb) - \alpha_{TE}\hat{H}_n^{(2)}(kb)}, \tag{3.82}$$

$$\alpha_{TM} = \frac{\tilde{\hat{H}}_n^{(1)\prime}(ka)}{\tilde{\hat{H}}_n^{(2)\prime}(ka)},$$

$$\alpha_{TE} = \frac{\hat{H}_n^{(1)}(ka)}{\hat{H}_n^{(2)}(ka)}.$$

3.2.4 Galerkin's Moment-Method Formulation

As we have discussed before, Galerkin's moment method is employed to solve the unknown surface current density on the patch. That is, the patch surface current density can be expanded into a linear combination of known basis functions, and we have

$$\tilde{J}(n) = \begin{bmatrix} \tilde{J}_U(n) \\ \tilde{J}_L(n) \end{bmatrix} = \sum_{i=1}^{N} I_i\tilde{J}_i(n)\,, \tag{3.83}$$

where I_i is the unknown coefficient for $\tilde{J}_i(n)$, the spectral amplitude of the ith expansion function. Then, by using the same set of basis functions as the weighting functions and after (3.83) is weighted by each weighting function, we can have an equation in matrix form for a source-free case written as

$$[0] = [Z][I] \tag{3.84}$$

The left-hand side of (3.84) is a zero matrix due to Parseval's theorem [5] and the boundary condition that the electric field and the patch surface current are complementary to each other at $r = b$. The elements in $[I]$ are unknown coefficients I_i, and the elements in $[Z]$ are given by

$$Z_{ij} = \sum_{n=m}^{\infty} \tilde{J}_i^*(n) S(n,m) \bar{\bar{\tilde{Y}}}^{-1} \tilde{J}_j(n) . \tag{3.85}$$

3.2.5 Basis Functions for Excited Patch Surface Current

As Galerkin's procedure is applied, a set of basis functions is needed as the expansion and weighting functions. In the case considered here, these basis functions should be orthogonal and can also be transformed into closed forms to simplify formulation and computation. Here the cavity-model theory is employed to derive the basis functions for the patch surface current distributions.

Based on cavity-model theory, the region between the circular patch and the ground metallic sphere is treated as a cavity bounded by a magnetic wall along the edge of the patch and by electric walls from above and below, and the field in the cavity is independent of the r-coordinate. For this cavity structure the potential functions ϕ_{TM} and ϕ_{TE} satisfy the following wave equation:

$$\left[\frac{1}{\sin\theta}\frac{\partial}{\partial\theta}\left(\sin\theta\frac{\partial}{\partial\theta}\right) + \frac{1}{\sin^2\theta}\frac{\partial^2}{\partial\phi^2} + (kb)^2\right]\begin{Bmatrix}\phi_{TM}\\ \phi_{TE}\end{Bmatrix} = 0 . \tag{3.86}$$

Equation (3.86) can be solved using the method of separation of variables, and we have the solutions given by

$$\begin{Bmatrix}\phi_{TM}\\ \phi_{TE}\end{Bmatrix} = e^{jm\phi} P_\nu^m(\cos\theta) , \tag{3.87}$$

with the condition

$$(kb)^2 = \nu(\nu+1) . \tag{3.88}$$

Also, due to the boundary condition

$$H_\phi|_{\theta=\theta_0} = 0 , \tag{3.89}$$

we have the following constraints: For $TM_{m\nu}$ modes,

$$P_\nu^{m'}(\cos\theta_0) = 0 , \tag{3.90}$$

and for $TE_{m\nu}$ modes,

$$P_\nu^m(\cos\theta_0) = 0 . \tag{3.91}$$

Note that there are only some special values of ν satisfying (3.90) and (3.91). Each value of ν corresponds to a specific operation mode or the ith mode in

(3.83). By further imposing the discontinuity boundary condition on the top electric wall, we have, for $TM_{m\nu}$ modes,

$$J_{TM_{m\nu}}(\theta) = \begin{cases} \begin{bmatrix} \dfrac{\partial P_\nu^m(\cos\theta)}{\partial\theta} \\ \dfrac{jm}{\sin\theta} P_\nu^m(\cos\theta) \end{bmatrix}, & \theta < \theta_0, \\ 0, & \theta > \theta_0, \end{cases} \tag{3.92}$$

and for $TE_{m\nu}$ modes,

$$J_{TE_{m\nu}}(\theta) = \begin{cases} \begin{bmatrix} \dfrac{jm}{\sin\theta} P_\nu^m(\cos\theta) \\ -\dfrac{\partial P_\nu^m(\cos\theta)}{\partial\theta} \end{bmatrix}, & \theta < \theta_0, \\ 0, & \theta > \theta_0. \end{cases} \tag{3.93}$$

Finally, from the definition of the vector Legendre transform, the spectral amplitudes of the $TM_{m\nu}$ and $TE_{m\nu}$ surface current distributions are, respectively, given by

$$\tilde{J}_{TM_{m\nu}} = \frac{1}{S(n,m)} \begin{bmatrix} \sin^2\theta_0 \dfrac{\nu(\nu+1)}{n(n+1)-\nu(\nu+1)} P_n^{m'}(\cos\theta_0) P_\nu^m(\cos\theta_0) \\ jm P_n^m(\cos\theta_0) P_\nu^m(\cos\theta_0) \end{bmatrix}, \tag{3.94}$$

$$\tilde{J}_{TE_{m\nu}} = \frac{1}{S(n,m)} \begin{bmatrix} 0 \\ \sin^2\theta_0 \dfrac{n(n+1)}{n(n+1)-\nu(\nu+1)} P_n^m(\cos\theta_0) P_\nu^{m'}(\cos\theta_0) \end{bmatrix}. \tag{3.95}$$

3.2.6 Resonance Characteristics

From considering (3.84), the existence of nontrivial solutions for $[I]$ requires that the determinant of the $[Z]$ matrix must be zero; that is,

$$\det[Z] = 0. \tag{3.96}$$

The condition of (3.96) is again found to be satisfied by the complex frequencies $f = f' + jf''$, which give the resonant frequency, f', and the half-power bandwidth, $2f''/f'$, for the microstrip structure.

Typical results for the fundamental mode of TM_{11} are calculated and shown in Figures 3.2 and 3.3. The three cases of positive uniaxial ($\epsilon_r = 2.47$, $\epsilon_\theta = 1.98$), isotropic ($\epsilon_r = \epsilon_\theta = 2.47$), and negative uniaxial ($\epsilon_r = 2.47$, $\epsilon_\theta = 2.96$) substrates are studied. Figure 3.2 shows the case for a ground sphere of $a = 5$ cm, and cases for $a = 3$ cm and 10 cm are plotted in Figure 3.3. From the results it is seen that both the resonant frequency and half-power bandwidth are increased due to the positive uniaxial anisotropy and decreased due to the negative uniaxial anisotropy.

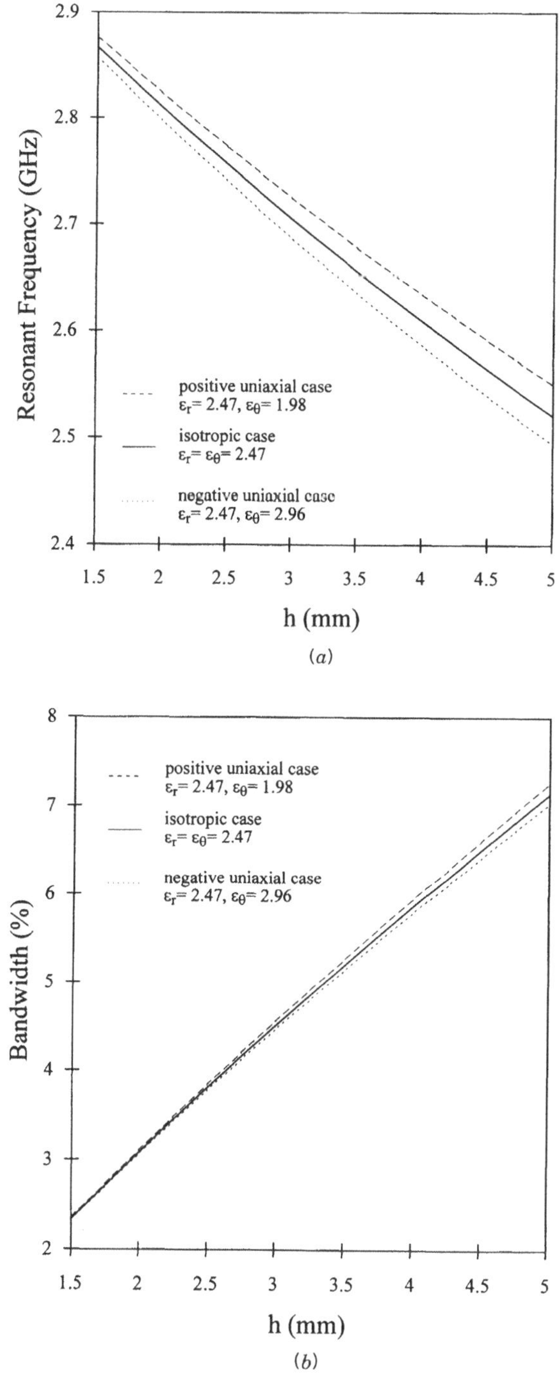

FIGURE 3.2 Dependence of (*a*) resonant frequency and (*b*) half-power bandwidth on substrate thickness for the cases of positive uniaxial, isotropic, and negative uniaxial substrates; $a = 5$ cm, $r_d = 1.88$ cm.

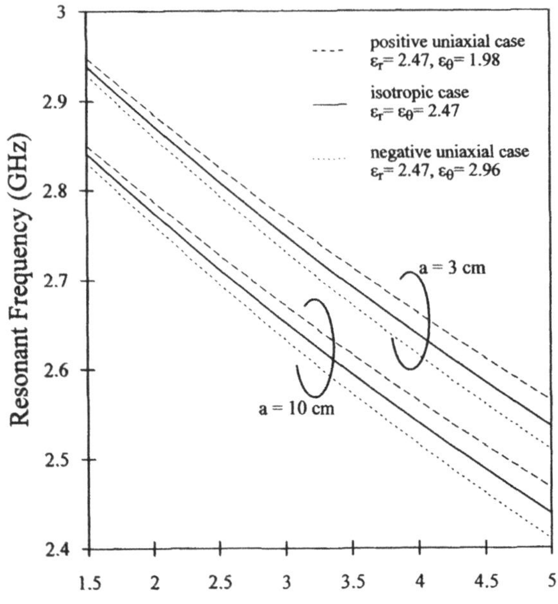

(*a*)

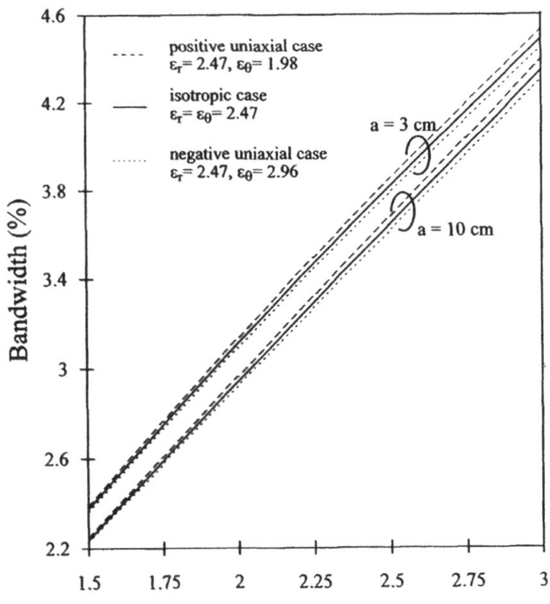

(*b*)

FIGURE 3.3 Dependence of the (*a*) resonant frequency and (*b*) half-power bandwidth on substrate thickness for $a = 3$ and 10 cm. Other parameters are the same as in Figure. 3.2.

The frequency shifts and bandwidth variations also increase with increasing substrate thickness.

3.2.7 Radiation Characteristics

Once the resonant frequency for the given operation mode is obtained, the radiation pattern for that operation mode can also be calculated under the resonant condition. That is, when the operating frequency is near the resonant frequency, we may assume that the patch current can be described dominantly by the cavity-model current function of the corresponding mode; that is,

$$\tilde{J}(n) = \begin{bmatrix} \tilde{J}_U(n) \\ \tilde{J}_L(n) \end{bmatrix} = \begin{cases} \tilde{J}_{\mathrm{TM}_{m\nu}}(n)\,, & \text{for } \mathrm{TM}_{m\nu} \text{ excitation}\,, \\ \tilde{J}_{\mathrm{TE}_{m\nu}}(n)\,, & \text{for } \mathrm{TE}_{m\nu} \text{ excitation}\,. \end{cases} \tag{3.97}$$

By substituting the resonant frequency and the dominant cavity-model current function into (3.79) and after some straightforward manipulation, the coefficients E_n and F_n in (3.74)–(3.75) in the air region can be derived as

$$E_n = \frac{j\sqrt{\mu_0\epsilon_0}\,b}{\hat{H}_n^{(2)\prime}(k_0 b)} \frac{\tilde{J}_U(n)}{Y_{11}(n)}\,, \tag{3.98}$$

$$F_n = \frac{\epsilon_0 b}{\hat{H}_n^{(2)}(k_0 b)} \frac{\tilde{J}_L(n)}{Y_{22}(n)}\,. \tag{3.99}$$

Then, by using the large-argument asymptotic formulas of

$$\hat{H}_n^{(2)}(k_0 r) \xrightarrow[k_0 r \to \infty]{} j^{n+1} e^{-jk_0 r} \tag{3.100}$$

and

$$\hat{H}_n^{(2)\prime}(k_0 r) \xrightarrow[k_0 r \to \infty]{} j^n e^{-jk_0 r}\,, \tag{3.101}$$

the far-zone electric fields can be written as

$$E_\theta^a = \frac{e^{-jk_0 r}}{r} \cos m\phi \sum_{n=m}^{\infty} \left[\frac{-j^{n+1}}{\sqrt{\mu_0\epsilon_0}} E_n \frac{\partial P_n^m(\cos\theta)}{\partial\theta} + \frac{j^n}{\epsilon_0} F_n \frac{m P_n^m(\cos\theta)}{\sin\theta} \right], \tag{3.102}$$

$$E_\phi^a = \frac{e^{-jk_0 r}}{r} \sin m\phi \sum_{n=m}^{\infty} \left[\frac{j^n}{\sqrt{\mu_0\epsilon_0}} E_n \frac{m P_n^m(\cos\theta)}{\sin\theta} + \frac{j^{n+1}}{\epsilon_0} F_n \frac{\partial P_n^m(\cos\theta)}{\partial\theta} \right]. \tag{3.103}$$

The far-zone radiation patterns calculated for the TM_{11} mode are shown in Figure 3.4. The operating frequency used for calculating the radiation patterns is

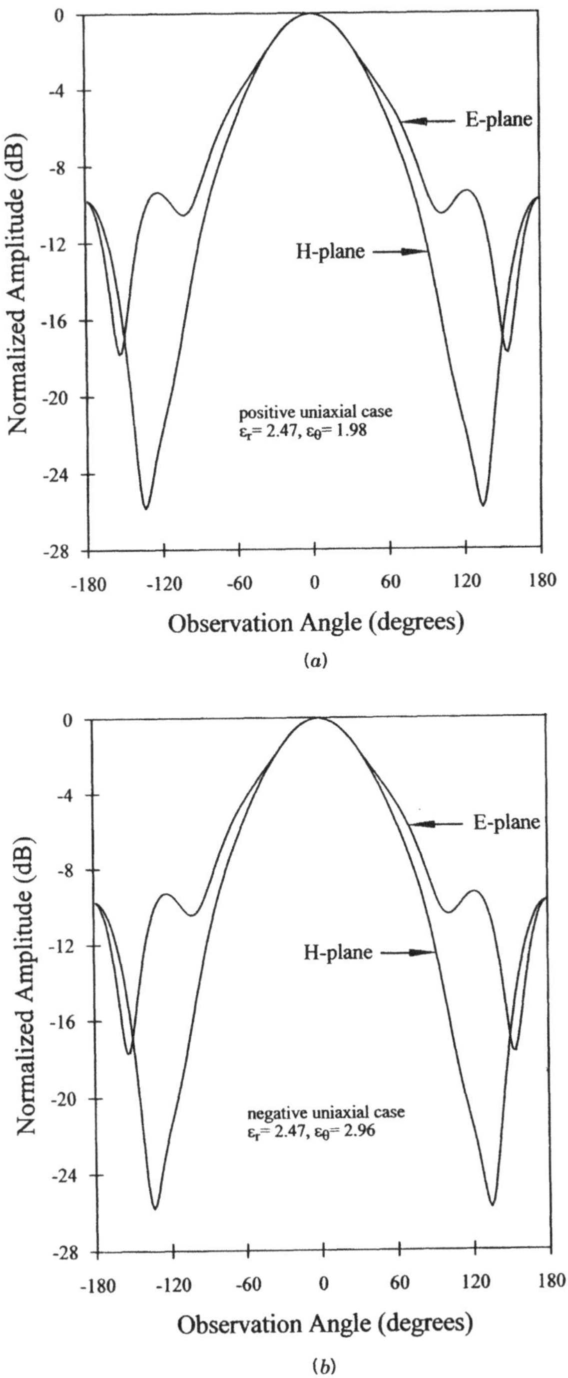

FIGURE 3.4 Radiation patterns of the spherical circular microstrip antenna with $a = 5$ cm, $r_d = 1.88$ cm, $h = 0.16$ cm. (*a*) positive uniaxial case; (*b*) negative uniaxial case.

the resonant frequency obtained in Section 3.2.6, and the current distribution on the circular patch is selected to be

$$\tilde{J}(n) = \begin{bmatrix} \tilde{J}_U(n) \\ \tilde{J}_L(n) \end{bmatrix} = \tilde{J}_{\mathrm{TM}_{11}}(n), \tag{3.104}$$

where $\tilde{J}_{\mathrm{TM}_{11}}(n)$ is as defined in (3.94). In Figure 3.4, both positive and negative uniaxial anisotropy are shown. It is observed that the radiation patterns are almost exactly the same for both. The patterns for the cases of other sphere radii are also calculated, again showing similar results. This indicates that the radiation patterns of the patch antenna are insensitive to the uniaxial-anisotropy variation in the substrate. As for the curvature effect on the radiation patterns, results are similar to those obtained in Chapter 2 for cylindrical microstrip structures.

3.2.8 Scattering Characteristics

The electromagnetic scattering from the present spherical circular microstrip structure has also been analyzed. Numerical results of the radar cross sections (RCSs) for various spherical radii have been calculated and reported [6]. The effects of the substrate's uniaxial anisotropy on the scattering behavior are also investigated.

For RCS calculation, we first assume a plane wave incident from the direction of θ_i in the x–z plane; that is, $\phi_i = 0°$, for theoretical simplicity. This assumption can be justified due to the symmetry of the circular patch. Thus we have

$$E^i = E_0(-\hat{x}\cos\theta_i + \hat{z}\sin\theta_i)e^{jk_0(x\sin\theta_i + z\cos\theta_i)}, \tag{3.105}$$

$$H^i = \hat{y}\frac{E_0}{\eta_0}e^{jk_0(x\sin\theta_i + z\cos\theta_i)}. \tag{3.106}$$

The incident plane wave illuminates the spherical microstrip structure and we have the boundary condition that the total electric fields tangential to the surface of the patch must be zero; that is, on the patch,

$$\hat{r} \times (\mathbf{E}^D + \mathbf{E}^i) = 0, \tag{3.107}$$

where $\mathbf{E}^D$ is generated by the disk current shown by (3.79) and $\mathbf{E}^i$ is due to the incident plane wave expressed as (3.72) with $r = b$. After applying Galerkin's moment method to (3.107), (3.84) is modified to be

$$[0] = [Z][I] + [V], \tag{3.108}$$

where the left-hand side of (3.108) becomes a zero matrix again due to Parseval's theorem [5] and the boundary condition of (3.107). The elements in $[I]$ and $[Z]$ have the same meanings as in (3.84) and the elements in $[V]$ are expressed as

$$V_i = \sum_{n=m}^{\infty} \tilde{J}_i^*(n) S(n, m) \tilde{E}^i . \tag{3.109}$$

From (3.108), the unknown coefficients of I_i in $[I]$ can be calculated and the total induced patch current can be expressed as in (3.83). Then the monostatic RCS can be calculated from the radiation field as

$$\sigma_{\theta\theta} = 4\pi r^2 |E_\theta|^2, \tag{3.110}$$

where E_θ is the radiation field and can be expressed as [7]

$$E_\theta(\theta, \phi) = b \frac{e^{-jk_0 r}}{r} \sum_{m=0}^{\infty} \cos m\phi \sum_{n=m}^{\infty} \left[\frac{-j^n \tilde{J}_U(n)}{Y_{11} \hat{H}_n^{(2)'}(k_0 b)} P_n^{m'}(\cos\theta) \sin\theta + \frac{j^{n+1} \tilde{J}_L(n)}{Y_{22} \hat{H}_n^{(2)}(k_0 b)} \frac{-jm P_n^m(\cos\theta)}{\sin\theta} \right]. \tag{3.111}$$

Typical RCS results are shown in Figures 3.5 to 3.7. The monostatic RCSs of the microstrip structure for various sphere radii are presented in Figure 3.5. The curve $\pi r_d^2 \cos\theta_i$ in the figure is the physical area of the patch seen by the incident wave. First note that the RCS resonant peak occurs at nearly the resonant frequency of the microstrip structure, and the curvature of the patch causes a shift of the RCS resonant peak to higher frequencies. It is also seen that the RCS resonant response peak increases with decreasing sphere radius. Figure 3.6 shows the monostatic RCS results for various substrate thicknesses. It is seen that the RCS resonant peak value is slightly affected by the substrate thickness. However, the frequency locations of the RCS resonant peaks move to lower frequencies for a greater substrate thickness. The uniaxial-anisotropy effect is shown in Figure 3.7.

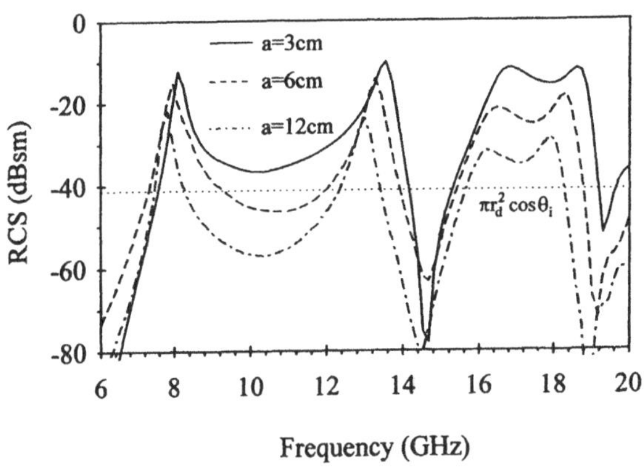

FIGURE 3.5 Radar cross section versus frequency for different sphere radii; $r_d = 7.1$ mm, $h = 0.7874$ mm, $\epsilon_r = \epsilon_\theta = 2.2$, $(\theta_i, \phi_i) = (63°, 0°)$. (From Ref. [6], © 1995 John Wiley & Sons, Inc., reprinted with permission.)

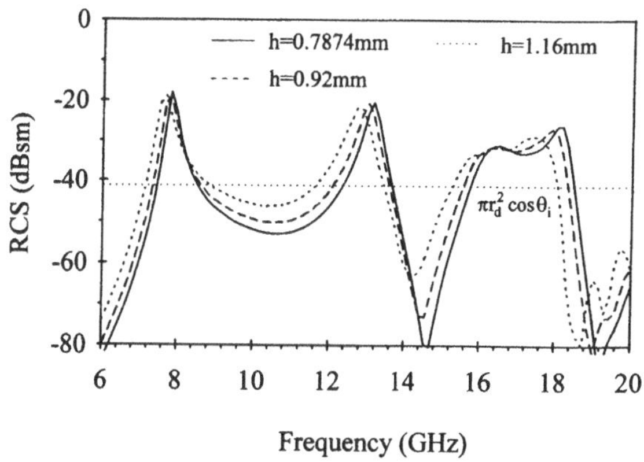

FIGURE 3.6 Radar cross section versus frequency for different substrate thicknesses, $a = 10$ cm, $r_d = 7.1$ mm, $\epsilon_r = \epsilon_\theta = 2.2$, $(\theta_i, \phi_i) = (63°, 0°)$. (From Ref. [6], © 1995 John Wiley & Sons, Inc., reprinted with permission.)

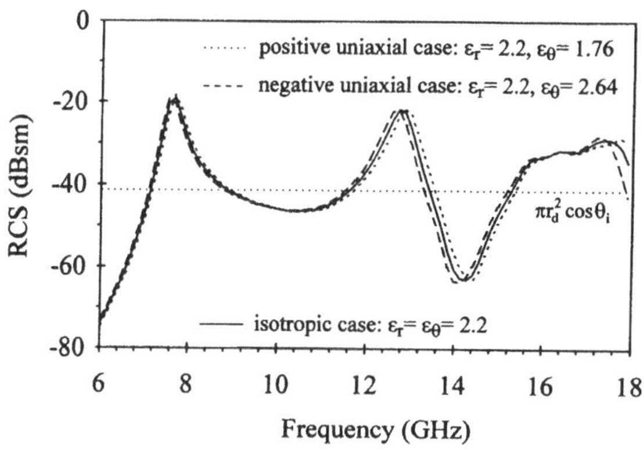

FIGURE 3.7 Radar cross section versus frequency for a uniaxial substrate with $a = 10$ cm, $r_d = 7.1$ mm, $\epsilon_r = 2.2$, $\epsilon_\theta = 1.76$, 2.2, 2.64 [anisotropic ratio $(=\epsilon_\theta/\epsilon_r) = 0.8$, 1.0, 1.2], $(\theta_i, \phi_i) = (63°, 0°)$. (From Ref. [6], © 1995 John Wiley & Sons, Inc., reprinted with permission.)

It is observed that the uniaxial anisotropy causes a shift of the resonant peak of the RCS values. The negative uniaxial anisotropy shifts the resonant RCS peak to lower frequencies and the positive uniaxial anisotropy moves the resonant peak to higher frequencies. It is also found that the RCS resonant peak value is almost independent of the uniaxial anisotropy.

3.3 SPHERICAL ANNULAR-RING MICROSTRIP PATCH

Theoretical calculations of complex resonant frequencies of the TM_{11} and TM_{12} modes for a spherical annular-ring microstrip structure are performed using a spectral-domain Green's function formulation and the Galerkin's moment-method procedure as described in Section 3.2. Convergent solutions of complex resonant frequencies are tested by using different numbers of cavity-model basis functions for the annular-ring microstrip patch. The effects of the sphere radius on the resonant frequency and quality factor (its inverse gives half-power bandwidth) of the microstrip structure are analyzed.

3.3.1 Theoretical Formulation

Consider the geometry of a spherical annular-ring microstrip structure shown in Figure 3.8. The substrate is assumed to be isotropic with a relative permittivity of ϵ_1. The annular-ring patch with inner radius r_1 $(=b\theta_1)$ and outer radius r_2 $(=b\theta_2)$ is printed on the substrate. By following the derivation procedure in Section 3.2.3, the relationship of the tangential electric field and the surface current density on the patch is first derived to be

$$\tilde{E}(n) = \bar{\bar{Y}}^{-1}\tilde{J}(n)\,, \tag{3.112}$$

with

$$\bar{\bar{Y}}(n) = \begin{bmatrix} Y_{11}(n) & 0 \\ 0 & Y_{22}(n) \end{bmatrix}, \tag{3.113}$$

$$Y_{11}(n) = j\sqrt{\frac{\epsilon_0}{\mu_0}}\frac{\hat{H}_n^{(2)}(k_0b)}{\hat{H}_n^{(2)\prime}(k_0b)} - j\sqrt{\frac{\epsilon_0\epsilon_1}{\mu_0}}\frac{\hat{H}_n^{(1)}(k_1b)\hat{H}_n^{(2)\prime}(k_1a) - \hat{H}_n^{(1)\prime}(k_1a)\hat{H}_n^{(2)}(k_1b)}{\hat{H}_n^{(1)\prime}(k_1b)\hat{H}_n^{(2)\prime}(k_1a) - \hat{H}_n^{(1)\prime}(k_1a)\hat{H}_n^{(2)\prime}(k_1b)}, \tag{3.114}$$

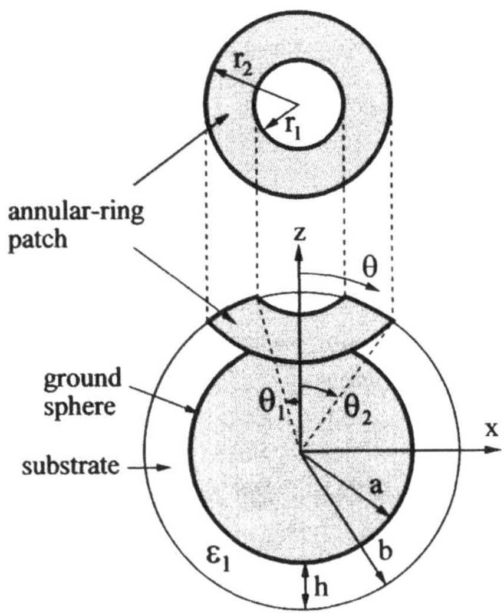

FIGURE 3.8 Geometry of a spherical annular-ring microstrip structure.

$$Y_{22}(n) = -j\sqrt{\frac{\epsilon_0}{\mu_0}}\frac{\hat{H}_n^{(2)'}(k_0b)}{\hat{H}_n^{(2)}(k_0b)} + j\sqrt{\frac{\epsilon_0\epsilon_1}{\mu_0}}\frac{\hat{H}_n^{(1)'}(k_1b)\hat{H}_n^{(2)}(k_1a) - \hat{H}_n^{(1)}(k_1a)\hat{H}_n^{(2)'}(k_1b)}{\hat{H}_n^{(1)}(k_1b)\hat{H}_n^{(2)}(k_1a) - \hat{H}_n^{(1)}(k_1a)\hat{H}_n^{(2)}(k_1b)} . \tag{3.115}$$

In (3.112), $k_1 = k_0\sqrt{\epsilon_1}$; $\tilde{E}(n)$ is again the spectral amplitude of the tangential electric field at $r = b$, and $\bar{\bar{Y}}^{-1}(n)$ is the Green's function in the spectral domain. We then solve (3.112) using Galerkin's moment-method procedure. First, the unknown spectral surface current density is expanded by a set of known basis functions, which again can be derived using cavity-model theory. From the cavity-model assumption, the TM fields between the annular-ring patch and the conducting sphere have only θ and ϕ variations, and the electric current on the annular-ring patch must have no component normal to the edge. Thus the TM fields satisfying the boundary condition $H_\phi(\theta = \theta_1) = H_\phi(\theta = \theta_2) = 0$ are given by

$$E_r = E_0 A_{\text{TM}}(\theta)e^{jm\phi} , \tag{3.116}$$

$$H_\theta = \frac{jmE_0}{\omega\mu_0 a \sin\theta} A_{\text{TM}}(\theta)e^{jm\phi} , \tag{3.117}$$

$$H_\phi = \frac{-E_0}{\omega\mu_0 a}\frac{\partial A_{\text{TM}}(\theta)}{\partial\theta} e^{jm\phi}, \tag{3.118}$$

with

$$A_{\text{TM}}(\theta) = P_\nu^{m'}(-\cos\theta_1)P_\nu^m(\cos\theta) + P_\nu^{m'}(\cos\theta_1)P_\nu^m(-\cos\theta) , \tag{3.119}$$

$$A'_{\text{TM}}(\theta_2) = P_\nu^{m'}(-\cos\theta_1)P_\nu^{m'}(\cos\theta_2) - P_\nu^{m'}(\cos\theta_1)P_\nu^{m'}(-\cos\theta_2) = 0 . \tag{3.120}$$

From (3.117) and (3.118), the surface current distribution of the $\text{TM}_{m\nu}$ mode on the annular-ring patch is written as [8]

$$J_{\text{TM}_{m\nu}} = \begin{bmatrix} J_\theta \\ J_\phi \end{bmatrix} = \begin{cases} \begin{bmatrix} \dfrac{\partial A_{\text{TM}}(\theta)}{\partial\theta} \\ \dfrac{jm}{\sin\theta} A_{\text{TM}}(\theta) \end{bmatrix}, & \theta_1 < \theta < \theta_2 , \\ 0 , & \text{elsewhere} . \end{cases} \tag{3.121}$$

By applying the vector Legendre transform, the spectral amplitude of the surface current distribution for the $\text{TM}_{m\nu}$ mode is derived to be

$$\tilde{J}_{\mathrm{TM}_{m\nu}} = \frac{1}{S(n,m)} \times \begin{bmatrix} \dfrac{\nu(\nu+1)}{n(n+1)-\nu(\nu+1)} [\sin^2\theta_2 P_n^{m'}(\cos\theta_2) A_{\mathrm{TM}}(\theta_2) - \sin^2\theta_1 P_n^{m'}(\cos\theta_1) A_{\mathrm{TM}}(\theta_1)] \\ jm[P_n^m(\cos\theta_2) A_{\mathrm{TM}}(\theta_2) - P_n^m(\cos\theta_1) A_{\mathrm{TM}}(\theta_1)] \end{bmatrix} . \tag{3.122}$$

Similarly, the $\mathrm{TE}_{m\nu}$-mode surface current distribution is given by

$$J_{\mathrm{TE}_{m\nu}} = \begin{bmatrix} J_\theta \\ J_\phi \end{bmatrix} = \begin{cases} \begin{bmatrix} \dfrac{jm}{\sin\theta} A_{\mathrm{TE}}(\theta) \\ -\dfrac{\partial A_{\mathrm{TE}}(\theta)}{\partial\theta} \end{bmatrix}, & \theta_1 < \theta < \theta_2 , \\ 0 , & \text{elsewhere} , \end{cases} \tag{3.123}$$

with

$$A_{\mathrm{TE}}(\theta) = P_\nu^m(-\cos\theta_1) P_\nu^m(\cos\theta) - P_\nu^m(\cos\theta_1) P_\nu^m(-\cos\theta) , \tag{3.124}$$

$$A_{\mathrm{TE}}(\theta_2) = P_\nu^m(-\cos\theta_1) P_\nu^m(\cos\theta_2) - P_\nu^m(\cos\theta_1) P_\nu^m(-\cos\theta_2) = 0 , \tag{3.125}$$

whose spectral amplitude is derived to be

$$\tilde{J}_{\mathrm{TE}_{m\nu}} = \frac{1}{S(n,m)} \times \begin{bmatrix} 0 \\ \dfrac{n(n+1)}{n(n+1)-\nu(\nu+1)} [\sin^2\theta_2 P_n^m(\cos\theta_2) A'_{\mathrm{TE}}(\theta_2) - \sin^2\theta_1 P_n^m(\cos\theta_1) A'_{\mathrm{TE}}(\theta_1)] \end{bmatrix} , \tag{3.126}$$

with

$$A'_{\mathrm{TE}}(\theta_i) = P_\nu^m(-\cos\theta_1) P_\nu^{m'}(\cos\theta_i) + P_\nu^m(\cos\theta_1) P_\nu^{m'}(-\cos\theta_i) , \qquad i = 1, 2 . \tag{3.127}$$

With (3.122) and (3.126) as the expansion functions, a similar expression of (3.84) can also be derived. Solving this homogeneous matrix equation of the microstrip structure, complex resonant frequencies are obtained. The results are discussed below.

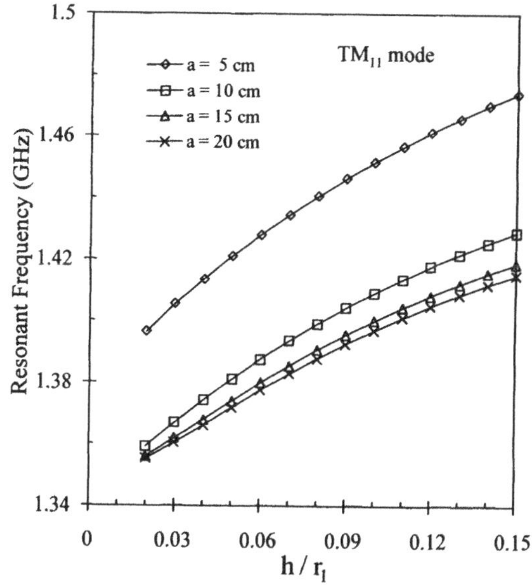

(*a*)

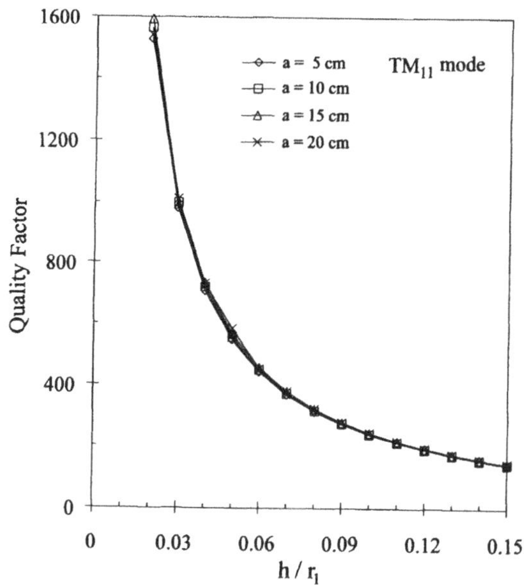

(*b*)

FIGURE 3.9 (*a*) Resonant frequency and (*b*) quality factor of the TM_{11} mode versus substrate thickness for different sphere radii; $r_1 = 1.5$ cm, $r_2 = 3$ cm, $\epsilon_1 = 2.65$. (From Ref. [8], © 1993 John Wiley & Sons, Inc.)

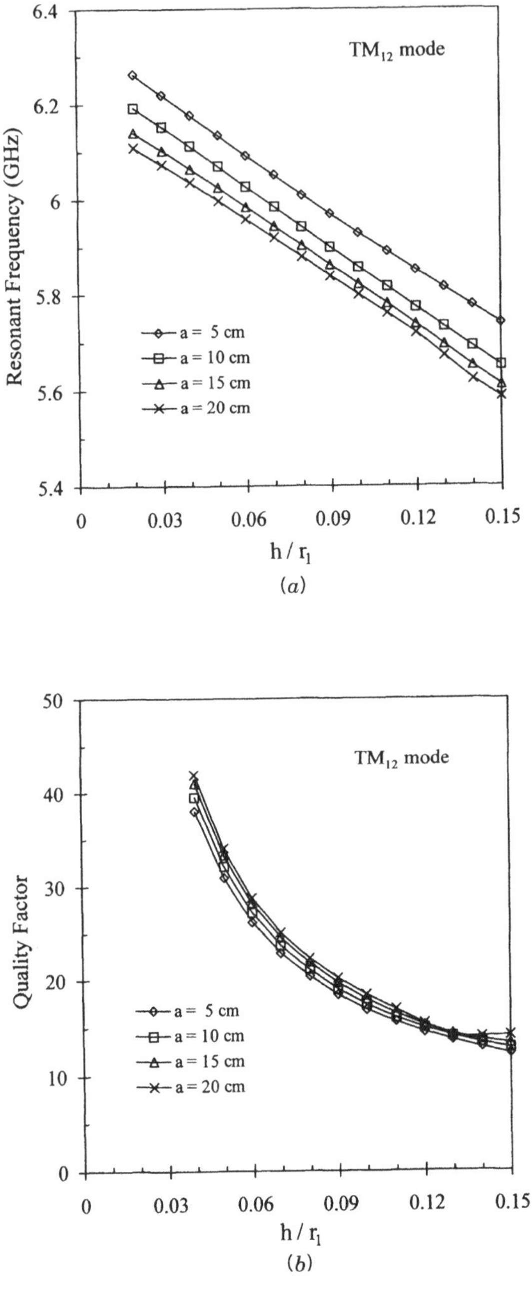

FIGURE 3.10 (*a*) Resonant frequency and (*b*) quality factor of the TM_{12} mode versus substrate thickness for different sphere radii; $r_1 = 1.5$ cm, $r_2 = 3$ cm, $\epsilon_1 = 2.65$. (From Ref. [8], © 1993 John Wiley & Sons, Inc.)

3.3.2 Complex Resonant Frequency Results

The convergence is first tested. It is observed that the convergent solutions can be reached by using five cavity-model basis functions ($N = 5$: TM_{11}, TE_{11}, TM_{12}, TE_{12}, and TM_{13} modes) and the imaginary parts of complex resonant frequencies converge faster than the real parts of complex resonant frequencies. The calculated resonant frequency and quality factor for the annular-ring patch excited in the TM_{11} and TM_{12} modes are, respectively, presented in Figures 3.9 and 3.10. The results for various sphere radii are shown. As the sphere radius decreases, the resonant frequencies for both modes are increased. Similar to the explanation given in Chapter 2 for a cylindrical microstrip structure, the effective radius of the annular-ring patch decreases as the curvature radius becomes smaller. It is clear that the TM_{12} mode has a lower Q factor than the TM_{11} mode. This indicates that the TM_{12} mode is an efficiently radiating mode for antenna application, whereas the TM_{11} mode is best suited for resonator applications. Also, the quality factor for the TM_{11} mode is varied only slightly with the sphere radii. On the other hand, variation in the quality factor is more sensitive to sphere radii for the TM_{12} mode. This suggests that a spherical annular-ring microstrip structure with a smaller sphere radius is more suitable as a radiator.

3.4 SPHERICAL MICROSTRIP PATCH WITH A SUPERSTRATE

Superstrate-loaded spherical circular and annular-ring microstrip structures have also been studied [9,10]. This dielectric superstrate layer can provide protection for the microstrip patch against heat, physical damage or environmental hazards, but it can also alter the resonant frequency of the microstrip structure, which may shift the operation frequency outside the operation band of interest. Various effects of the superstrate layer on the resonance of the spherical microstrip structures are discussed below.

3.4.1 Circular Microstrip Patch

Figure 3.11 shows the geometry under consideration. The superstrate layer has a thickness of t and a relative permittivity of ϵ_2. The substrate is assumed to be isotropic with a relative permittivity of ϵ_1. Other parameters are as given in Figure 3.1. By applying a full-wave formulation similar to that described in Section 3.2.3, and after some lengthy manipulation, we have the elements Y_{11} and Y_{22} in (3.80) written as

$$Y_{11} = j\alpha_0 \frac{\alpha_1[\hat{H}_n^{(1)}(k_2b) - X_1\sqrt{\epsilon_{12}}\hat{H}_n^{(1)'}(k_2b)] - \alpha_2[\hat{H}_n^{(2)}(k_2b) - X_1\sqrt{\epsilon_{12}}\hat{H}_n^{(2)'}(k_2b)]}{\alpha_1\hat{H}_n^{(1)'}(k_2b) - \alpha_2\hat{H}_n^{(2)'}(k_2b)}, \tag{3.128}$$

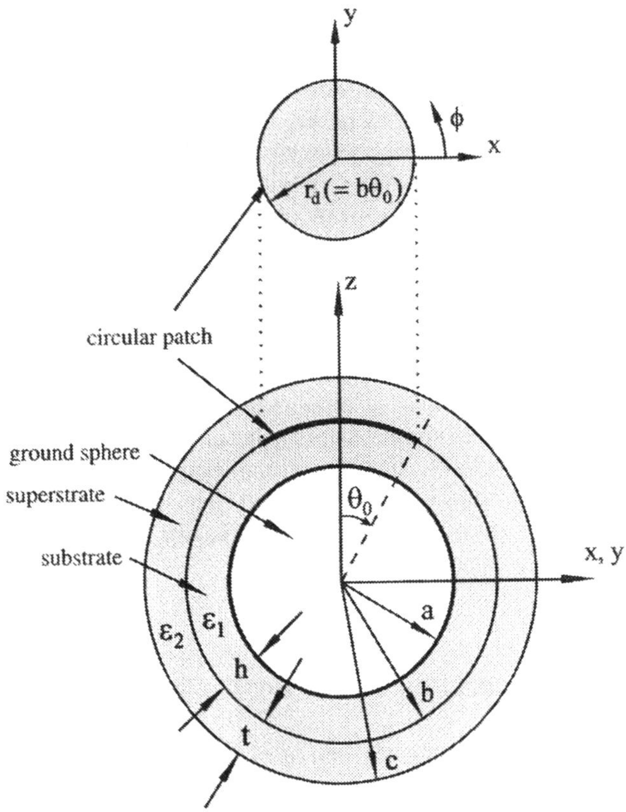

FIGURE 3.11 Geometry of a superstrate-loaded spherical circular microstrip structure.

$$Y_{22} = j\alpha_0 \frac{\alpha_3[\hat{H}_n^{(1)'}(k_2 b) - X_2\sqrt{\epsilon_{12}}\hat{H}_n^{(1)}(k_2 b)] - \alpha_4[\hat{H}_n^{(2)'}(k_2 b) - X_2\sqrt{\epsilon_{12}}\hat{H}_n^{(2)}(k_2 b)]}{\alpha_3\hat{H}_n^{(1)}(k_2 b) - \alpha_4\hat{H}_n^{(2)}(k_2 b)}, \tag{3.129}$$

with

$$\alpha_0 = \sqrt{\frac{\epsilon_0\epsilon_2}{\mu_0}}, \qquad \epsilon_{12} = \frac{\epsilon_1}{\epsilon_2}, \tag{3.130}$$

$$\alpha_1 = \frac{\hat{H}_n^{(2)}(k_2 c)}{\hat{H}_n^{(2)}(k_0 c)} - \frac{1}{\sqrt{\epsilon_2}}\frac{\hat{H}_n^{(2)'}(k_2 c)}{\hat{H}_n^{(2)'}(k_0 c)}, \tag{3.131}$$

$$\alpha_2 = \frac{\hat{H}_n^{(1)}(k_2 c)}{\hat{H}_n^{(2)}(k_0 c)} - \frac{1}{\sqrt{\epsilon_2}}\frac{\hat{H}_n^{(1)'}(k_2 c)}{\hat{H}_n^{(2)'}(k_0 c)}, \tag{3.132}$$

$$\alpha_3 = \frac{\hat{H}_n^{(2)}(k_2 c)}{\hat{H}_n^{(2)}(k_0 c)} - \sqrt{\epsilon_2}\,\frac{\hat{H}_n^{(2)'}(k_2 c)}{\hat{H}_n^{(2)'}(k_0 c)}, \tag{3.133}$$

$$\alpha_4 = \frac{\hat{H}_n^{(1)}(k_2 c)}{\hat{H}_n^{(2)}(k_0 c)} - \sqrt{\epsilon_2}\,\frac{\hat{H}_n^{(1)'}(k_2 c)}{\hat{H}_n^{(2)'}(k_0 c)}, \tag{3.134}$$

$$X_1 = \frac{\hat{H}_n^{(2)'}(k_1 a)\hat{H}_n^{(1)}(k_1 b) - \hat{H}_n^{(2)}(k_1 b)\hat{H}_n^{(1)'}(k_1 a)}{\hat{H}_n^{(2)'}(k_1 a)\hat{H}_n^{(1)'}(k_1 b) - \hat{H}_n^{(2)'}(k_1 b)\hat{H}_n^{(1)'}(k_1 a)}, \tag{3.135}$$

$$X_2 = \frac{\hat{H}_n^{(2)}(k_1 a)\hat{H}_n^{(1)'}(k_1 b) - \hat{H}_n^{(2)'}(k_1 b)\hat{H}_n^{(1)}(k_1 a)}{\hat{H}_n^{(2)}(k_1 a)\hat{H}_n^{(1)}(k_1 b) - \hat{H}_n^{(2)}(k_1 b)\hat{H}_n^{(1)}(k_1 a)}, \tag{3.136}$$

$$k_1 = k_0\sqrt{\epsilon_1}, \qquad k_2 = k_0\sqrt{\epsilon_2}.$$

By substituting (3.128)–(3.129) into (3.85) for the evaluation of elements in the [Z] matrix and solving the determinant of [Z], the resonance problem of the superstrate-loaded spherical circular microstrip structure is solved. Figure 3.12 presents the calculated resonant frequency results for various sphere radii, and Figure 3.13 shows the superstrate permittivity effects on the resonant frequency. From the results it is seen that the resonant frequency decreases monotonically with increased superstrate thickness. The larger the superstrate permittivity, the greater the decrease in resonant frequency. In Figure 3.12 we also show the resonant frequency of a superstrate-loaded planar circular microstrip structure [11] (see also Appendix C) for comparison. The results are calculated using a curve-

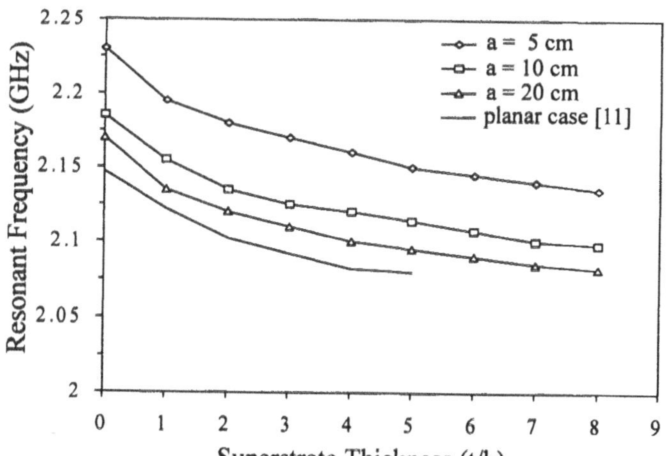

FIGURE 3.12 Resonant frequency versus superstrate thickness for various sphere radii; $\epsilon_1 = \epsilon_2 = 2.5$, $r_d = 2.5$ cm, $h = 1.588$ mm. (From Ref. [9], © 1993 IEEE, reprinted with permission.)

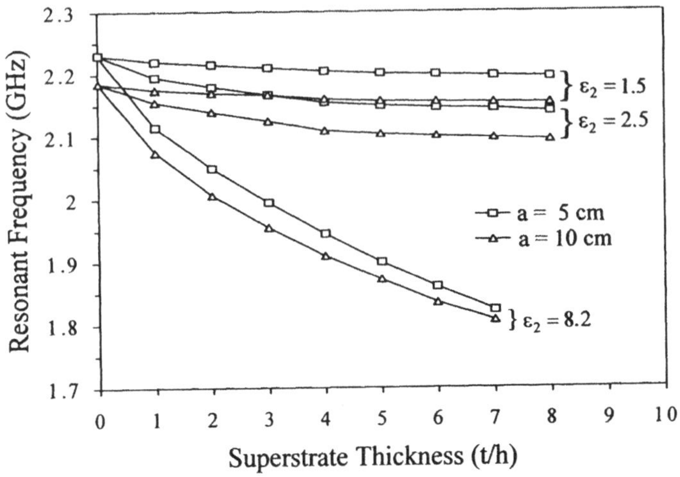

FIGURE 3.13 Resonant frequency versus superstrate thickness for various superstrate permittivities; $\epsilon_1 = 2.5$, $r_d = 2.5$ cm, $h = 1.588$ mm. (From Ref. [9], © 1993 IEEE, reprinted with permission.)

fitting formula, which is a form of the multivariable polynomial developed using a database generated from a full-wave approach incorporating Galerkin's moment method. This curve-fitting formula makes possible fast, accurate determination of the complex resonant frequencies of the planar circular microstrip structure. Details of the curve-fitting formulas are listed in Appendix C. It is observed that when the sphere radius increases, the curve calculated for the spherical microstrip structure approaches that of the planar structure.

Figure 3.14 shows half-power bandwidth versus superstrate thickness for various sphere radii and superstrate permittivities. It is observed that the superstrate loading greatly increases the half-power bandwidth of the microstrip structure. For the case of $\epsilon_2 = 1.5$ and 2.5 in Figure 3.14*b*, the bandwidth varies slightly with increasing superstrate thickness. As for the case of $\epsilon_2 = 8.2$, the bandwidth increases significantly, especially for larger superstrate thicknesses.

The radiation pattern of the microstrip structure at resonance is also analyzed. By applying a formulation similar to that described in Section 3.2.7, the far-zone radiated fields are derived as

$$E(\theta, \phi) = \begin{bmatrix} E_\theta \\ E_\phi \end{bmatrix} = e^{jm\phi} \frac{be^{-jk_0 r}}{r} j^n \sum_{n=m}^{\infty} \tilde{\bar{L}}(n, m, \theta)\bar{\bar{G}}\bar{\bar{Y}}^{-1}\tilde{J}, \tag{3.137}$$

with

$$\bar{\bar{G}} = \begin{bmatrix} G_{11} & 0 \\ 0 & G_{22} \end{bmatrix}, \tag{3.138}$$

$$G_{11} = \frac{1}{\hat{H}_n^{(2)\prime}(k_0 c)} \frac{\hat{H}_n^{(1)\prime}(k_2 c) - S_1 \hat{H}_n^{(2)\prime}(k_2 c)}{\hat{H}_n^{(1)\prime}(k_2 b) - S_1 \hat{H}_n^{(2)\prime}(k_2 b)}, \tag{3.139}$$

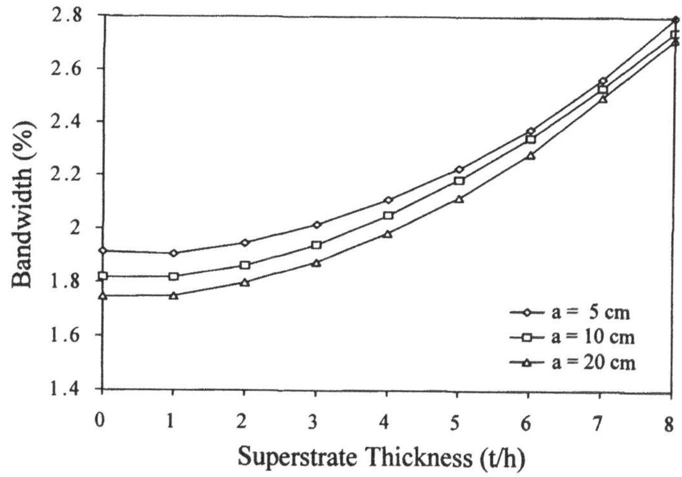

(*a*)

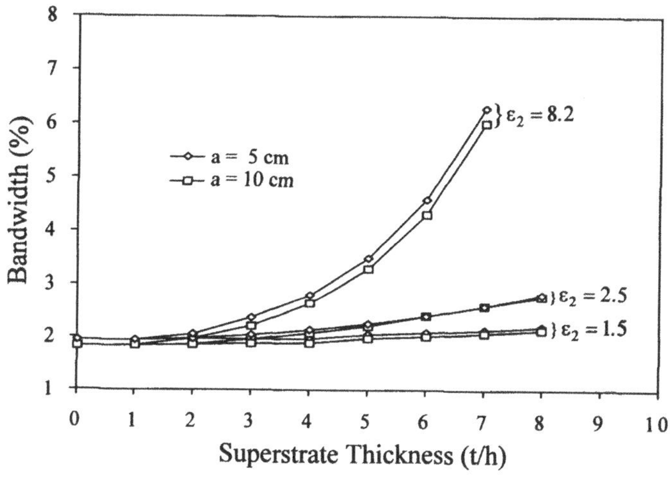

(*b*)

FIGURE 3.14 Half-power bandwidth versus superstrate thickness for various (*a*) sphere radii in Figure 3.12 and (*b*) superstrate permittivities in Figure 3.13; $\epsilon_1 = 2.5$, $r_d = 2.5$ cm, $h = 1.588$ mm. (From Ref. [9], © 1993 IEEE, reprinted with permission.)

$$G_{22} = \frac{j}{\hat{H}_n^{(2)}(k_0 c)} \frac{\hat{H}_n^{(1)}(k_2 c) - S_2 \hat{H}_n^{(2)}(k_2 c)}{\hat{H}_n^{(1)}(k_2 b) - S_2 \hat{H}_n^{(2)}(k_2 b)}, \tag{3.140}$$

$$S_1 = \frac{\alpha_2}{\alpha_1}, \qquad S_2 = \frac{\alpha_4}{\alpha_3}, \tag{3.141}$$

where the α_i are defined in (3.131)–(3.134). Note that this expression for the far-zone electric field can also be reduced to the special case without a superstrate layer when setting $c = b$, $k_2 = k_0$.

E- and *H*-plane radiation patterns for different superstrate thicknesses are plotted in Figure 3.15. It is seen that the 3-dB beamwidth of the *E*-plane pattern decreases with increasing superstrate thickness. For the H-plane pattern it is observed that the superstrate loading effect is small. The effects of sphere-radius

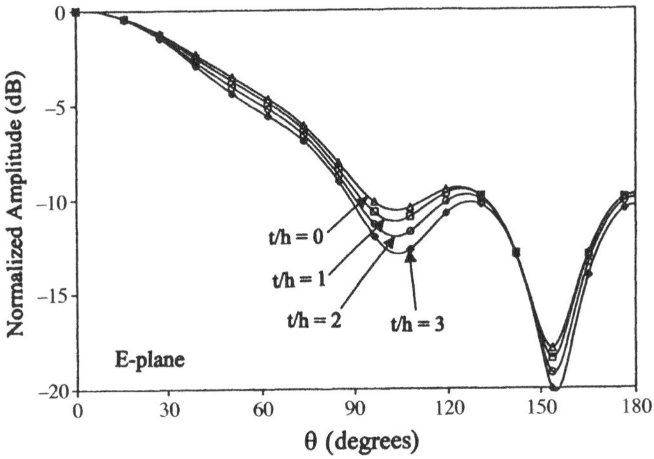

(*a*)

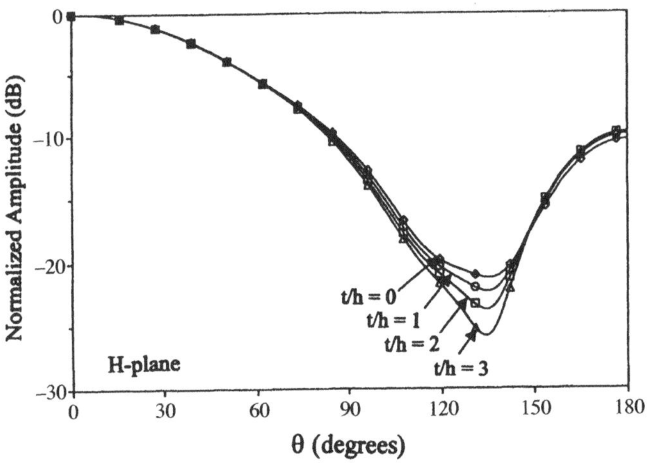

(*b*)

FIGURE 3.15 (*a*) *E*-plane and (*b*) *H*-plane radiation patterns with $a = 5$ cm, $h = 1.6$ mm, $r_d = 1.88$ cm, $\epsilon_1 = 2.47$, $\epsilon_2 = 2.32$, $t/h = 0, 1, 2, 3$. (From Ref. [9], © 1993 IEEE, reprinted with permission.)

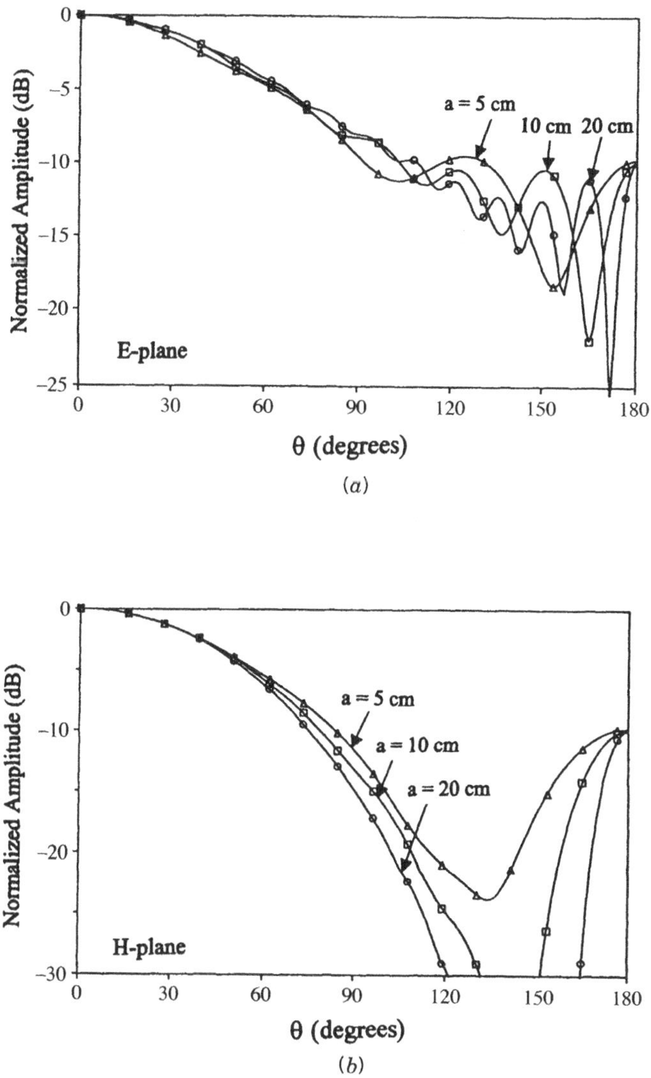

FIGURE 3.16 (*a*) *E*-plane and (*b*) *H*-plane radiation patterns with $t/h = 1$, $h = 1.6$ mm, $r_d = 1.88$ cm, $\epsilon_1 = 2.47$, $\epsilon_2 = 2.32$, $a = 5$, 10, 20 cm.

variation on the patterns are shown in Figure 3.16. The side-lobe level and backward radiation are seen to increase with decreasing sphere radius.

3.4.2 Annular-Ring Microstrip Patch

The geometry under consideration is shown in Figure 3.17. By expanding the surface current density on the annular-ring patch using (3.122) and (3.126) and applying the same theoretical formulation as in Section 3.4.1, the resonance

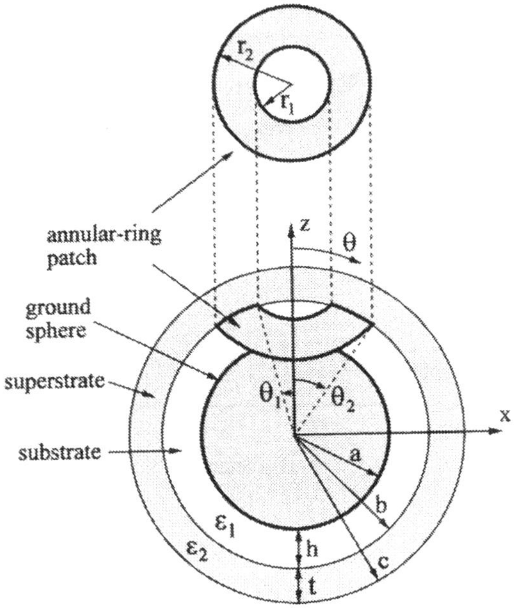

FIGURE 3.17 Geometry of a superstrate-loaded spherical annular-ring microstrip structure.

problem of a superstrate-loaded spherical annular-ring microstrip structure is solved. Both the TM_{12} and TM_{11} modes are studied. The results for the TM_{12} mode are shown in Figures 3.18 and 3.19. Seven cavity-model basis functions (TE_{1n}, $n = 1, 2, 3$ and TM_{1n}, $n = 1, 2, 3, 4$) are used in the theoretical calculation. From the results in Figure 3.18, the resonant frequency again is seen to decrease with increasing superstrate thickness and is smaller for a larger sphere radius. The quality factor also decreases significantly with increasing superstrate thickness. This indicates that adding a superstrate layer will greatly increase the radiation loss of the microstrip structure.

Results showing the effects of superstrate permittivity on the resonant frequency and quality factor are presented in Figure 3.19. The cases of $r_2 = 6$ cm, $r_1 = 3$ cm and $r_2 = 6$ cm, $r_1 = 1.5$ cm are shown. It is observed that the lower the superstrate permittivity, the less the decrease in resonant frequency and quality factor. Also, the quality factor is smaller for a structure with a larger inner radius, which reduces the energy stored under the annular-ring patch. This suggests that an annular-ring microstrip structure with a larger inner radius excited in the TM_{12} mode is a better radiator. However, the resonant frequency for the case with a larger inner radius is more sensitive to superstrate loading.

Figure 3.20 shows the results for TM_{11}-mode excitation. Again, the resonant frequency decreases with increasing superstrate thickness. However, the variation is smaller than that in Figure 3.18. Also, the quality factor varies slightly with superstrate loading. This is probably because, as a resonator, the wave energy is

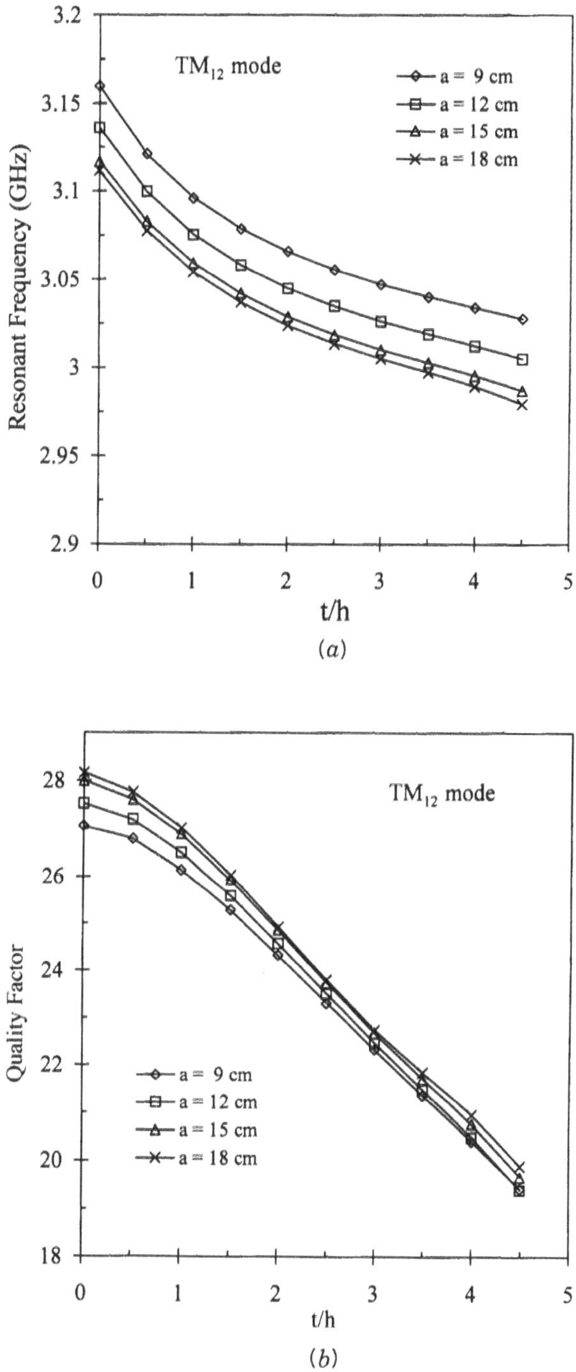

FIGURE 3.18 (*a*) Resonant frequency and (*b*) quality factor of the TM_{12} mode versus normalized superstrate thickness; $r_1 = 3$ cm, $r_2 = 6$ cm, $\epsilon_1 = \epsilon_2 = 2.47$. (From Ref. [10], © 1994 John Wiley & Sons, Inc.)

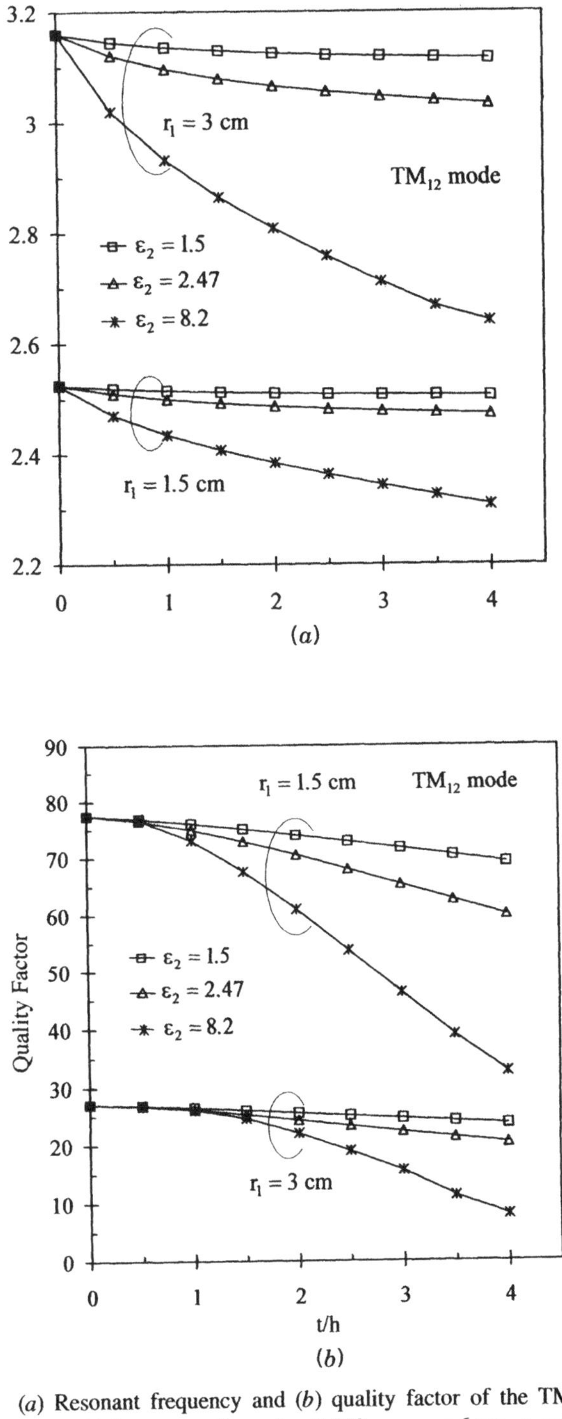

FIGURE 3.19 (*a*) Resonant frequency and (*b*) quality factor of the TM_{12} mode versus normalized superstrate thickness; $a = 9$ cm, $h = 0.159$ cm, $r_2 = 6$ cm, $\epsilon_1 = 2.47$. (From Ref. [10], © 1994 John Wiley & Sons, Inc.)

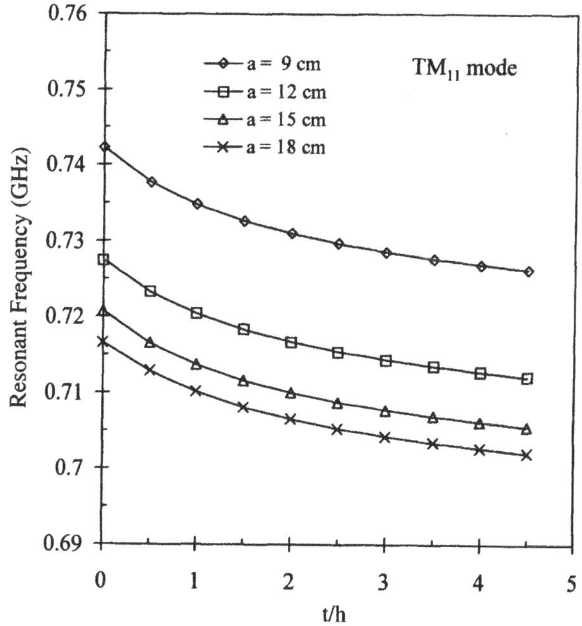

(*a*)

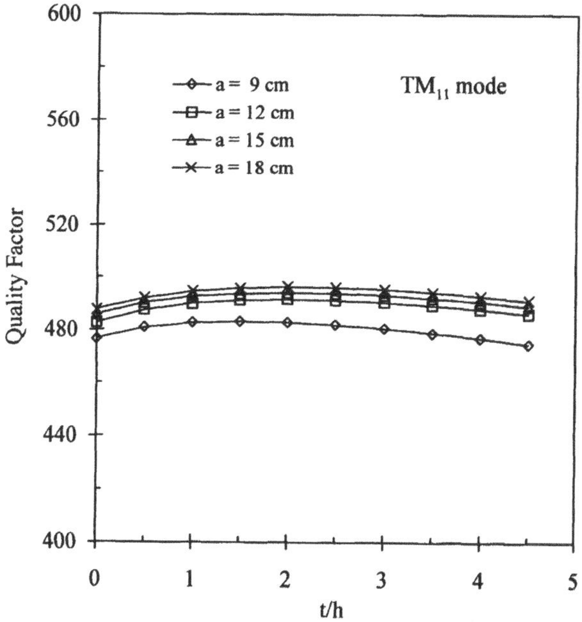

(*b*)

FIGURE 3.20 (*a*) Resonant frequency and (*b*) quality factor of the TM_{11} mode versus normalized superstrate thickness. Parameters are the same as in Figure 3.18. (From Ref. [10], © 1994 John Wiley & Sons, Inc.)

confined almost entirely inside the structure, and thus adding a superstrate layer will have very little effect on radiation loss in the structure.

3.5 SPHERICAL MICROSTRIP PATCH WITH AN AIR GAP

For the analysis of a spherical microstrip structure with an air gap [12,13], the effects of an air gap on bandwidth enhancement are the major concern. Spherical circular and annular-ring microstrip structures with an air gap are discussed below.

3.5.1 Circular Microstrip Patch

Given the geometry in Figure 3.21, an air gap ($\epsilon_s = 1.0$) with a thickness of S is placed between the substrate layer and the ground sphere. The theoretical treatment is similar to that described in Section 3.2, and the complex resonant

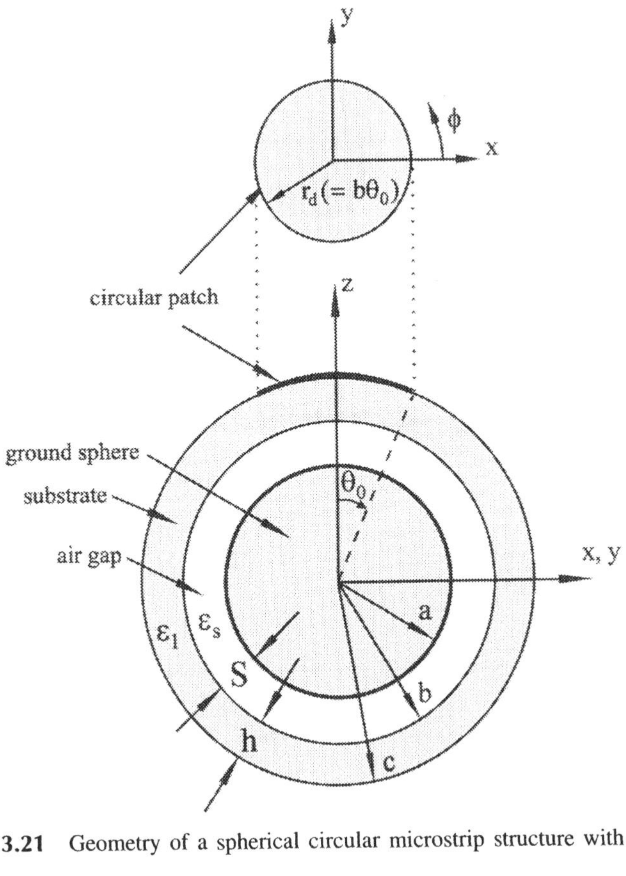

FIGURE 3.21 Geometry of a spherical circular microstrip structure with an air gap.

frequencies are calculated by replacing Y_{11} and Y_{22} in (3.80) with the following expressions [12]:

$$Y_{11}(n) = -R_2(n) + R_1(n)\,, \tag{3.142}$$

$$Y_{22}(n) = T_2(n) - T_1(n), \tag{3.143}$$

where

$$R_i(n) = -j\sqrt{\frac{\epsilon_0\epsilon_i}{\mu_0}}\,\frac{\hat{J}_n(k_i r) + P_i(n)\hat{Y}_n(k_i r)}{\hat{J}'_n(k_i r) + P_i(n)\hat{Y}'_n(k_i r)}\,, \qquad i = 0, 1, 2\,, \tag{3.144}$$

$$T_i(n) = -j\sqrt{\frac{\epsilon_0\epsilon_i}{\mu_0}}\,\frac{\hat{J}'_n(k_i r) + Q_i(n)\hat{Y}'_n(k_i r)}{\hat{J}_n(k_i r) + Q_i(n)\hat{Y}_n(k_i r)}\,, \qquad i = 0, 1, 2\,, \tag{3.145}$$

with

$$P_0 = -\frac{\hat{J}'_n(k_0 a)}{\hat{Y}'_n(k_0 a)}\,, \qquad Q_0 = -\frac{\hat{J}_n(k_0 a)}{\hat{Y}_n(k_0 a)}\,, \tag{3.146}$$

$$P_1(n) = -\frac{R_0(n)\hat{J}'_n(k_1 b) + j\hat{J}_n(k_1 b)\sqrt{\epsilon_0\epsilon_1/\mu_0}}{R_0(n)\hat{Y}'_n(k_1 b) + j\hat{Y}_n(k_1 b)\sqrt{\epsilon_0\epsilon_1/\mu_0}}\,, \tag{3.147}$$

$$Q_1(n) = -\frac{T_0(n)\hat{J}_n(k_1 b) + j\hat{J}'_n(k_1 b)\sqrt{\epsilon_0\epsilon_1/\mu_0}}{T_0(n)\hat{Y}_n(k_1 b) + j\hat{Y}'_n(k_1 b)\sqrt{\epsilon_0\epsilon_1/\mu_0}}\,, \tag{3.148}$$

$$P_2 = Q_2 = -j\,, \tag{3.149}$$

$$k_1 = k_0\sqrt{\epsilon_1} = \sqrt{\mu_0\epsilon_0\epsilon_1}\,.$$

Real and imaginary parts of the calculated complex resonant frequencies in the TM_{11} mode are shown in Figure 3.22. The cases of $\epsilon_1 = 2.32$, $\epsilon_s = 1.0$ (an air gap is present) and $\epsilon_1 = \epsilon_s = 2.32$ (no air gap is present) are presented for comparison. For the latter case the substrate thickness is now $S + h$. It is seen that when an air gap is present, the resonant frequency increases with increasing air-gap thickness. This behavior is quite different from that for the case of $\epsilon_1 = \epsilon_s = 2.32$, where the resonant frequency decreases as the substrate thickness $(S + h)$ increases. This is due mainly to the effective permittivity of the region under the patch being lowered with the presence of an air gap. A smaller sphere radius also increases the resonant frequency. As for the imaginary resonant frequencies shown in Figure 3.22*b*, it is seen that the case of $\epsilon_1 = 2.32$, $\epsilon_s = 1.0$ has a larger value than that of $\epsilon_1 = \epsilon_s = 2.32$. This indicates that radiation loss of the microstrip structure increases with the presence of an air gap. A spherical structure with a smaller radius is also seen to be a more efficient radiator. From the results of Figure 3.22, the half-power bandwidth of the microstrip structure can also readily be calculated.

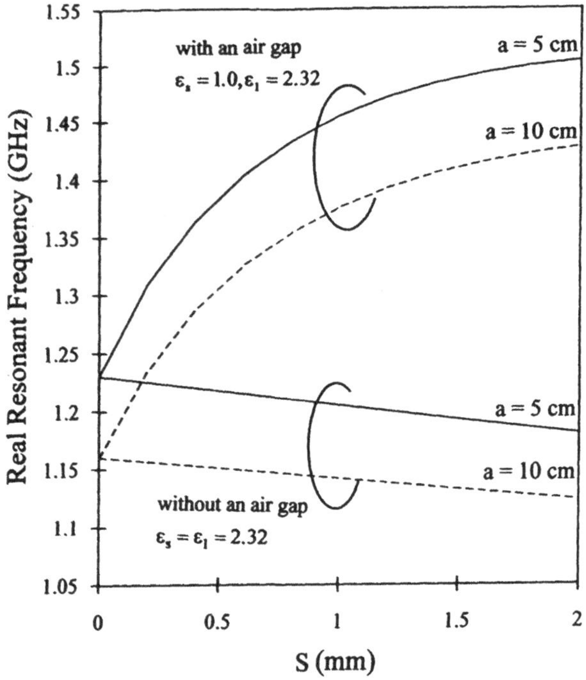

(*a*)

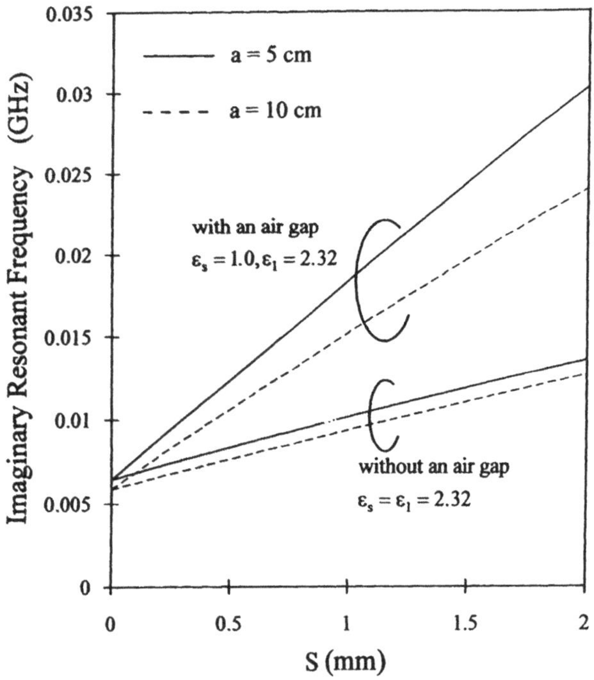

(*b*)

FIGURE 3.22 (*a*) Real and (*b*) imaginary parts of complex resonant frequencies in the TM_{11} mode with $\epsilon_1 = 2.32$, $r_d = 5$ cm, $h = 1.59$ mm, $\epsilon_s = 1.0$ (air gap) or 2.32. (From Ref. [12], © 1993 IEEE, reprinted with permission.)

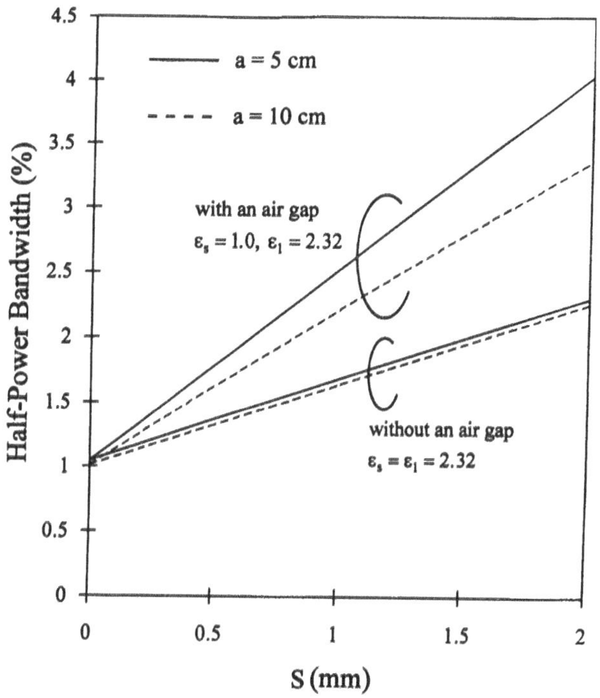

FIGURE 3.23 Half-power bandwidth of the microstrip structure shown in Figure 3.22. (From Ref. [12], © 1993 IEEE, reprinted with permission.)

The results are shown in Figure 3.23. The bandwidth is seen to be considerably increased due to the presence of an air gap, and a spherical structure with a smaller radius also has a larger bandwidth.

3.5.2 Annular-Ring Microstrip Patch

The geometry of a spherical annular-ring microstrip structure with an air gap is depicted in Figure 3.24. By expanding the surface current density on the annular-ring patch using (3.122) and (3.126), replacing Y_{11} and Y_{22} in (3.80) using (3.142) and (3.143), and applying a theoretical formulation in Section 3.2, complex resonant frequencies of the annular-ring microstrip structure with an air gap are obtained. The TM_{11} mode is studied first. In this case, six cavity-model basis functions of TE_{1n}, TM_{1n}, $n = 1, 2, 3$ [each value of n represents a specific ν in (3.120) and (3.125)], are enough to obtain convergent solutions. The real and imaginary parts of calculated complex resonant frequencies are shown in Figure 3.25. It is seen that when the inner radius of the annular-ring patch approaches zero (i.e., becomes a circular patch), the results agree with the data obtained in [12] for a spherical circular microstrip antenna. The resonant frequency decreases

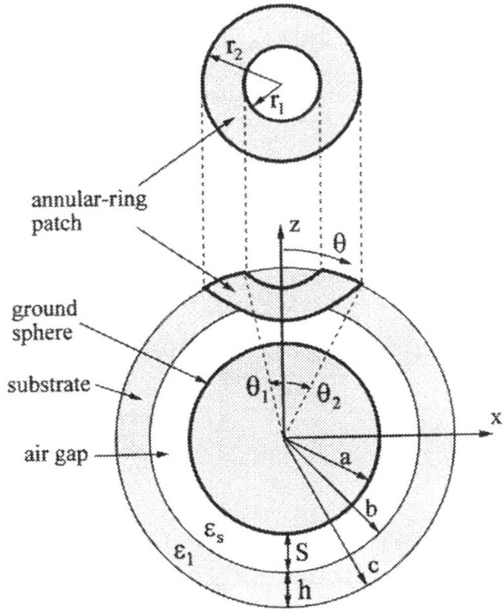

FIGURE 3.24 Geometry of a spherical annular-ring microstrip structure with an air gap.

when the inner radius increases. This behavior can be used for the design of a reduced-size printed antenna. However, it is also noted that the imaginary resonant frequency decreases with increasing inner radius. That is, the radiation loss of the TM_{11} mode operation decreases as the inner radius increases. This is because the fields from the inner and outer edges of the annular-ring patch interfere destructively for the TM_{11} mode and the radiation loss is thus reduced [14].

Since the TM_{11} mode is a poor radiation mode, the effects of introducing an air gap are discussed for the TM_{12} mode, which is a relatively better radiation mode. The results are shown in Figures 3.26 and 3.27. Various annular-ring patches of $r_2 = 5$ cm, $r_1 = 3.34$ cm and $r_2 = 5$ cm, $r_1 = 2.5$ cm are studied. It is seen that the resonant frequency is smaller for the patch with the smaller inner radius, which is contrary to the TM_{11} mode. With the presence of an air gap, the resonant frequency also increases, as seen for the circular patch case. However, for larger air-gap thickness, the resonant frequency starts to decrease slightly. This is probably because in this case the effective relative permittivity of the region under the annular-ring patch varies slightly and the thickness of the region starts to dominate the effects, which reduces the resonant frequency. As for the imaginary resonant frequency, it indicates that the radiation loss of the structure increases with increasing air-gap thickness. Figure 3.27 shows the bandwidth for the case in Figure 3.26, and the bandwidth is seen to increase with increasing air-gap thickness and inner radius. However, the effect of the air gap on the bandwidth enhancement is reduced for the larger inner-radius case. This is probably because for the TM_{12} mode, the fields from the inner edge of the patch interfere

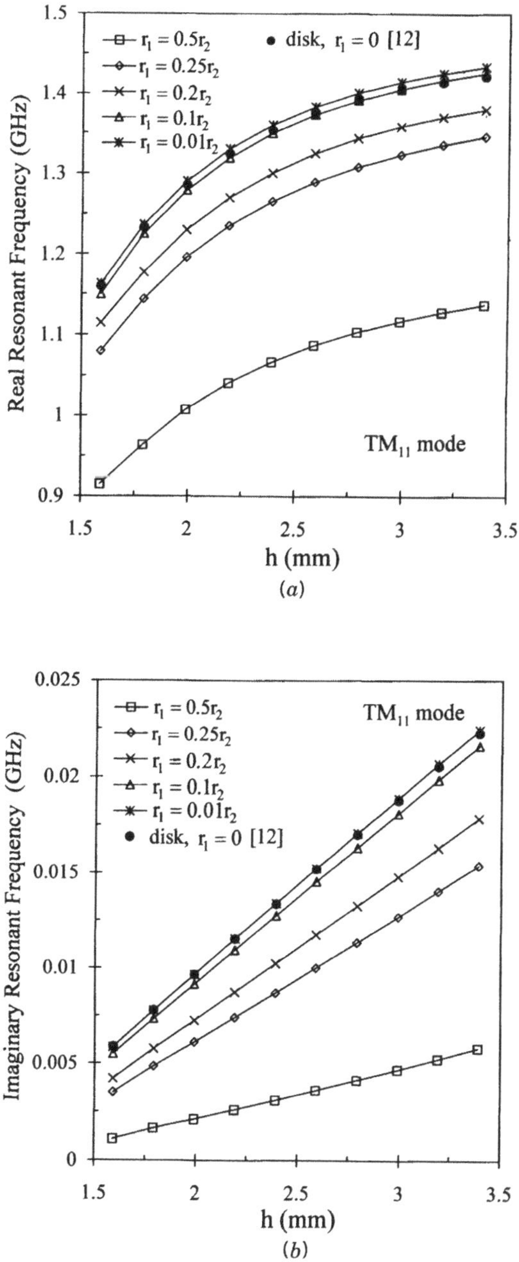

FIGURE 3.25 (*a*) Real and (*b*) imaginary parts of complex resonant frequencies in the TM_{11} mode with $a = 10$ cm, $\epsilon_1 = 2.32$, $S = 0$, $r_2 = 5$ cm. (From Ref. [13], © 1994 John Wiley & Sons, Inc.)

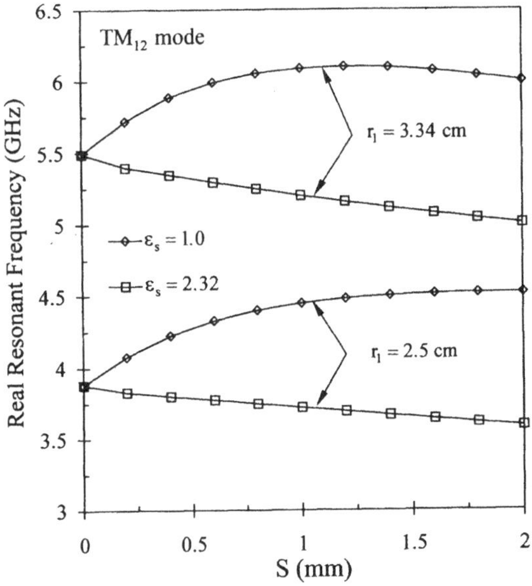

(*a*)

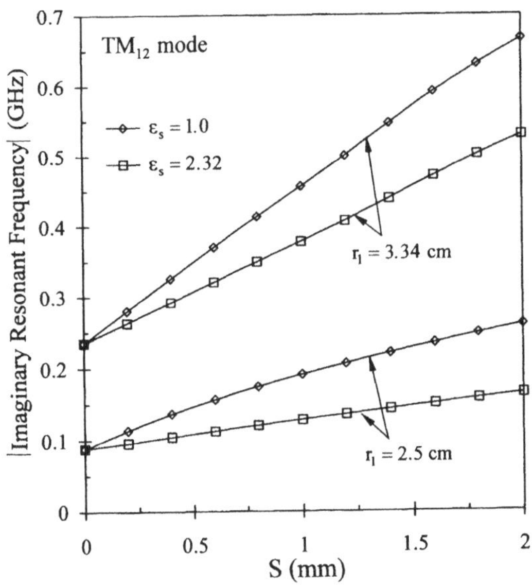

(*b*)

FIGURE 3.26 (*a*) Real and (*b*) imaginary parts of complex resonant frequencies in the TM_{12} mode with $a = 10$ cm, $h = 0.159$ cm, $\epsilon_1 = 2.32$, $r_2 = 5$ cm, $\epsilon_s = 1.0$ (air gap) or 2.32. (From Ref. [13], © 1994 John Wiley & Sons, Inc.)

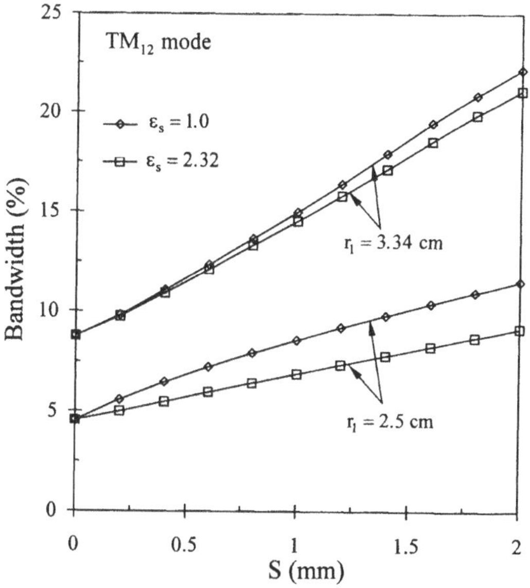

FIGURE 3.27 Half-power bandwidth for the case shown in Figure 3.26. (From Ref. [13], © 1994 John Wiley & Sons, Inc.)

constructively with the fields from the outer edge, and with increasing inner radius, the fields from the inner edge increase and start to dominate the radiation loss of the structure, which reduces the air-gap effect on bandwidth enhancement.

REFERENCES

1. N. G. Alexopoulos, "Integrated-circuit structures on anisotropic substrates," *IEEE Trans. Microwave Theory Tech.*, vol. 33, pp. 847–881, Oct. 1985.
2. K. L. Wong and H. T. Chen, "Electromagnetic scattering by a uniaxially anisotropic sphere," *IEE Proc.*, pt. H, vol. 139, pp. 314–318, Aug. 1992.
3. R. F. Harrington, *Time-Harmonic Electromagnetic Fields*, McGraw-Hill, New York, 1961, Chap. 6.
4. T. Uwaro and T. Itoh, in T. Itoh, ed., *Numerical Techniques for Microwave and Millimeter-Wave Passive Structures*, John Wiley & Sons, New York, 1989, Chap. 5.
5. W. Y. Tam and K. M. Luk, "Resonance in spherical-circular microstrip structure," *IEEE Trans. Microwave Theory Tech.*, vol. 39, pp. 700–704, Apr. 1991.
6. T. J. Chang and H. T. Chen, "Full-wave analysis of scattering from a spherical-circular microstrip antenna," *Microwave Opt. Technol. Lett.*, vol. 10, pp. 49–52, Sept. 1995.
7. H. T. Chen and K. L. Wong, "Cross polarization characteristics of a probe-fed spherical-circular microstrip patch antenna," *Microwave Opt. Technol. Lett.*, vol. 6, pp. 705–710, Sept. 20, 1993.

8. K. L. Wong and H. D. Chen, "Resonance in a spherical annular-ring microstrip structure," *Microwave Opt. Technol. Lett.*, vol. 6, pp. 852–856, Dec. 5, 1993.
9. K. L. Wong, S. F. Hsiao, and H. T. Chen, "Resonance and radiation of a superstrate-loaded spherical-circular microstrip patch antenna," *IEEE Trans. Antennas Propagat.*, vol. 41, pp. 686–690, May 1993.
10. H. D. Chen and K. L. Wong, "Resonance frequency of a superstrate-loaded annular-ring microstrip structure on a spherical body," *Microwave Opt. Technol. Lett.*, vol. 7, pp. 364–367, June 5, 1994.
11. I. S. Chang, Curve-Fitting Formulas for Fast Determination of Accurate Resonant Frequency of Circular Microstrip Patches with Superstrate, M.S. thesis, National Sun Yat-Sen University, Kaohsiung, Taiwan, 1993.
12. K. L. Wong and H. T. Chen, "Resonance in a spherical-circular microstrip structure with an airgap," *IEEE Trans. Microwave Theory Tech.*, vol. 41, pp. 1466–1468, Aug. 1993.
13. H. D. Chen and K. L. Wong, "Analysis of spherical annular-ring microstrip structures with an airgap," *Microwave Opt. Technol. Lett.*, vol. 7, pp. 205–207, Mar. 1994.
14. W. C. Chew, "A broad-band annular-ring microstrip antenna," *IEEE Trans. Antennas Propagat.*, vol. 30, pp. 918–922, Sept. 1982.

CHAPTER FOUR

Characteristics of Cylindrical Microstrip Antennas

4.1 INTRODUCTION

In this chapter we describe the characteristics of cylindrical microstrip antennas excited by a coax feed or through a coupling slot fed by a microstrip feed line. Typical types of rectangular, triangular, circular, and annular-ring microstrip antennas are analyzed. Characterization of curvature effects on the input impedance and radiation characteristics is of major concern. Calculated solutions obtained from various theoretical techniques, such as the full-wave approach, cavity-model analysis, and GTLM theory, are shown and discussed. Some experimental results are also presented for comparison.

4.2 PROBE-FED CASE: FULL-WAVE SOLUTION

Solutions obtained using the full-wave approach are first addressed. The problem of cylindrical rectangular and triangular microstrip antennas with a probe feed as shown in Figure 4.1 is solved in this section. The case with a protective superstrate layer is also treated. The configuration is similar to that shown in Figure 2.1 and the same parameters have the same meanings in both cases. The microstrip patch antenna is excited by a probe feed located at (ϕ_p, z_p), pointing along the $\hat{\rho}$ direction and extending from $\rho = a$ to $\rho = b$. Because the radius of the excitation probe is usually a very small fraction of the operating wavelength, the probe feed can be approximately treated as a line source with a current density written as

$$\mathbf{J}(\boldsymbol{\rho}') = \frac{\hat{\rho} I_0 \delta(\phi' - \phi_p)\delta(z' - z_p)}{\rho'}, \qquad a \leq \rho' \leq b, \tag{4.1}$$

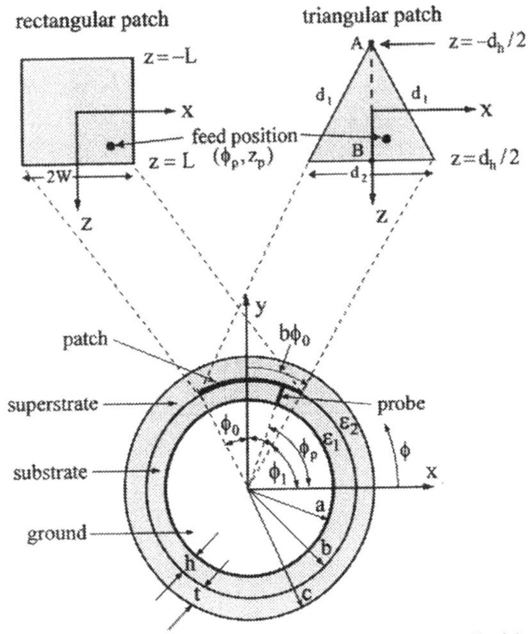

FIGURE 4.1 Configurations of probe-fed superstrate-loaded cylindrical rectangular and triangular microstrip antennas.

where I_0 is the amplitude of the input current. To solve the unknown patch surface current excited through the probe feed, the boundary condition that the total electric field tangential to the patch surface must be zero is applied; that is, on the patch,

$$\mathbf{E}^D(\phi, z) + \mathbf{E}^P(\phi, z) = 0\,, \tag{4.2}$$

where $\mathbf{E}^D(\phi, z)$ is the electric field due to the patch current and $\mathbf{E}^P(\phi, z)$ is the electric field due to the probe current with the patch being absent. To derive $\mathbf{E}^D(\phi, z)$, the theoretical formulation technique described in Section 2.2 can be used, and an expression similar to that in (2.58) can be derived.

As for the derivation of $\mathbf{E}^P(\phi, z)$, field expressions of the transverse electric and magnetic fields resulting from the probe current in a layered medium are first solved. By assuming that the patch is absent and the probe current is an ideal line source described by (4.1), the $\hat{z}$-directed electric and magnetic fields due to a point source inside the substrate layer at (ρ', ϕ_p, z_p) are written as

$$E_z^P(\rho, \phi, z) = \frac{jI_0\, d\rho'}{8\pi} \sum_{n=-\infty}^{\infty} e^{jn(\phi-\phi_p)} \int_{-\infty}^{\infty} dk_z\, e^{jk_z(z-z_p)}$$

$$
\times\begin{cases}
\dfrac{k_z k_{\rho 1}}{\omega\epsilon_0\epsilon_1} H_n^{(1)\prime}(k_{1\rho}\rho')J_n(k_{1\rho}\rho) + C_{1e}H_n^{(1)}(k_{1\rho}\rho) + D_{1e}J_n(k_{1\rho}\rho)\,, & a \le \rho < \rho'\,,\\
\dfrac{k_z k_{\rho 1}}{\omega\epsilon_0\epsilon_1} H_n^{(1)}(k_{1\rho}\rho)J_n'(k_{1\rho}\rho') + C_{1e}H_n^{(1)}(k_{1\rho}\rho) + D_{1e}J_n(k_{1\rho}\rho)\,, & \rho' \le \rho < b\,,\\
C_{2e}H_n^{(1)}(k_{2\rho}\rho) + D_{2e}J_n(k_{2\rho}\rho)\,, & b \le \rho < c\,,\\
C_{3e}H_n^{(1)}(k_{3\rho}\rho)\,, & c \le \rho\,,
\end{cases}
\tag{4.3}
$$

$$
H_z^P(\rho, \phi, z) = \frac{I_0 d\rho'}{8\pi} \sum_{n=-\infty}^{\infty} e^{jn(\phi-\phi_p)} \int_{-\infty}^{\infty} dk_z\, e^{jk_z(z-z_p)}
$$

$$
\times\begin{cases}
\dfrac{n}{\rho'} H_n^{(1)}(k_{1\rho}\rho')J_n(k_{1\rho}\rho) + C_{1h}H_n^{(1)}(k_{1\rho}\rho) + D_{1h}J_n(k_{1\rho}\rho)\,, & a \le \rho < \rho'\,,\\
\dfrac{n}{\rho'} H_n^{(1)}(k_{1\rho}\rho)J_n(k_{1\rho}\rho') + C_{1h}H_n^{(1)}(k_{1\rho}\rho) + D_{1h}J_n(k_{1\rho}\rho)\,, & \rho' \le \rho < b\,,\\
C_{2h}H_n^{(1)}(k_{2\rho}\rho) + D_{2h}J_n(k_{2\rho}\rho)\,, & b \le \rho < c\,,\\
C_{3h}H_n^{(1)}(k_{3\rho}\rho)\,, & c \le \rho\,.
\end{cases}
\tag{4.4}
$$

From the $\hat{z}$-directed electric and magnetic fields given above, the transverse components (E_ρ, E_ϕ, H_ρ, H_ϕ) can readily be obtained by applying (2.3)–(2.6). By further imposing boundary conditions on E_ρ and E_ϕ at the perfectly conducting ground cylinder ($\rho = a$) and matching the continuous conditions of the transverse electric and magnetic fields across the boundaries at $\rho = b$ and c, the coefficients C_{ix} and D_{ix}, $x = e, h$, $i = 1, 2, 3$, can be obtained. After some straightforward manipulation and summing up all the field contributions from the point sources along the line source shown by (4.1), expressions for the transverse fields of $\mathbf{E}^P(\phi, z)$ on the surface of $\rho = b$ are written as

$$
\left.\begin{bmatrix} E_\phi^P \\ E_z^P \end{bmatrix}\right|_{\rho=b} = \frac{1}{2\pi} \sum_{n=-\infty}^{\infty} e^{jn(\phi-\phi_p)} \int_{-\infty}^{\infty} dk_z\, e^{-jk_z(z-z_p)} \begin{bmatrix} T_{11} & T_{12} \\ T_{21} & T_{22} \end{bmatrix} \begin{bmatrix} R_1 \\ R_2 \end{bmatrix}, \tag{4.5}
$$

where the elements in the $[T]$ and $[R]$ matrices are given in [1]. Next, by substituting the derived transverse fields of $\mathbf{E}^D(\phi, z)$ and $\mathbf{E}^P(\phi, z)$ into (4.2), we obtain the following integral equation on the patch:

$$
\begin{aligned}
&\sum_{n=-\infty}^{\infty} e^{jn\phi} \int_{-\infty}^{\infty} dk_z\, e^{jk_z z} \begin{bmatrix} \tilde{G}_{\phi\phi} & \tilde{G}_{\phi z} \\ \tilde{G}_{z\phi} & \tilde{G}_{zz} \end{bmatrix} \begin{bmatrix} \tilde{J}_\phi(n, k_z) \\ \tilde{J}_z(n, k_z) \end{bmatrix} \\
&\quad = -\sum_{n=-\infty}^{\infty} e^{jn(\phi-\phi_p)} \int_{-\infty}^{\infty} dk_z e^{jk_z(z-z_p)} \begin{bmatrix} T_{11} & T_{12} \\ T_{21} & T_{22} \end{bmatrix} \begin{bmatrix} R_1 \\ R_2 \end{bmatrix}.
\end{aligned}
\tag{4.6}
$$

The integral equation above is then solved by applying Galerkin's moment method, described in Section 2.2.2. After choosing suitable basis functions for expansion of the excited surface current density on the microstrip patch, taking spectral amplitudes of the chosen basis functions, and substituting the results into (4.6), we have

$$\sum_{n=-\infty}^{\infty} e^{jn\phi} \int_{-\infty}^{\infty} dk_z\, e^{jk_z z} \begin{bmatrix} \tilde{G}_{\phi\phi} & \tilde{G}_{\phi z} \\ \tilde{G}_{z\phi} & \tilde{G}_{zz} \end{bmatrix} \begin{bmatrix} \sum_{q=1}^{Q} I_{\phi q} \tilde{J}_{\phi q} \\ \sum_{p=1}^{P} I_{zp} \tilde{J}_{zp} \end{bmatrix}$$

$$= -\sum_{n=-\infty}^{\infty} e^{jn(\phi-\phi_p)} \int_{-\infty}^{\infty} dk_z\, e^{jk_z(z-z_p)} \begin{bmatrix} T_{11} & T_{12} \\ T_{21} & T_{22} \end{bmatrix} \begin{bmatrix} R_1 \\ R_2 \end{bmatrix}. \tag{4.7}$$

By further using the chosen basis functions as testing functions and integrating (4.7) over the patch area, we can have the following matrix equation:

$$\begin{bmatrix} (Z_{lq}^{\phi\phi})_{Q\times Q} & (Z_{lp}^{\phi z})_{Q\times P} \\ (Z_{kq}^{z\phi})_{P\times Q} & (Z_{kp}^{zz})_{P\times P} \end{bmatrix} \begin{bmatrix} (I_{\phi q})_{Q\times 1} \\ (I_{zp})_{P\times 1} \end{bmatrix} = \begin{bmatrix} (V_{\phi q})_{Q\times 1} \\ (V_{zp})_{P\times 1} \end{bmatrix}, \tag{4.8}$$

where the elements in the [Z] matrix have been written in (2.73)–(2.76), and $I_{\phi n}$ and I_{zm} are again the unknown coefficients to be solved; the elements in [V] matrix are given as

$$V_{\phi q} = \sum_{n=-\infty}^{\infty} e^{-jn\phi_p} \int_{-\infty}^{\infty} dk_z e^{-jk_z z_p} \tilde{J}_{\phi q}(-n, -k_z)[T_{11}R_1 + T_{12}R_2], \tag{4.9}$$

$$V_{zp} = \sum_{n=-\infty}^{\infty} e^{-jn\phi_p} \int_{-\infty}^{\infty} dk_z\, e^{-jk_z z_p} \tilde{J}_{zp}(-n, -k_z)[T_{21}R_1 + T_{22}R_2]. \tag{4.10}$$

By solving (4.8), the unknown patch surface current density can be obtained and the input impedance is readily evaluated from

$$Z_{in} = -\frac{1}{I_0^2} \int_a^b E_\rho^D(\rho, \phi_p, z_p)\, d\rho, \tag{4.11}$$

with

$$E_\rho^D(\rho, \phi, z) = \frac{1}{2\pi} \sum_{n=-\infty}^{\infty} e^{jn\phi} \int_{-\infty}^{\infty} dk_z\, e^{jk_z z} \bar{\bar{\tilde{G}}}{}_\rho^{\mathrm{T}}(n, k_z) \begin{bmatrix} \tilde{J}_\phi(n, k_z) \\ \tilde{J}_z(n, k_z) \end{bmatrix}, \tag{4.12}$$

where $\bar{\bar{G}}_\rho = [G^D_{\rho\phi}, G^D_{\rho z}]$ relates the $\hat{\rho}$-directed component of the electric field inside the substrate layer due to the patch current and the superscript T denotes a transpose of the matrix. The expressions of $\bar{\bar{G}}_\rho$ in the spectral domain have been derived as

$$\bar{\bar{G}}_\rho(n, k_z) = \frac{1}{X_{11}X_{22} - X_{12}X_{21}} \begin{bmatrix} a_{31}X_{22} - a_{32}X_{21} & c_{31}X_{22} - c_{32}X_{12} \\ a_{32}X_{11} - a_{31}X_{12} & c_{32}X_{11} - c_{31}X_{12} \end{bmatrix} \times \begin{bmatrix} \dfrac{jk_z[H_n^{(1)\prime}(k_{1\rho}\rho) - \gamma_1 J_n'(k_{1\rho}\rho)]}{k_{1\rho}} \\ -\dfrac{n\omega\mu_0[H_n^{(1)}(k_{1\rho}\rho) - \gamma_2 J_n(k_{1\rho}\rho)]}{\rho k_{1\rho}^2} \end{bmatrix}, \tag{4.13}$$

with

$$\gamma_1 = \frac{H_n^{(1)}(k_{1\rho}a)}{J_n(k_{1\rho}a)}, \qquad \gamma_2 = \frac{H_n^{(1)\prime}(k_{1\rho}a)}{J_n'(k_{1\rho}a)}, \tag{4.14}$$

$$a_{31} = \frac{Y_{21}}{H_n^{(1)}(k_{1\rho}b) - \gamma_1 J_n(k_{1\rho}b)}, \tag{4.15}$$

$$a_{32} = \frac{a_{31}Y_{22}}{Y_{21}}, \tag{4.16}$$

$$c_{31} = \frac{(jnk_z k_{1\rho}/b\omega\mu_0)(1/k_{1\rho}^2 - 1/k_{2\rho}^2)Y_{21} + (k_{1\rho}/k_{2\rho})[c_{21}H_n^{(1)\prime}(k_{2\rho}b) + d_{21}J_n'(k_{2\rho}b)]}{G_1}, \tag{4.17}$$

$$c_{32} = \frac{(jnk_z k_{1\rho}/b\omega\mu_0)(1/k_{1\rho}^2 - 1/k_{2\rho}^2)Y_{22} + (k_{1\rho}/k_{2\rho})[c_{22}H_n^{(1)\prime}(k_{2\rho}b) + d_{22}J_n'(k_{2\rho}b)]}{G_1}, \tag{4.18}$$

$$c_{21} = -\frac{jnk_z k_{2\rho}}{\omega\mu_0 c}\left(\frac{1}{k_{2\rho}^2} - \frac{1}{k_{3\rho}^2}\right)\frac{H_n^{(1)}(k_{3\rho}c)J_n(k_{2\rho}c)}{G_2}, \tag{4.19}$$

$$c_{22} = \frac{H_n^{(1)}(k_{3\rho}c)J_n'(k_{2\rho}c) - (k_{2\rho}/k_{3\rho})H_n^{(1)\prime}(k_{3\rho}c)J_n(k_{2\rho}c)}{G_2}, \tag{4.20}$$

$$d_{21} = -\frac{H_n^{(1)}(k_{2\rho}c)}{J_n(k_{2\rho}c)} d_{22}, \tag{4.21}$$

$$d_{22} = -\frac{H_n^{(1)}(k_{3\rho}c)H_n^{(1)\prime}(k_{2\rho}c) - (k_{2\rho}/k_{3\rho})H_n^{(1)\prime}(k_{3\rho}c)H_n^{(1)}(k_{2\rho}c)}{G_2}, \tag{4.22}$$

$$G_1 = H_n^{(1)\prime}(k_{1\rho}b) - \gamma_2 J_n'(k_{1\rho}b), \tag{4.23}$$

$$G_2 = H_n^{(1)}(k_{2\rho}c)J_n'(k_{2\rho}c) - H_n^{(1)\prime}(k_{2\rho}c)J_n(k_{2\rho}c). \tag{4.24}$$

In addition to the expressions listed above, X_{ij} and Y_{ij} are elements in matrices $[X]$ and $[Y]$ given in (2.33), whose expressions are written in (2.35)–(2.42).

As for the far-zone radiated fields in spherical coordinates, they can be given approximately by

$$\begin{bmatrix} E_\theta \\ E_\phi \end{bmatrix} \simeq \frac{1}{\sin\theta}\begin{bmatrix} -1 & 0 \\ 0 & \eta_0 \end{bmatrix}\begin{bmatrix} E_z^D \\ H_z^D \end{bmatrix}, \tag{4.25}$$

with

$$\begin{bmatrix} E_z^D \\ H_z^D \end{bmatrix} = \frac{1}{2\pi}\sum_{n=-\infty}^{\infty} e^{jn\phi}\int_{-\infty}^{\infty} dk_z\, e^{jk_z z}\begin{bmatrix} a_{31} & a_{32} \\ c_{31} & c_{32} \end{bmatrix} \begin{bmatrix} X_{11} & X_{12} \\ X_{21} & X_{22} \end{bmatrix}^{-1}\begin{bmatrix} \tilde{J}_\phi(n, k_z) \\ \tilde{J}_z(n, k_z) \end{bmatrix} H_n^{(1)}(k_{3\rho}\rho), \tag{4.26}$$

where η_0 in (4.25) is free-space intrinsic impedance, and E_z^D and H_z^D are again the $\hat{z}$-directed components of the electric and magnetic fields due to the patch current. For large values of $k_{3\rho}\rho$, the integral in (4.26) can be evaluated using the asymptotic formulation, which yields

$$\begin{bmatrix} E_z^D \\ H_z^D \end{bmatrix} \simeq \frac{-je^{jk_0 r}}{\pi r}\sum_{n=-\infty}^{\infty}(-j)^m e^{jn\phi}\begin{bmatrix} a_{31} & a_{32} \\ c_{31} & c_{32} \end{bmatrix}\begin{bmatrix} X_{11} & X_{12} \\ X_{21} & X_{22} \end{bmatrix}^{-1}\begin{bmatrix} \tilde{J}_\phi(n, k_z) \\ \tilde{J}_z(n, k_z) \end{bmatrix}, \quad m = |n|. \tag{4.27}$$

4.2.1 Rectangular Patch

For a rectangular patch, the cavity-model basis functions of (2.61)–(2.62) are selected for moment-method computation. To test the numerical convergence of the results calculated, 12 ($N = 6$, $M = 6$) basis functions are required to obtain good convergent solutions of the input impedance. A microstrip patch antenna excited at the TM_{01} mode (i.e., the patch current is excited primarily in the $\hat{z}$ direction) is studied first. Excellent agreement between the results calculated and measured for two different feed positions is shown in Figure 4.2. The resonant frequencies, determined from the zero crossing of the reactance curve, are found to be only very slightly affected by curvature variations. It is also shown that the

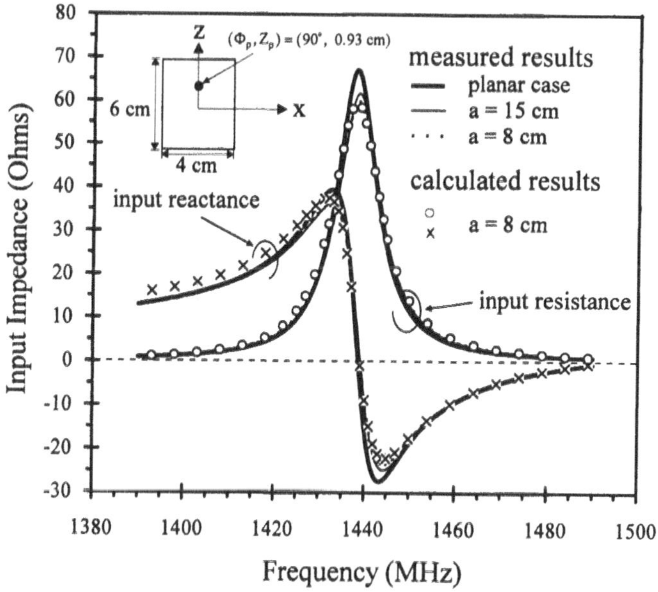

(a)

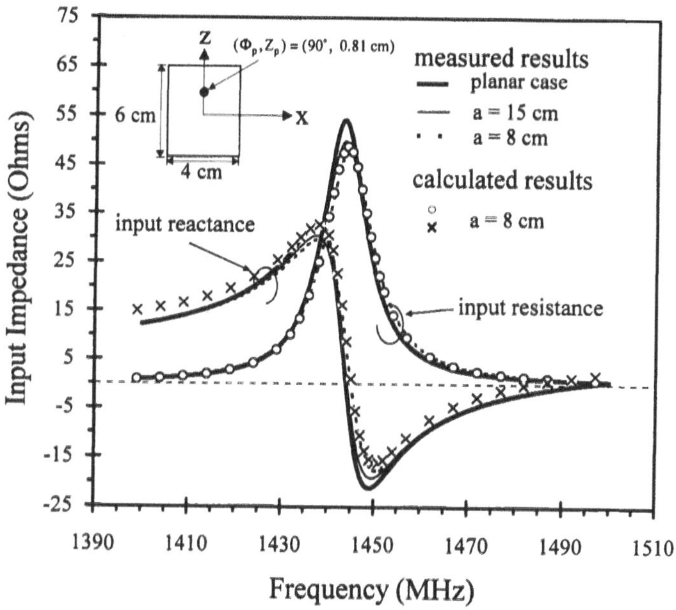

(b)

FIGURE 4.2 Input impedance measured at the TM_{01} mode versus frequency; $\epsilon_1 = 2.98$, $h = 0.762$ mm, $t = 0$, $2L = 6$ cm, $2b\phi_0 = 4$ cm. (*a*) $(\phi_p, z_p) = (90°, 0.93$ cm); (*b*) $(\phi_p, z_p) = (90°, 0.81$ cm). The results calculated for $a = 8$ cm are full-wave solutions. (From Ref. [2], © 1996 John Wiley & Sons, Inc.)

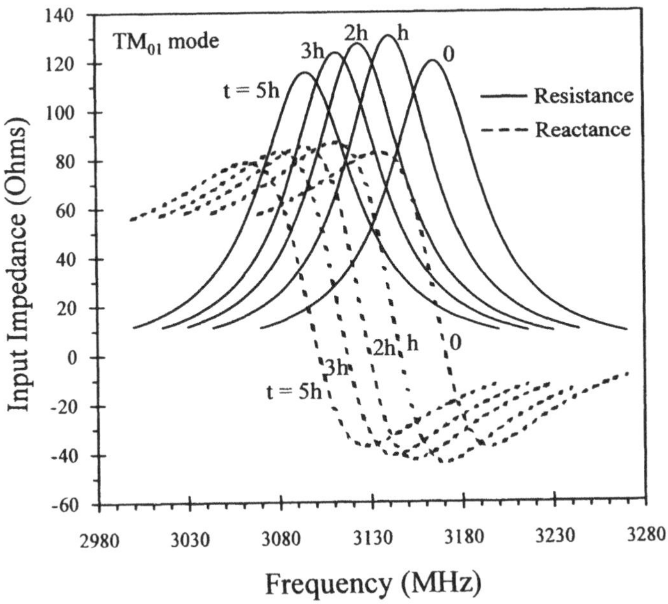

FIGURE 4.3 Input impedance calculated at the TM_{01} mode versus frequency with a superstrate presence; $a = 5\,\text{cm}$, $\epsilon_1 = \epsilon_2 = 2.32$, $h = 0.795\,\text{mm}$, $2L = 3\,\text{cm}$, $2b\phi_0 = 4\,\text{cm}$, $(\phi_p, z_p) = (90°, 1.0\,\text{cm})$. (From Ref. [1], © 1994 John Wiley & Sons, Inc.)

resonant input resistance decreases when the cylinder radius decreases. A case with superstrate loading is shown in Figure 4.3, where again the resonant frequency decreases with increasing superstrate thickness. However, the resonant input resistance increases initially with superstrate loading until it reaches a maximum value, and then decreases with increasing superstrate thickness.

Results for TM_{10} mode excitation (the patch is excited in the $\hat{\phi}$ direction) are shown in Figure 4.4. In contrast to the results shown in Figure 4.2, the resonant frequency is very sensitive to curvature variation and increases with decreasing cylinder radius. This difference in the curvature effect for the TM_{01} and TM_{10} modes is due primarily to the different excitation directions of the patch current for each mode.

Figure 4.5 shows radiation patterns in the x–y plane (H plane) for various cylinder radii; the two cases $t = 0$ and $t = 3h$ are presented. To confirm the correctness of the theoretical results, the convergent solutions are compared with the measured data. The results demonstrate that when the microstrip antenna is excited at the TM_{01} mode, the H-plane pattern is broadened with decreasing cylinder radius. On the other hand, the H-plane pattern is relatively insensitive to superstrate loading. Radiation patterns in the y–z plane (E plane) are presented in Figure 4.6. In this case, superstrate loading causes the 3-dB beamwidth to decrease with increasing superstrate thickness and cylinder radius.

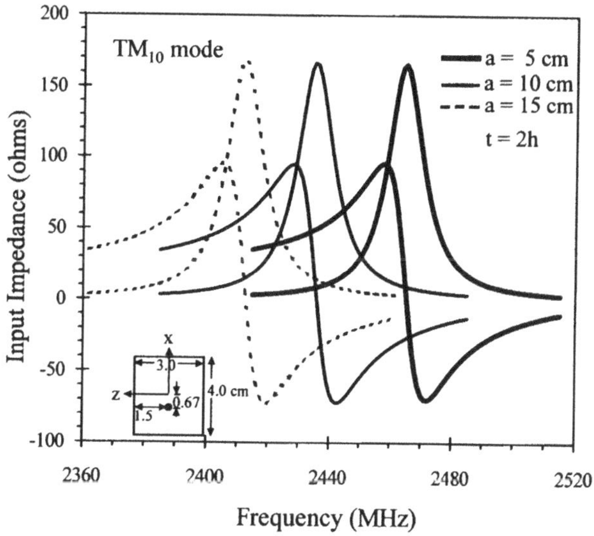

FIGURE 4.4 Input impedance calculated at the TM_{10} mode for various cylinder radii; $\epsilon_1 = \epsilon_2 = 2.32$, $h = 0.795$ mm, $t = 2h$, $2L = 3$ cm, $2b\phi_0 = 4$ cm. (From Ref. [1], © 1994 John Wiley & Sons, Inc.)

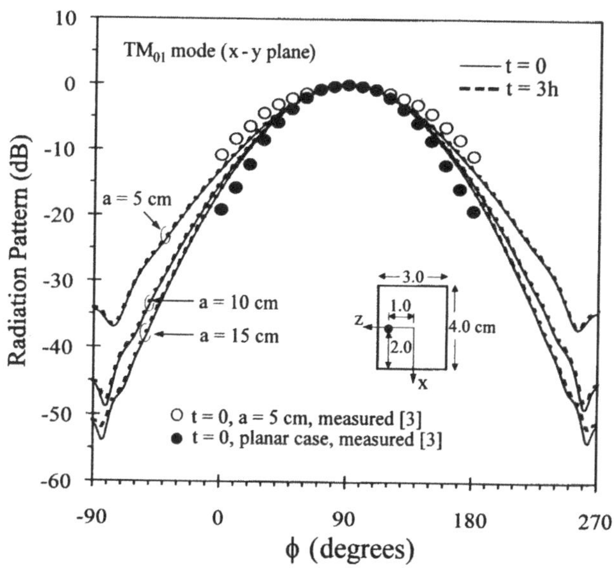

FIGURE 4.5 Radiation pattern in the x–y plane (H plane) as a function of superstrate thickness and cylinder radius for the antenna parameters shown in Figure 4.3.

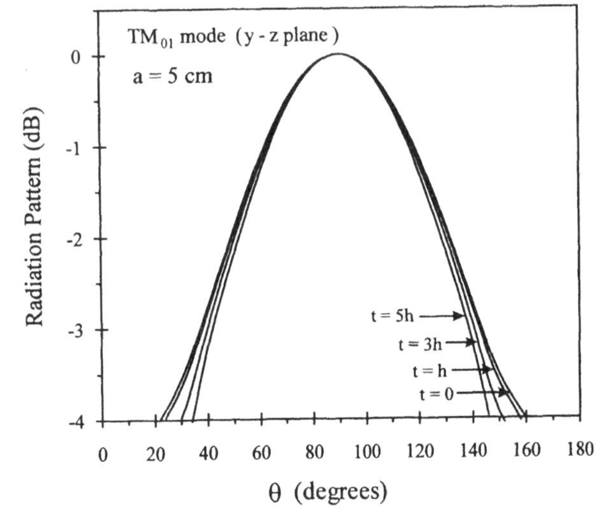

(a)

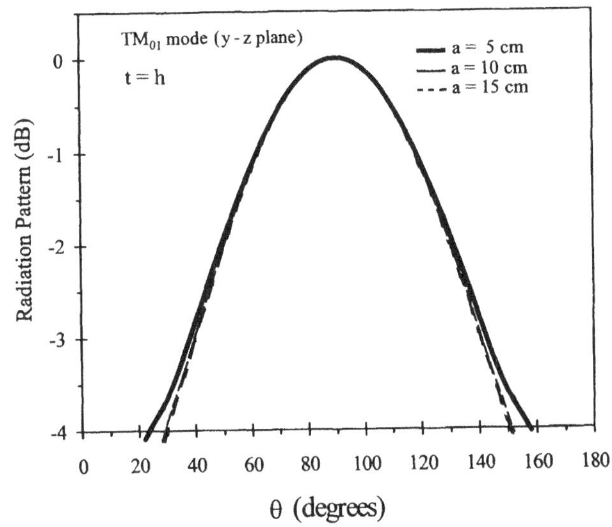

(b)

FIGURE 4.6 Radiation pattern in the y–z plane (E plane) for various (a) superstrate thicknesses and (b) cylinder radii; antenna parameters are as in Figure 4.3.

4.2.2 Triangular Patch

For a triangular patch, the basis functions of (2.99) and (2.100) are chosen for moment-method calculation. Figure 4.7 shows the input impedance calculated and measured at the TM_{10} mode (excited in the $\hat{z}$ direction) for a triangular microstrip antenna mounted on a ground cylinder of $a = 15$ cm. Although only two ($N = 1$,

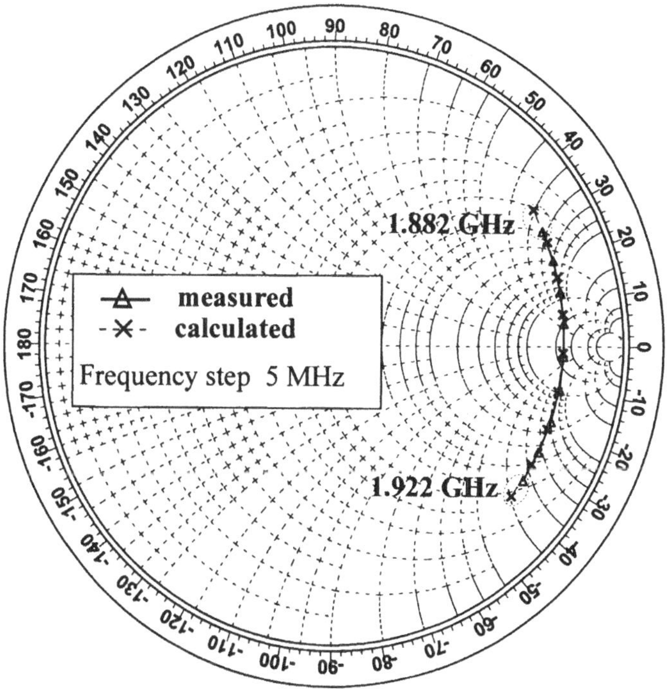

FIGURE 4.7 Input impedance measured and calculated versus frequency for a triangular patch with $a = 15$ cm; $\epsilon_1 = 3.0$, $h = 0.762$ mm, $d_1 = d_2 = 6$ cm, $\phi_p = 90°$, $z_p = 0.9$ cm. (From Ref. [4], © 1997 John Wiley & Sons, Inc.)

$M = 1$) basis functions are used, good agreement between the results is obtained. This is primarily because the resonant frequency of the fundamental mode (TM_{10} mode) of the triangular patch is far below that of other higher-order modes, as discussed in Section 2.6. The results for antennas with various cylinder radii are presented in Figure 4.8. It is seen that the resonant frequency increases with decreasing cylinder radius; however, the resonant input resistance decreases when the cylinder radius is decreased. The results calculated for the radiation patterns, shown in Figure 4.9, indicate that with decreasing cylinder radius, both E- and H-plane patterns are slightly broadened, and radiation is increased in the backward direction. For comparison, radiation patterns measured in the upper half-plane are plotted in Figure 4.10. From the results it can be seen that the results calculated agree in general with the data measured.

4.3 PROBE-FED CASE: CAVITY-MODEL SOLUTION

Given the configuration in Figure 4.1, a thin substrate is assumed. Due to the limitation of cavity-model analysis, only the case without a superstrate ($t = 0$) is

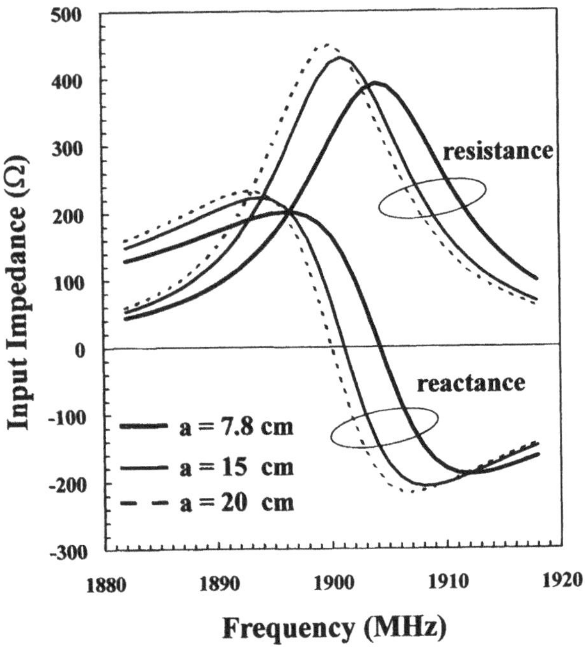

FIGURE 4.8 Input impedance calculated versus frequency for a triangular patch with various cylinder radii; antenna parameters are as in Figure 4.7. (From Ref. [4], © 1997 John Wiley & Sons, Inc.)

considered here. Based on the observations described in Section 1.2.2, the region between the patch and the ground cylinder can be treated as a cavity bounded by a magnetic wall around the cavity and two electric walls on the top and bottom of the cavity. When considering cylindrical coordinates, we can have a source-free field satisfying the wave equation

$$\frac{1}{\rho^2}\frac{\partial^2(\rho E_\phi)}{\partial\rho\,\partial\phi}-\frac{1}{\rho^2}\frac{\partial^2 E_\rho}{\partial\phi^2}-\frac{\partial^2 E_\rho}{\partial z^2}+\frac{\partial^2 E_z}{\partial z\,\partial\rho}-k^2E_\rho=0\,, \tag{4.28}$$

where $k^2=\omega^2\mu_0\epsilon_0\epsilon_1$. Since the electric field is assumed to have only an E_ρ component independent of ρ, (4.28) thus becomes

$$\frac{1}{\rho^2}\frac{\partial^2 E_\rho}{\partial\phi^2}+\frac{\partial^2 E_\rho}{\partial z^2}+k^2E_\rho=0\,. \tag{4.29}$$

By further assuming that the substrate thickness is much less than the cylinder radius and using the approximation of $\rho=b$, eigenfunctions of E_ρ $(=\hat{\rho}\psi_{mn})$ and eigenvalues of k $(=k_{mn})$ satisfying the boundary conditions at the magnetic walls around the cavity can be obtained. The expressions of ψ_{mn} and k_{mn} represent, respectively, the field distribution and wavenumber of TM_{mn} excitation.

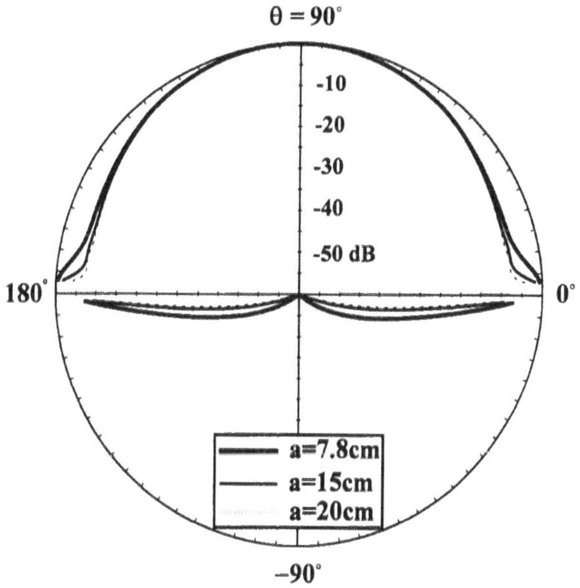

(*a*)

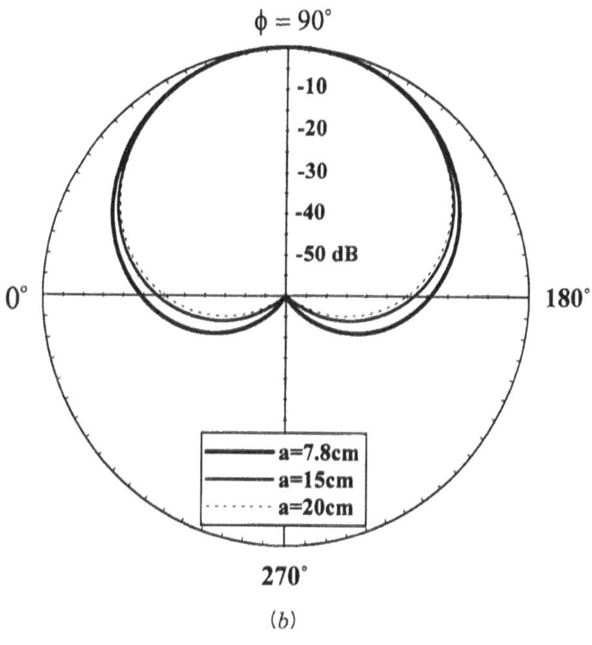

(*b*)

FIGURE 4.9 Radiation patterns calculated at resonance for a triangular patch; antenna parameters are as in Figure 4.7. (*a*) *E*-plane pattern; (*b*) *H*-plane pattern. (From Ref. [4], © 1997 John Wiley & Sons, Inc.)

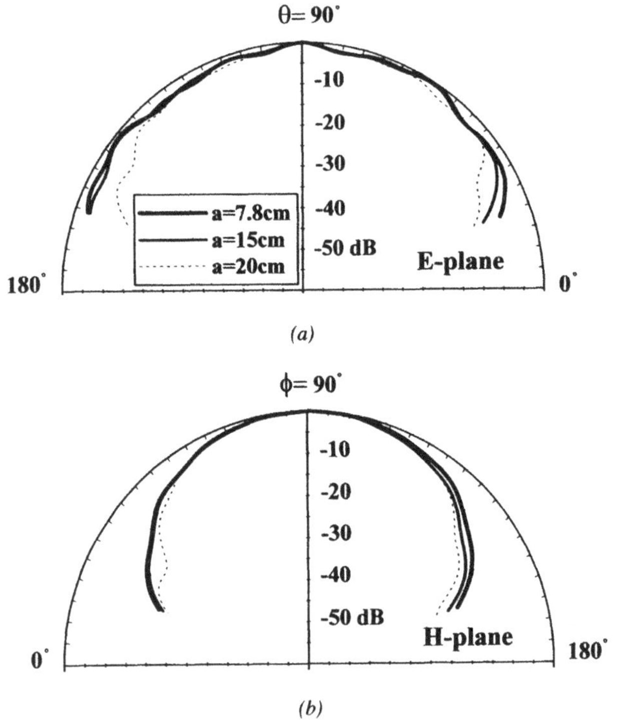

FIGURE 4.10 Radiation patterns measured at resonance for a triangular patch; antenna parameters are as in Figure 4.7. (*a*) *E*-plane pattern; (*b*) *H*-plane pattern. (From Ref. [4], © 1997 John Wiley & Sons, Inc.)

Once ψ_{mn} is derived, the equivalent magnetic currents along the cavity can be obtained from

$$\begin{aligned}\mathbf{M} &= M_\phi \hat{\phi} + M_z \hat{z} \\ &= \hat{\rho}\psi_{mn} \times \hat{n} ,\end{aligned} \tag{4.30}$$

where $\hat{n}$ is a unit vector pointing outward at the boundary around the cavity. This equivalent magnetic current accounts for the radiation of the cylindrical microstrip patch antenna. By utilizing the theoretical treatment for the problem of magnetic current radiation in the presence of a cylindrical body [5], the far-zone radiated fields can be expressed as

$$E_\theta = E_0 \frac{e^{-jk_0 r}}{\pi r} \sin\theta \sum_{p=-\infty}^{\infty} e^{jp\phi} j^{p+1} f_p(-k_0 \cos\theta) , \tag{4.31}$$

$$E_\phi = E_0 \frac{k_0 e^{-jk_0 r}}{\pi \omega \mu_0 r} \sin\theta \sum_{p=-\infty}^{\infty} e^{jp\phi} j^{p+1} g_p(-k_0 \cos\theta)\,, \tag{4.32}$$

where

$$f_p(u) = \frac{j\omega\epsilon_0 \tilde{M}_\phi(p,u)}{(k_0^2 - u^2) H_p^{(2)\prime}(a\sqrt{k_0^2 - u^2})}\,, \tag{4.33}$$

$$g_p(u) = \frac{1}{\sqrt{k_0^2 - u^2}\, H_p^{(2)\prime}(a\sqrt{k_0^2 - u^2})} \left[-\tilde{M}_z(p,u) + \frac{pu\tilde{M}_\phi(p,u)}{a(k_0^2 - u^2)} \right]. \tag{4.34}$$

The functions $\tilde{M}_\phi$ and $\tilde{M}_z$ are the Fourier transforms of M_ϕ and M_z, respectively. Once the cavity and radiation fields are determined, the time-average radiated power can be calculated by

$$P_{\text{rad}} = \frac{1}{240\pi} \int_0^{\pi} \int_0^{2\pi} (E_\theta E_\theta^* + E_\phi E_\phi^*) r^2 \sin\theta \, d\phi \, d\theta\,. \tag{4.35}$$

The time-average copper loss of the microstrip patch is

$$P_{\text{cu}} = \frac{1}{2} \sqrt{\frac{\omega\mu_0}{2\sigma_c}} \int_{\text{patch}} |\mathbf{H}|^2 \, dS, \tag{4.36}$$

where $\mathbf{H}$ is the magnetic field in the cavity; σ_c is the conductivity of the copper. The time-average dielectric loss of the substrate is defined by

$$P_d = \frac{\omega\epsilon_0\epsilon_1 \tan\delta}{2} \int_{\text{cavity}} |E_\rho|^2 \, dV\,, \tag{4.37}$$

where V is the volume of the cavity under the patch and $\tan\delta$ is the loss tangent of the substrate. As for the time-average electric energy stored inside the cavity, we have

$$W_e = \frac{\epsilon_0\epsilon_1}{4} \int_{\text{cavity}} |E_\rho|^2 \, dV. \tag{4.38a}$$

The time-average magnetic energy stored inside the cavity is written as

$$W_m = \frac{\mu_0}{4} \int_{\text{cavity}} |\mathbf{H}|^2 \, dV. \tag{4.38b}$$

From these losses, the input impedance can be expressed as

$$Z_{\text{in}} = \frac{2[P + j2\omega(W_m - W_e)]}{|I|^2}, \tag{4.39}$$

where $P = P_{\text{rad}} + P_{\text{cu}} + P_d$; I is the total input current. It is also noted that in the formulation above, the complex wavenumber k should be replaced by

$$k_{\text{eff}} = k_0\sqrt{\epsilon_r(1 - j\delta_{\text{eff}})}, \qquad k_0 = \omega\sqrt{\mu_0\epsilon_0}, \tag{4.40}$$

where δ_{eff} is the effective loss tangent defined as

$$\delta_{\text{eff}} = \frac{P_{\text{rad}} + P_{\text{cu}} + P_d}{2\omega W_e}. \tag{4.41}$$

To derive the input impedance, the electric field inside the cavity is modified by incorporating a unit input current density at the feed point; that is,

$$\mathbf{E} = \hat{\rho}j\omega\mu_0 \sum_{m,n} \frac{\psi_{mn}}{k_{\text{eff}}^2 - k_{mn}^2} \frac{\langle J_\rho, \psi_{mn}\rangle}{\langle \psi_{mn}, \psi_{mn}\rangle}, \tag{4.42}$$

with

$$\langle J_\rho, \psi_{mn}\rangle = \int_v J_\rho \psi_{mn}^* \, dV, \tag{4.43}$$

$$\langle \psi_{mn}, \psi_{mn}\rangle = \int_v \psi_{mn}\psi_{mn}^* \, dV, \tag{4.44}$$

where J_ρ is the input probe current and is usually modeled as a unit-amplitude current ribbon with an effective width of w_p given to be about 4.5 times the radius of the actual probe feed. With the result of (4.42), the input impedance of the antenna can be derived from

$$Z_{\text{in}} = \frac{V}{I} = -\frac{h}{w_p^2} \int_{w'-w_p/2}^{w'+w_p/2} E_\rho \, dw'. \tag{4.45}$$

Based on the cavity-model theory above, the expressions for the far-zone radiated fields and the input impedance for various cylindrical microstrip antennas with rectangular, triangular, circular, and annular-ring patches are shown. The results calculated are presented and discussed below.

4.3.1 Rectangular Patch

For a cylindrical rectangular patch having a straight dimension of $2L$ and a curved dimension of $2b\phi_0$ $(=2W)$, the resonant frequency of the TM_{mn} mode is given as

$$f_{mn} = \frac{c}{2\sqrt{\epsilon_1}}\left[\left(\frac{m}{2b\phi_0}\right)^2 + \left(\frac{n}{2L}\right)^2\right]. \tag{4.46}$$

For the source-free case, ψ_{mn} and k_{mn} are derived as

$$\psi_{mn} = E_0 \cos\left[\frac{m\pi}{2\phi_0}(\phi - \phi_1)\right]\cos\left[\frac{n\pi}{2L}(z - L)\right], \tag{4.47}$$

$$k_{mn} = \left[\left(\frac{m\pi}{2W}\right)^2 + \left(\frac{n\pi}{2L}\right)^2\right]^{1/2}, \tag{4.48}$$

and the far-zone radiated fields are expressed as [6]

$$E_\theta = \frac{E_0 h e^{-jk_0 r}}{2\pi^2 r \sin\theta}[1 - (-1)^n e^{-j2k_0 L\cos\theta}] \sum_{p=-\infty}^{\infty} \frac{e^{jp\phi} j^{p+1} e^{-jp\phi_1} I(\phi_0, m, -p)}{H_p^{(2)}(k_0 a \sin\theta)}, \tag{4.49}$$

$$E_\phi = -j\frac{E_0 h}{2\pi^2 r} e^{-jk_0 r} I(L, n, -k_0\cos\theta) \sum_{p=-\infty}^{\infty} \frac{e^{jp\phi} j^{p+1} e^{-jp\phi_1}}{H_p^{(2)\prime}(k_0 a \sin\theta)}[1 - (-1)^m e^{-j2p\phi_0}]$$
$$-j\frac{E_0 h e^{-jk_0 r}\cos\theta}{2\pi^2 a k_0 r \sin^2\theta}[1 - (-1)^n e^{-j2k_0 L\cos\theta}] \sum_{p=-\infty}^{\infty} \frac{p e^{jp\phi} j^{p+1} e^{-jp\phi_1} I(\phi_0, m, -p)}{H_p^{(2)\prime}(k_0 a \sin\theta)}, \tag{4.50}$$

where

$$I(L, n, u) = \int_{-L}^{L} \cos\left[\frac{n\pi}{2L}(z - L)\right] e^{-juz}\, dz, \tag{4.51}$$

$$I(\phi_0, m, -p) = \int_0^{2\phi_0} \cos\frac{m\pi\phi}{2\phi_0} e^{-jp\phi}\, d\phi. \tag{4.52}$$

By further assuming the patch to be fed by a coax at (ϕ_p, z_p), the eigenfunction of (4.47) is modified as [6]

$$\psi_{mn} = j\omega\mu_0 \sum_{m,n} C_{mn} \cos\left[\frac{m\pi}{2\phi_0}(\phi - \phi_1)\right]\cos\left[\frac{n\pi}{2L}(z - L)\right], \tag{4.53}$$

where

$$C_{mn} = \frac{w_p}{k_{\text{eff}}^2 - k_{mn}^2}\frac{e_m e_n}{4a\phi_0 L}\cos\left[\frac{m\pi}{2\phi_0}(\phi_p - \phi_1)\right]\cos\left[\frac{n\pi}{2L}(z_p - L)\right] j_0\left(\frac{m\pi w_p}{4a\phi_0}\right), \tag{4.54}$$

$$j_0(x) = \frac{\sin x}{x}, \tag{4.55}$$

$$e_m = \begin{cases} 1, & m=0, \\ 2, & m \neq 0, \end{cases} \qquad e_n = \begin{cases} 1, & n=0, \\ 2, & n \neq 0. \end{cases} \tag{4.56}$$

The far-zone radiated fields of (4.49) and (4.50) are modified as [7]

$$E_\theta = \frac{A}{\sin\theta} \sum_{p=0}^{\infty} \frac{e_p j^p}{H_p^{(2)}(k_0 a \sin\theta)} \frac{p}{(m\pi/2\phi_0)^2 - p^2}$$

$$\times \begin{cases} \cos p(\phi - \pi/2) \sin p\phi_0, & m = 0, 2, 4, \ldots, \\ \sin p(\phi - \pi/2) \cos p\phi_0, & m = 1, 3, 5, \ldots, \end{cases} \tag{4.57}$$

$$E_\phi = \frac{A k_0 \cos\theta}{a[(n\pi/a)^2 - k_0^2 \cos^2\theta]} \sum_{p=0}^{\infty} \frac{e_p j^p}{H_p^{(2)\prime}(k_0 a \sin\theta)} \begin{cases} \sin p(\phi - \pi/2)(-\sin p\phi_0) \\ \cos p(\phi - \pi/2)\cos p\phi_0 \end{cases}$$

$$- \frac{A\cos\theta}{k_0 a \sin^2\theta} \sum_{p=0}^{\infty} \frac{e_p j^p}{H_p^{(2)\prime}(k_0 a \sin\theta)} \frac{p^2}{(m\pi/2\phi_0)^2 - p^2}$$

$$\times \begin{cases} \sin p(\phi - \pi/2)(-\sin p\phi_0), & m = 0, 2, 4, \ldots, \\ \cos p(\phi - \pi/2)\cos p\phi_0, & m = 1, 3, 5, \ldots, \end{cases} \tag{4.58}$$

with

$$A = \frac{\omega\mu_0 C_{mn} h}{\pi^2 r} e^{-jk_0 r}[1 - (-1)^n e^{-j2k_0 L\cos\theta}], \tag{4.59}$$

$$e_p = \begin{cases} 1, & p=0, \\ 2, & p \neq 0. \end{cases} \tag{4.60}$$

In (4.59), C_{mn} is a function of the feed position (ϕ_p, z_p) and is given by (4.54). As for the input impedance seen at the feed position, we have [6]

$$Z_{\text{in}} = \sum_{m=0}^{\infty} \sum_{n=0}^{\infty} \frac{h e_m e_n}{8\pi\epsilon_0\epsilon_1 a L\phi_0} \cos^2\left[\frac{n\pi}{2L}(z_p - L)\right] \cos^2\left[\frac{m\pi}{2\phi_0}(\phi_p - \phi_1)\right]$$

$$\times j_0^2 \left(\frac{m\pi w_p}{4a\phi_0}\right) \frac{\delta_{\text{eff}} f^3 - jf(f^2 - f_{mn}^2)}{\delta_{\text{eff}}^2 f^4 + (f^2 - f_{mn}^2)^2}. \tag{4.61}$$

Typical results of the input impedance and radiation pattern calculated using the formulation above have been presented in [3,6]. Some results are presented in

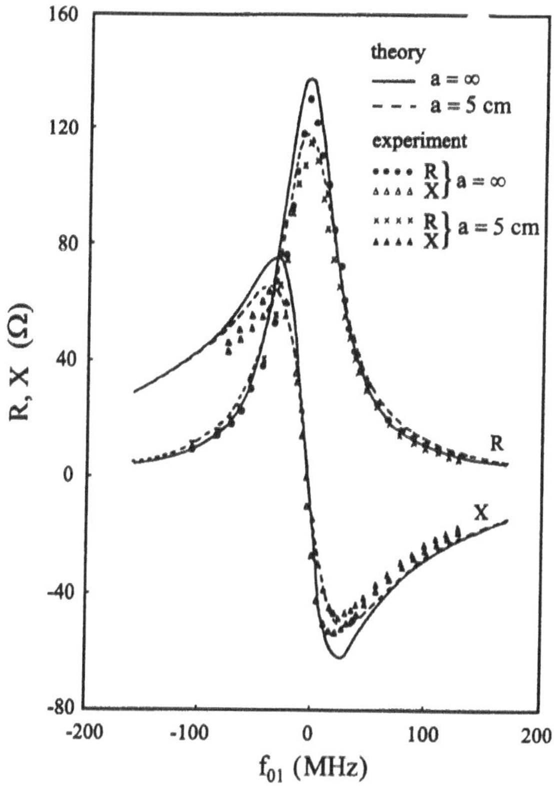

FIGURE 4.11 Input impedance calculated and measured for a rectangular patch at the TM_{01} mode; $2L = 3$ cm, $2b\phi_0 = 4$ cm, $\epsilon_1 = 2.32$, $t = 0$, $h = 0.795$ mm, $z_p = 1.0$ cm, $\phi_p = 90°$; measured $f_{01} = 3170$ MHz; calculated $f_{01} = 3232$ MHz. (From Ref. [3], © 1987 IEE, reprinted with permission.)

Figures 4.11 and 4.12. It is observed that due to the cavity-model approximation in the formulation, the resonant frequency calculated for the TM_{01} mode is not affected by the curvature variation and has a difference of about 2% compared to the data measured. However, the resonant input resistance is in excellent agreement with the measured data, and the radiation patterns calculated agree with the measured results. The results obtained, especially for the resonant input resistance and radiation characteristics, make cavity-model analysis useful in cylindrical microstrip antenna designs.

4.3.2 Triangular Patch

For the consideration of a triangular patch, the equilateral case is discussed. In this case, referring to the geometry shown in Figure 4.1, the triangular patch is assumed to be equilateral with a side length of $d_1(=d_2)$. By solving the

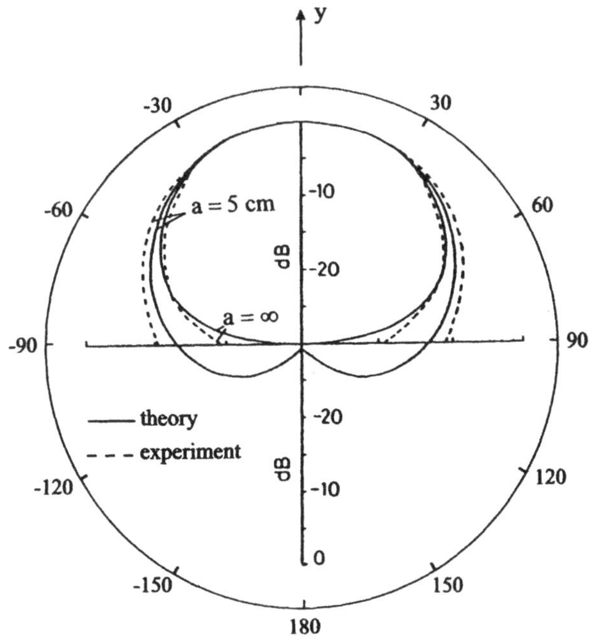

FIGURE 4.12 Radiation patterns calculated and measured in the x–y plane (H plane) for the TM_{01} mode; antenna parameters are as in Figure 4.11. (From Ref. [3], © 1987 IEE, reprinted with permission.)

homogeneous Helmholtz equation with a boundary condition for a cylindrical triangular cavity, we can derive the field distribution and wavenumber for TM_{mn}-mode excitation, expressed as [8]

$$\psi_{mn} = E_0\left\{\cos\left[2\pi l\left(\frac{z}{\sqrt{3}\,d_1} - \frac{1}{3}\right)\right]\cos\frac{\pi(m-n)(\phi-\pi/2)}{3\phi_0}\right.$$

$$+\cos\left[2\pi m\left(\frac{z}{\sqrt{3}\,d_1} - \frac{1}{3}\right)\right]\cos\frac{\pi(n-l)(\phi-\pi/2)}{3\phi_0}$$

$$\left.+\cos\left[2\pi n\left(\frac{z}{\sqrt{3}\,d_1} - \frac{1}{3}\right)\right]\cos\frac{\pi(l-m)(\phi-\pi/2)}{3\phi_0}\right\}, \quad (4.62)$$

$$k_{mn} = \frac{4\pi}{3d_1}(m^2 + mn + n^2), \quad (4.63)$$

where $l + m + n = 0$. Then, by following the theoretical formulation in Section 4.3, the far-zone radiated fields can be written as [8]

$$E_\theta = \sum_{p=-\infty}^{\infty} \frac{C_p}{H_p^{(2)}(ak_0 \sin\theta)} \{[A_p(m-n)B(l) + A_p(n-l)B(m) + A_p(l-m)B(n)]$$

$$- 2e^{-jd_1k_0\cos\theta/2\sqrt{3}}[(-1)^l A_p(m-n) + (-1)^m A_p(n-l) + (-1)^n A_p(l-m)]\},$$

$$(4.64)$$

$$E_\phi = \sum_{p=-\infty}^{\infty} \frac{j\sqrt{3}\, C_p \sin\theta}{H_p^{(2)\prime}(ak_0 \sin\theta)} \Big\{[A_p^{(2)}(m-n)B(l) + A_p^{(2)}(n-l)B(m) + A_p^{(2)}(l-m)B(n)]$$

$$- [A_p^{(1)}(m-n)B(l) + A_p^{(1)}(n-l)B(m) + A_p^{(1)}(l-m)B(n)]$$

$$- \frac{p\cos\theta}{\sqrt{3}\, ak_0 \sin\theta}[A_p(m-n)B(l) + A_p(n-l)B(m) + A_p(l-m)B(n)]$$

$$+ \frac{2p\cos\theta}{\sqrt{3}\, ak_0 \sin^2\theta} e^{-jd_1k_0\cos\theta/2\sqrt{3}}[(-1)^l A_p(m-n) + (-1)^m A_p(n-l)$$

$$+ (-1)^n A_p(l-m)]\Big\}, \qquad (4.65)$$

with

$$A_p^{(1)}(x) = \int_0^{\phi_0} \cos\frac{\pi x(\phi - \phi_0)}{3\phi_0} e^{-jp\phi}\, d\phi, \qquad (4.66)$$

$$A_p^{(2)}(x) = \int_{\phi_0}^{2\phi_0} \cos\frac{\pi x(\phi - \phi_0)}{3\phi_0} e^{-jp\phi}\, d\phi, \qquad (4.67)$$

$$A_p(x) = A_p^{(1)}(x) + A_p^{(2)}(x), \qquad (4.68)$$

$$B(x) = \int_{-d_1/2\sqrt{3}}^{d_1/2\sqrt{3}} \cos\left[\left(\frac{z}{\sqrt{3}\, d_1} - \frac{7}{12}\right)2\pi x\right] e^{jk_0 z\cos\theta} dz,$$

$$(4.69)$$

$$C_p = \frac{E_0 h e^{-jk_0 r} e^{jp(\phi-\phi_0)} j^{p+1}}{4\pi^2 r \sin\theta}. \qquad (4.70)$$

Based on (4.45), the input impedance is derived as [8]

$$Z_{\text{in}} = \sum_{m=0}^{\infty} \sum_{n=0}^{\infty} \left\{ \frac{ah}{2\pi\epsilon_0\epsilon_1 \langle \psi_{mn}, \psi_{mn} \rangle} \left[\sin\frac{2\pi l z_p}{\sqrt{3}\, d_1} \Delta(l) + \sin\frac{2\pi m z_p}{\sqrt{3}\, d_1} \Delta(m) \right.\right.$$

$$\left.\left. + \sin\frac{2\pi n z_p}{\sqrt{3}\, d_1} \Delta(n) \right]^2 \times \frac{f^3\delta_{\text{eff}} - jf(f^2 - f_{mn}^2)}{f^4\delta_{\text{eff}}^2 + (f^2 - f_{mn}^2)^2} \right\}, \tag{4.71}$$

where

$$\Delta(x) = \frac{\sin(\pi w_p x/\sqrt{3}\, d_1)}{\pi w_p x/\sqrt{3}\, d_1}. \tag{4.72}$$

From the formulation above, cavity-model solutions of the input impedance for the cylindrical triangular microstrip antenna with various cylinder radii are first calculated. The results are presented in Figure 4.13. Measured input impedances are also presented, and the cavity-model solutions agree in general with the data measured. The deviation in resonant frequency between theory and experiment is within 0.6% in the case shown here. Also, from the results it is seen that the input impedance level decreases with decreasing cylinder radius. This behavior agrees with the prediction of the full-wave solutions described in Section 4.2.2. The input resistance at resonant frequency as a function of feed position for various cylinder radii is also presented in Figure 4.14. The feed position is chosen from line section $\overline{AB}$ of length d_h (see Figure 4.1). From the results it is also seen that there exists a null resistance at $z_p/d_h \sim 0.17$ (i.e., at a position away from the tip of the triangular patch a distance of two-thirds of $\overline{AB}$). At this position, the excited field inside the cavity for the TM_{10} mode is zero.

Figure 4.15 shows cavity-model solutions for the radiation patterns. When the cylinder radius decreases, both the *E*- and *H*-plane patterns are seen to be broadened and backward radiation is increased. This characteristic is similar to that for a rectangular patch and agrees with the full-wave solutions and the data discussed in Section 4.2.2.

4.3.3 Circular Patch

Figure 4.16 shows the geometry of a probe-fed cylindrical circular microstrip antenna. The circular patch has a radius of r_d. Since it is inconvenient to describe the field expression inside the cavity under the patch using the cylindrical coordinates, new coordinates of $(\hat{\rho}, \hat{l}, \hat{\beta})$, as shown in the figure, are adopted to solve the problem. The relations between the components of a vector in new coordinates and cylindrical coordinates can be written in matrix form as

$$\begin{bmatrix} A_\rho \\ A_l \\ A_\beta \end{bmatrix} = \begin{bmatrix} 1 & 0 & 0 \\ 0 & \cos\beta & \sin\beta \\ 0 & -\sin\beta & \cos\beta \end{bmatrix} \begin{bmatrix} A_\rho \\ A_\phi \\ A_z \end{bmatrix}. \tag{4.73}$$

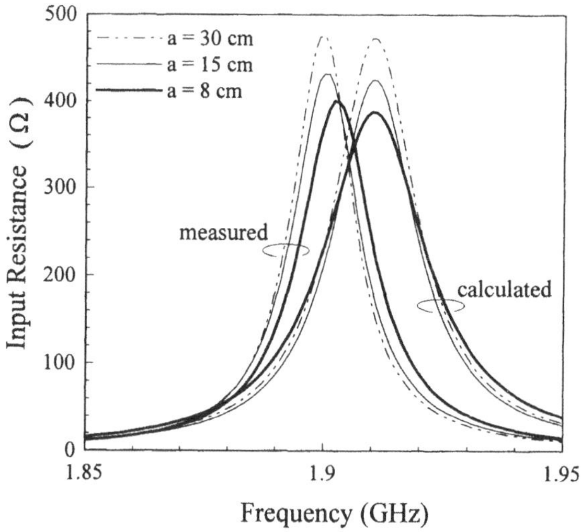

(*a*)

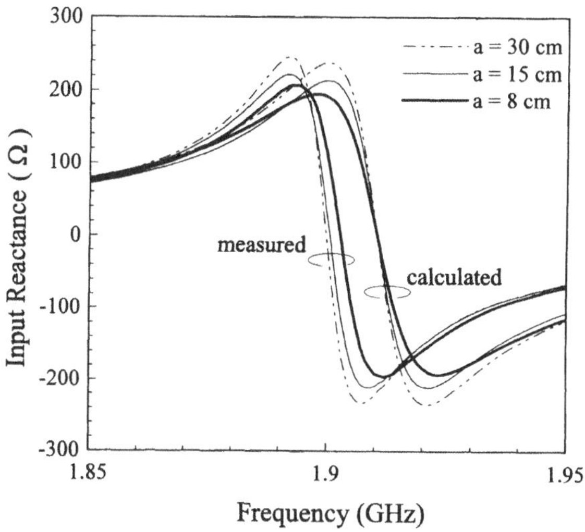

(*b*)

FIGURE 4.13 (*a*) Input resistance and (*b*) input reactance calculated and measured as a function of cylinder radius; $\epsilon_1 = 3.0$, $h = 0.762$ mm, $t = 0$, $d_1 = d_2 = 6$ cm, $\phi_p = 90°$, $z_p = -0.902$ cm. (From Ref. [8], © 1997 John Wiley & Sons, Inc.)

Using the new coordinates, the feed position of the circular patch can be assumed to be at (l_p, β_p). For the coordinates, the electric field E_ρ satisfies the following wave equation:

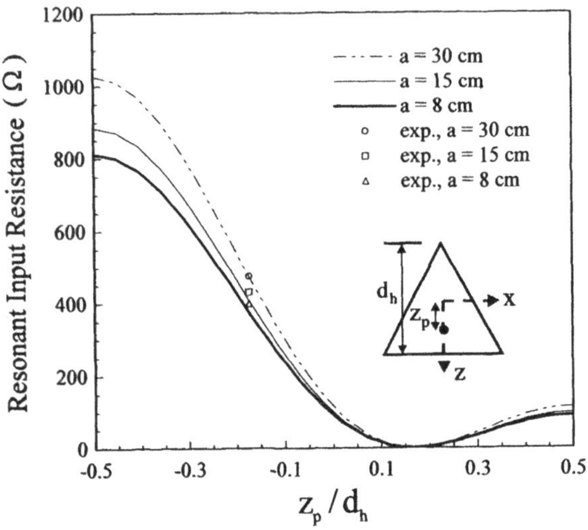

FIGURE 4.14 Dependence of input resistance at resonance frequency on the feed position; antenna parameters are as in Figure 4.13. (From Ref. [8], © 1997 John Wiley & Sons, Inc.)

$$\left(l^2 \frac{\partial^2}{\partial l^2} + l \frac{\partial}{\partial l} + \frac{\partial^2}{\partial \beta^2} + k^2 l^2 \right) E_\rho = 0 \,. \tag{4.74}$$

Solving the equation above with the boundary conditions associated with a cylindrical circular cavity, we have [9]

$$E_\rho = E_0 J_m(k_{mn} l) \cos[m(\beta - \beta_p)] \,, \tag{4.75}$$

where k_{mn} are the roots of $J'_m(k_{mn} r_d) = 0$. The resonant frequency is given by

$$f_{mn} = \frac{k_{mn} c r_d}{2\pi r_{de} \sqrt{\epsilon_1}} \,, \tag{4.76}$$

where c is the speed of light and r_{de} is the effective radius, given by

$$r_{de} = r_d \left[1 + \frac{2h}{\pi r_d \epsilon_1} \left(\ln \frac{\pi r_d}{2h} + 1.7726 \right) \right]^{1/2} . \tag{4.77}$$

Then, evaluating (4.31) and (4.32), we have the far-zone radiated fields written as

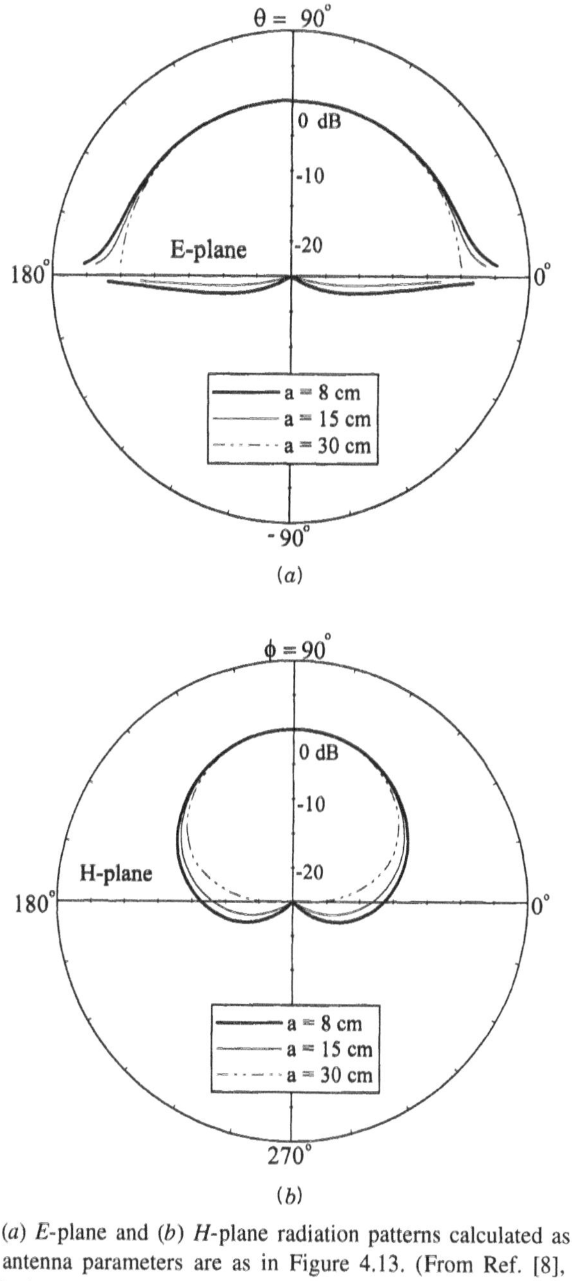

FIGURE 4.15 (*a*) *E*-plane and (*b*) *H*-plane radiation patterns calculated as a function of cylinder radius; antenna parameters are as in Figure 4.13. (From Ref. [8], © 1997 John Wiley & Sons, Inc.)

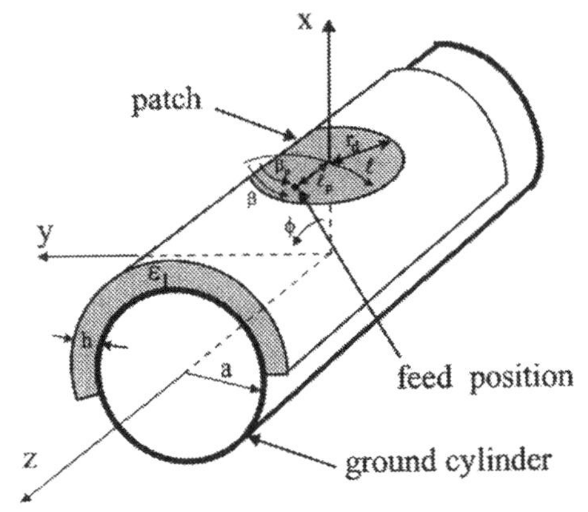

FIGURE 4.16 Geometry of a probe-fed cylindrical circular microstrip antenna.

$$E_\theta = \frac{r_d h E_0 J_m(k_{mn} r_d)}{2\pi^2 a} \frac{e^{-jk_0 r}}{r \sin\theta} \sum_{p=-\infty}^{\infty} \frac{e^{jp\phi} j^{p+1}}{H_p^{(2)}(ak_0 \sin\theta)} I_s(p, -k_0 \cos\theta), \quad (4.78)$$

$$E_\phi = \frac{j r_d h E_0 J_m(k_{mn} r_d) e^{-jk_0 r}}{2\pi^2 a r} \sum_{p=-\infty}^{\infty} \frac{e^{jp\phi} j^{p+1}}{H_p^{(2)\prime}(ak_0 \sin\theta)}$$

$$\times \left[I_c(p, -k_0 \cos\theta) - \frac{p \cos\theta}{ak_0 \sin^2\theta} I_s(p, -k_0 \cos\theta) \right], \quad (4.79)$$

where

$$I_s(p, u) = \int_0^{2\pi} \sin\beta \cos[m(\beta - \beta_p)] e^{-j(pr_d/a)\cos\beta - jur_d \sin\beta} d\beta, \quad (4.80)$$

$$I_c(p, u) = \int_0^{2\pi} \cos\beta \cos[m(\beta - \beta_p)] e^{-j(pr_d/a)\cos\beta - jur_d \sin\beta} d\beta. \quad (4.81)$$

When considering a circular patch excited by a probe feed at (l_p, β_p), which is again modeled by a unit-amplitude current ribbon of effective length w_p, (4.75) is modified to be

$$E_\rho = j\omega\mu_0 \sum_{m=0}^{\infty} \sum_{n=1}^{\infty} C_{mn} J_m(k_{mn} l) \cos[m(\beta - \beta_p)], \quad (4.82)$$

where

$$C_{mn} = e_m \frac{(-1)^m k_{mn}^2}{k_{\mathrm{eff}}^2 - k_{mn}^2} \frac{w_p J_m(k_{mn} l_p) j_0(m w_p/2)}{(k_{mn}^2 r_d^2 - m^2) J_m^2(k_{mn} r_d)}. \quad (4.83)$$

The input impedance is derived as

$$Z_{\text{in}} = \sum_{m=0}^{\infty} \sum_{n=1}^{\infty} \frac{h e_m k_{mn}^2}{2\pi^2 \epsilon_0 \epsilon_1} \frac{J_m^2(k_{mn} l_p) j_0^2(m w_p/2)}{J_m^2(k_{mn} r_d)(k_{mn}^2 r_d^2 - m^2)} \frac{\delta_{\text{eff}} f^3 - jf(f^2 - f_{mn}^2)}{\delta_{\text{eff}}^2 f^4 + (f^2 - f_{mn}^2)^2}. \quad (4.84)$$

From the expressions above, cavity-model solutions for a cylindrical circular microstrip antenna at the TM_{11} mode can be obtained. Related results have been reported in [9]. It is found that the results depend on the position of the feed with respect to the cylinder axis. When the circular patch is fed in the x–y plane ($\beta_p = 0°$), the characteristics are similar to those of the TM_{10} mode (excited in the $\hat{\phi}$ direction) of a cylindrical rectangular patch antenna. On the other hand, when the feed position is in the x–z plane ($\beta_p = 90°$), the results are similar to those of the TM_{01} mode (excited in the $\hat{z}$ direction) of a cylindrical rectangular patch antenna.

4.3.4 Annular-Ring Patch

In the probe-fed cylindrical annular-ring microstrip antenna depicted in Figure 4.17, the annular-ring patch has an inner radius r_1 and an outer radius r_2. To solve

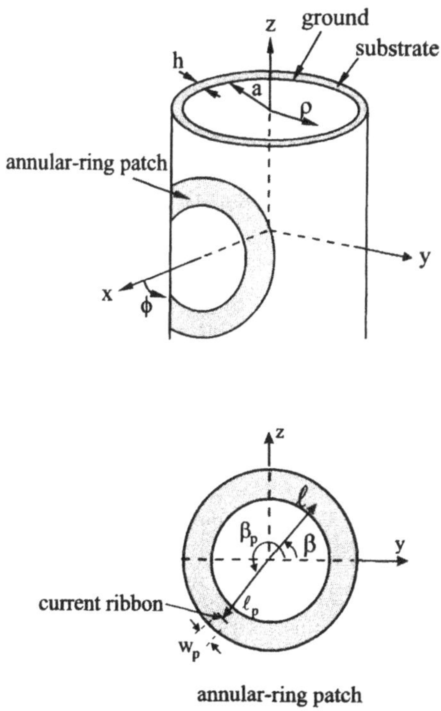

FIGURE 4.17 Geometry of a probe-fed cylindrical annular-ring microstrip antenna.

the problem, the coordinates of (4.73) are adopted. The feed position is assumed to be at (l_p, β_p) and the probe current is modeled as a $\hat{\rho}$-directed current ribbon of effective arc length w_p. With the probe current given by

$$\mathbf{J} = \frac{\hat{\rho} J_\rho(\beta)\delta(l - l_p)}{l_p}, \tag{4.85}$$

where

$$J_\rho(\beta) = \begin{cases} 1, & \pi + \beta_p - w_p/2 < \beta < \pi + \beta_p + w_p/2, \\ 0, & \text{elsewhere}, \end{cases} \tag{4.86}$$

the excited field inside the cylindrical annular-ring cavity satisfies the following wave equation:

$$\left(\frac{\partial^2}{\partial l^2} + \frac{1}{l}\frac{\partial}{\partial l} + \frac{1}{l^2}\frac{\partial^2}{\partial \beta^2} + k^2 \right) \mathbf{E} = j\omega\mu_0 \mathbf{J}. \tag{4.87}$$

Solving (4.87), we have [10]

$$\mathbf{E} = \hat{\rho} E_\rho = \hat{\rho} j\omega\mu_0 \sum_{m=0}^{\infty} \sum_{n=1}^{\infty} C_{mn} \Delta_{mn}(l) \cos[m(\beta - \beta_p)], \tag{4.88}$$

with

$$\Delta_{mn}(l) = J_m(k_{mn} l) - \frac{J'_m(k_{mn} r_{1e})}{Y'_m(k_{mn} r_{1e})} Y_m(k_{mn} l), \tag{4.89}$$

$$C_{mn} = \frac{w_p(-1)^m j_0(m w_p/2)\Delta_{mn}(l_p)}{(k_{\text{eff}}^2 - k_{mn}^2) N_{mn}^2}, \tag{4.90}$$

$$N_{mn} = \frac{\sqrt{\pi}}{\delta_m k_{mn}} [(k_{mn}^2 r_{2e}^2 - m^2)\Delta_{mn}^2(r_{2e}) - (k_{mn}^2 r_{1e}^2 - m^2)\Delta_{mn}^2(r_{1e})]^{1/2}, \tag{4.91}$$

where k_{mn} are the roots of the equation

$$J'_m(k_{mn} r_{2e}) - \frac{J'_m(k_{mn} r_{1e})}{Y'_m(k_{mn} r_{1e})} Y'_m(k_{mn} r_{2e}) = 0, \tag{4.92}$$

with

$$r_{1e} = R_1 \sqrt{1 - A_1}, \tag{4.93}$$

$$r_{2e} = R_2 \sqrt{1 + A_2}, \tag{4.94}$$

$$A_i = \frac{2h}{\pi r_i \epsilon_1}\left[\ln\frac{r_i}{2h} + 1.41\epsilon_1 + 1.77 + \frac{h}{r_i}(0.268\epsilon_1 + 1.65)\right], \qquad i = 1, 2 . \tag{4.95}$$

In the equations above, r_{1e} and r_{2e} are the effective inner and outer radii of the cylindrical annular-ring patch, whose expressions are obtained by accounting for the fringing field effects.

Next, from a consideration of the equivalent magnetic current radiating in the presence of a cylindrical surface [5], we can obtain the far-zone electric fields expressed as

$$E_\theta = E_{\theta 1} + E_{\theta 2}, \tag{4.96}$$

$$E_\phi = E_{\phi 1} + E_{\phi 2}, \tag{4.97}$$

with

$$E_{\theta i} = \frac{(-1)^i h r_{ie}\left[\sum_{m=0}^{\infty}\sum_{n=1}^{\infty} C_{mn}\Delta_{mn}(r_{ie})\right]e^{-jk_0 r}}{2ar\pi^2 \sin\theta} \sum_{p=-\infty}^{\infty} \frac{e^{jp\phi}j^{p+1}}{H_p^{(2)}(ak_0 \sin\theta)} I_{si}(p, -k_0\cos\theta), \tag{4.98}$$

$$E_{\phi i} = \frac{j(-1)^i h r_{ie}\left[\sum_{m=0}^{\infty}\sum_{n=1}^{\infty} C_{mn}\Delta_{mn}(r_{ie})\right]e^{-jk_0 r}}{2ar\pi^2} \sum_{p=-\infty}^{\infty} \frac{e^{jp\phi}j^{p+1}}{H_p^{(2)\prime}(ak_0 \sin\theta)}$$
$$\times\left[I_{ci}(p, -k_0\cos\theta) - \frac{p\cos\theta}{ak_0\sin^2\theta} I_{si}(p, -k_0\cos\theta)\right], \tag{4.99}$$

$$I_{si}(p, u) = I_s(p, u)\big|_{r_d = r_{ie}}, \qquad i = 1, 2, \tag{4.100}$$

$$I_{ci}(p, u) = I_c(p, u)\big|_{r_d = r_{ie}}, \qquad i = 1, 2, \tag{4.101}$$

where the expressions of $I_s(p, u)$ and $I_c(p, u)$ are given in (4.80) and (4.81), respectively. From (4.96)–(4.101), the radiation patterns of the cylindrical annular-ring patch antenna can be obtained. To determine the input impedance, we use

$$Z_{in} = \frac{-j\omega\mu_0 h}{w_p} \sum_{m=0}^{\infty}\sum_{n=1}^{\infty} C_{mn}(-1)^m j_0\left(\frac{m w_p}{2}\right)\Delta_{mn}(l_p), \tag{4.102}$$

where Δ_{mn} and C_{mn} are as given in (4.89) and (4.90), respectively.

From the formulation above, typical solutions of the input impedance for the TM_{12} mode of a cylindrical annular-ring microstrip antenna are as presented in

Figure 4.18, where the feed positions in the x–y plane ($\beta_p = 0°$) and x–z plane ($\beta_p = 90°$) are shown. Since for an annular-ring patch antenna, several adjacent modes are usually excited with excitation of the TM_{12} mode, and these undesired modes may contribute significantly to the input impedance of the antenna, a total of 30 modes of TM_{mn}, $n = 0$ to 5, $m = 1$ to 5, is included in the theoretical calculation. The results obtained using more than 30 modes are found to vary very slightly with the results obtained using 30 modes. From these results, it is seen that

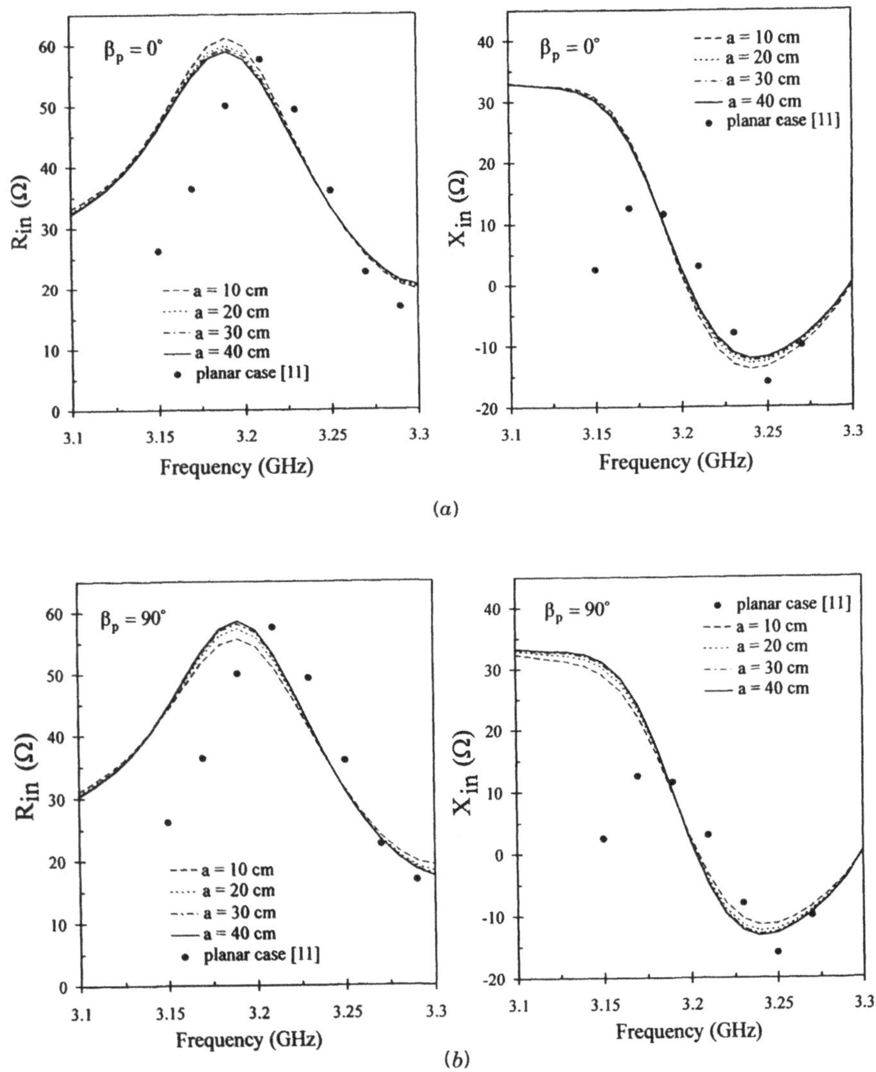

FIGURE 4.18 Input impedance versus frequency for an annular-ring patch at the TM_{12} mode; $\epsilon_1 = 2.2$, $h = 1.59$ mm, $r_1 = 3$ cm, $r_2 = 6$ cm, $l_p = 3.4$ cm. (a) $\beta_p = 0°$; (b) $\beta_p = 90°$. (From Ref. [10], © 1995 John Wiley & Sons, Inc.)

the resonant frequency is almost unaffected by the curvature variation. This behavior is due to the approximation adopted in cavity-model theory. However, it is observed that the resonant input resistance of an annular-ring patch antenna with a large cylinder radius agrees with that of the corresponding planar case [11]. For the case $\beta_p = 0°$, the resonant input resistance is also found to decrease with increasing cylinder radius, whereas the resonant input resistance increases with increasing cylinder radius for the case $\beta_p = 90°$. This behavior resembles that observed for a cylindrical circular patch antenna in the TM_{11} mode [9]. The radiation patterns calculated are plotted in Figures 4.19 and 4.20. Both the E- and H-plane patterns are shown. From the results it is seen that the back radiation increases with decreasing cylinder radius and is also more significant for the feed position placed in the x–y plane ($\beta_p = 0°$), similar to that of a cylindrical circular patch antenna [9].

4.4 PROBE-FED CASE: GENERALIZED TRANSMISSION-LINE MODEL SOLUTION

To apply GTLM theory [12], a microstrip patch antenna is modeled as sections of transmission lines taken in the direction that joins the radiating edges and loaded with wall admittance at the radiating edges. Each section of the transmission line is then replaced by an equivalent π network, whose circuit elements are derived as functions of antenna parameters. The wall admittances at the radiating edges are contributed from the self-admittance at the radiating edge and the mutual admittance between the radiating edges. Inclusion of mutual admittance in the analysis increases the accuracy of the GTLM solutions. From these derived equivalent lumped-circuit elements, an equivalent circuit for the microstrip patch antenna can be constructed and the input impedance of the antenna can then be evaluated from the simple circuit theory. Various GTLM formulations and the corresponding equivalent circuits for cylindrical microstrip patch antennas with various patch shapes are described in the following sections.

4.4.1 Rectangular Patch

A. Equivalent-Circuit Representation The transmission line is taken along the line joining the radiating apertures that are radiating the major portion of power. The effect of other apertures is considered as leakage from the transmission line. The analysis is based on the transmission-line mode with a single index, denoted by the TM_m mode. Here m stands for an angular eigenvalue. In a cavity the resonant modes are denoted by the double index TM_{mn}, the index n being associated with the eigenvalue in the $\hat{z}$ direction. The single-index transmission-line mode is equivalent to the sum of double-index cavity modes summed over the second index (i.e., $TM_m = \Sigma_n TM_{mn}$). Therefore, analysis of the TM_m mode takes into account the effect of the resonant modes TM_{mn} with $n = 0, 1, 2, \ldots$. The characteristics of a specific resonant mode, TM_{mn}, are then obtained from the

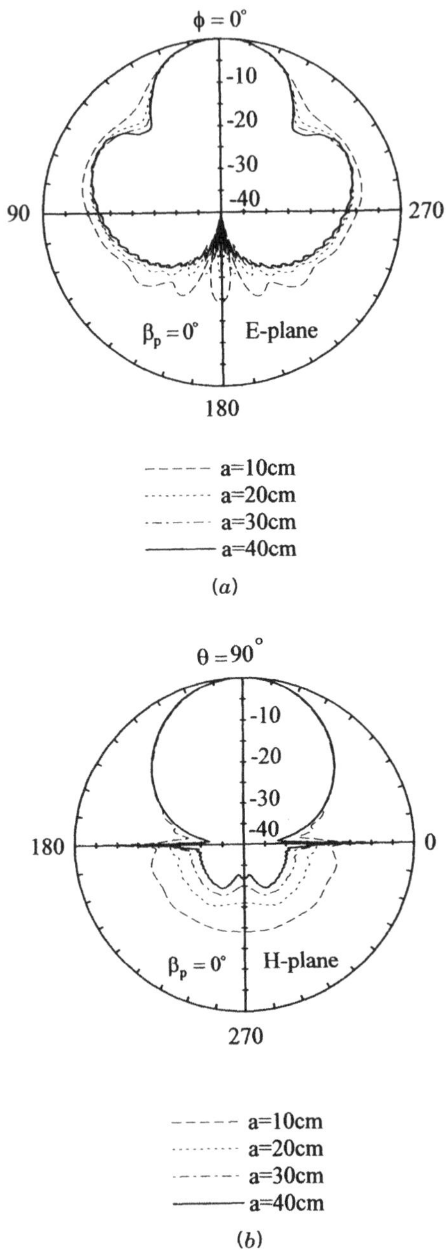

FIGURE 4.19 Radiation patterns for an annular-ring patch at the TM_{12} mode; antenna parameters with $\beta_p = 0°$ are as in Figure 4.18. (*a*) *E*-plane pattern; (*b*) *H*-plane pattern. (From Ref. [10], © 1995 John Wiley & Sons, Inc.)

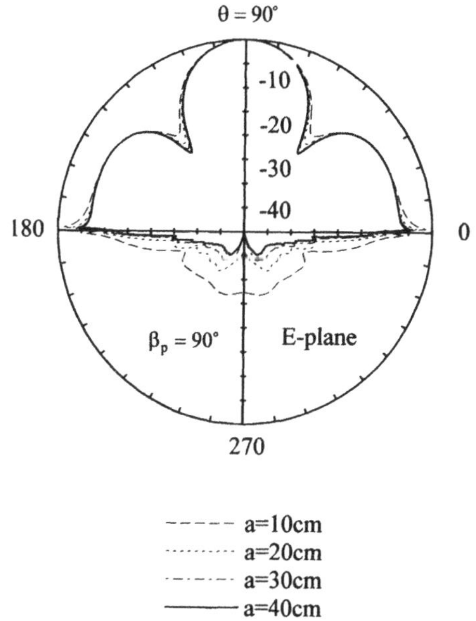

(*a*)

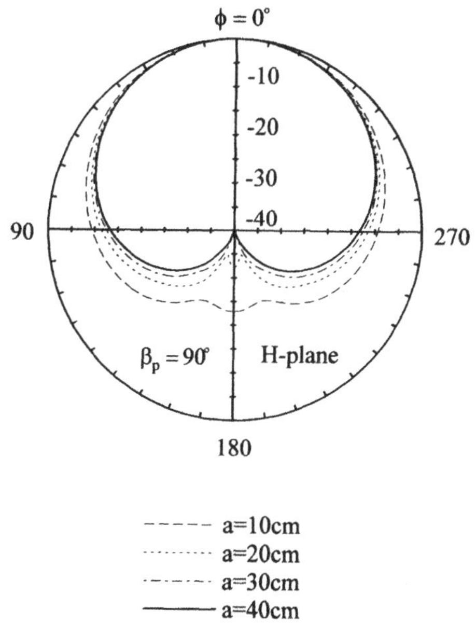

(*b*)

FIGURE 4.20 Radiation patterns for an annular-ring patch at the TM_{12} mode; antenna parameters with $\beta_p = 90°$ are as in Figure 4.18. (*a*) *E*-plane pattern; (*b*) *H*-plane pattern. (From Ref. [10], © 1995 John Wiley & Sons, Inc.)

characteristics of the TM_m mode near the neighborhood of the resonant frequency of such a specific mode.

To begin with, we refer to the geometry in Figure 4.1 (without the superstrate presence) for a cylindrical rectangular patch antenna and adopt the assumptions used in cavity-model analysis. The electric field inside the grounded substrate under the patch in this case has only the radial component E_ρ, which satisfies the wave equation of (4.29) and is independent of ρ. For the patch excited at the TM_{01} mode, the general solution for E_ρ can be written as

$$E_\rho = E_0(e^{jk_\phi(\phi-\phi_1)} + B_1 e^{-jk_\phi(\phi-\phi_1)})(e^{jk_z z} + B_2 e^{-jk_z z}), \tag{4.103}$$

with

$$k^2 = \omega^2 \mu_0 \epsilon_0 \epsilon_1 = \frac{k_\phi^2}{\rho^2 + k_z^2}, \tag{4.104}$$

where E_0, B_1, and B_2 are arbitrary constants. Then, since the rectangular patch is treated as a leaky transmission line along the $\hat{z}$ direction, the propagation constant k_z can be derived from the following impedance boundary condition:

$$y_t = \frac{H_z}{E_\rho}\bigg|_{\phi=\phi_1} = -\frac{H_z}{E_\rho}\bigg|_{\phi=\phi_1+2\phi_0}, \tag{4.105}$$

where y_t is the transverse wall admittance. When H_z is obtained by substituting (4.103) into Maxwell's equation $(-\nabla \times \mathbf{E})/j\omega\mu_0$, the magnetic field in the $\hat{z}$ direction can be written as

$$H_z = \frac{k_\phi E_0}{\omega\mu_0\rho}(e^{jk_\phi(\phi-\phi_1)} - B_1 e^{-jk_\phi(\phi-\phi_1)})(e^{jk_z z} + B_2 e^{-jk_z z}). \tag{4.106}$$

Substituting (4.103) and (4.106) into (4.105), we obtain

$$y_t = -\frac{k_\phi}{\omega\mu_0\rho}\frac{1-B_1}{1+B_1} = -\frac{k_\phi}{\omega\mu_0\rho}\frac{e^{jk_0 2\phi_0} - B_1 e^{-jk_0 2\phi_0}}{e^{jk_0 2\phi_0} + B_1 e^{-jk_0 2\phi_0}}. \tag{4.107}$$

By considering TM_{01}-mode excitation (i.e, the maximum radiation takes place from the edges at $z=-L$ and $z=L$, and the edges at $\phi=\phi_1$ and $\phi=\phi_1+2\phi_0$ radiate very little power), we can take $B_1 \cong 1$ as a first-order approximation. In this case, (4.107) becomes

$$y_t = -\frac{jk_\phi}{\omega\mu_0\rho}\tan(k_\phi 2\phi_0) \approx -\frac{2j\phi_0 k_\phi^2}{\omega\mu_0\rho}, \tag{4.108}$$

which yields

$$k_\phi^2 = \frac{j\omega\mu_0 \rho y_t}{2\phi_0}. \tag{4.109}$$

By substituting (4.109) into (4.104) and knowing that the imaginary part of y_t is very small due to the very small loss in nonradiating edges, k_z can be expressed approximately as

$$k_z \simeq k - \frac{j377 g_t}{4b\phi_0\sqrt{\epsilon_1}}, \tag{4.110}$$

where g_t is the real part of y_t at $\phi = \phi_1$ and $\phi = \phi_1 + 2\phi_0$, whose expression is derived later in this section.

With k_z derived, the rectangular patch is considered further as two sections of $\hat{z}$-directed transmission line separated by the feed position (ϕ_p, z_p), and each section of the transmission line can be replaced by an equivalent π network. The equivalent circuit is shown in Figure 4.21, where A, B, and C denote the positions $z = L$, $z = z_p$, and $z = -L$, respectively; g_1, g_2, and g_3 are the circuit elements of an equivalent π network replacing the section of transmission line between $z = z_p$ and $z = -L$; g_1', g_2', and g_3' are the circuit elements for the section of transmission line between $z = L$ and $z = z_p$. In this model we define the modal voltage as E_ρ and the modal current as $\pm bH_\phi$. The admittance at any position z is thus given as

$$Y(z) = \begin{cases} \dfrac{bH_\phi(z)}{E_\rho(z)}, & L > z > z_p, \\[2ex] \dfrac{-bH_\phi(z)}{E_\rho(z)}, & z_p > z > -L, \end{cases} \tag{4.111}$$

where E_ρ and H_ϕ can be expressed as

$$E_\rho = E_0(e^{jk_z z} + Ce^{-jk_z z}) \cos k_\phi(\phi - \phi_0), \tag{4.112}$$

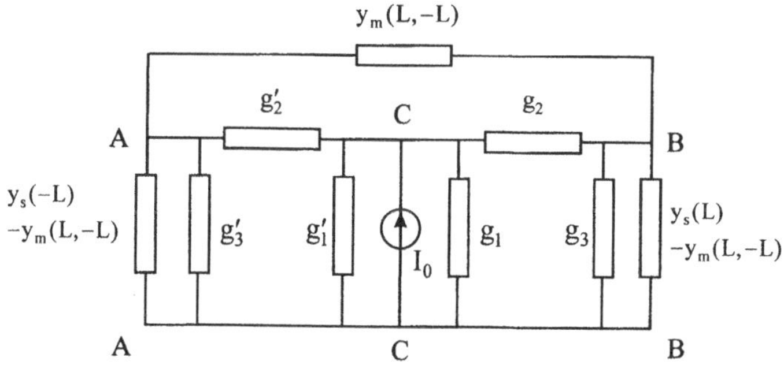

FIGURE 4.21 Equivalent circuit of a probe-fed cylindrical rectangular microstrip antenna, as shown in Figure 4.1, for GTLM analysis.

$$H_\phi = \frac{-k_z}{\omega\mu_0} E_0(e^{jk_z z} - Ce^{-jk_z z}) \cos k_\phi(\phi - \phi_0) . \tag{4.113}$$

The coefficient C in (4.106) and (4.107) is a constant to be determined. From the definition of (4.111), the wall admittance at $z = L$ is given by

$$y_w(L) = \frac{bH_\phi(L)}{E_\rho(L)} = \frac{b[H_\phi(L) + H_\phi(-L, L)]}{E_\rho(L)} . \tag{4.114}$$

By writing $y_s(L) = bH_\phi(L)/E_\rho(L)$ and $y_m(-L, L) = -bH_\phi(-L, L)/E_\rho(-L)$, (4.114) becomes

$$y_w(L) = y_s(L) - \frac{E_\rho(-L)}{E_\rho(L)} y_m(-L, L) , \tag{4.115}$$

where $H_\phi(L)$ is the magnetic field at $z = L$ generated by the magnetic current at $z = L$ and $H_\phi(-L, L)$ is the magnetic field at $z = L$ generated by the magnetic current at $z = -L$. These magnetic currents are equivalent currents due to the electric field distribution at $z = L$ and $z = -L$, respectively. Equation (4.115) can be rearranged further as

$$y_w(L) = y_s(L) - y_m(-L, L) + \frac{E_\rho(L) - E_\rho(-L)}{E_\rho(L)} y_m(-L, L) , \tag{4.116}$$

where y_s and y_m are defined as the self- and mutual admittance at the radiating edge. Similarly, for the radiating aperture at $z = -L$, we have the wall admittance expressed as

$$y_w(-L) = y_s(-L) - y_m(L, -L) + \frac{E_\rho(-L) - E_\rho(L)}{E_\rho(-L)} y_m(L, -L) . \tag{4.117}$$

Equations (4.116) and (4.117) explain the arrangement of the equivalent circuit shown in Figure 4.21.

Also, it should be noted that in Figure 4.21, I_0 is the feed current corresponding to the TM_{01} mode. From the condition of the magnetic field (and hence the modal current) discontinuity at $z = z_p$, we have

$$I_0 = \rho[H_\phi(z = z_p^+) - H_\phi(z = z_p^-)] . \tag{4.118}$$

By further extracting the zeroth term (i.e., for the TM_{01} mode) in the Fourier expansion of the total surface current at $\rho = b$, supplied by the total feed current I_t, it gives

$$I_0 = -\frac{I_t}{2\phi_0} . \tag{4.119}$$

Next, from the admittance matrix defined as

$$\begin{bmatrix} I_B \\ I_C \end{bmatrix} = \begin{bmatrix} Y_{11} & Y_{12} \\ Y_{21} & Y_{22} \end{bmatrix} \begin{bmatrix} V_B \\ V_C \end{bmatrix}, \tag{4.120}$$

with

$$I_B = bH_\phi(B), \qquad I_C = -bH_\phi(C),$$

$$V_B = E_\rho(B), \qquad V_C = E_\rho(C), \tag{4.121}$$

the elements Y_{ij} in the $\bar{\bar{Y}}$ matrix by definition are given as

$$Y_{11} = \left.\frac{I_B}{V_B}\right|_{V_C=0} = \left|\frac{bH_\phi(B)}{E_\rho(B)}\right|_{V_C=0}, \tag{4.122}$$

$$Y_{12} = \left.\frac{I_B}{V_C}\right|_{V_B=0} = \left.\frac{bH_\phi(B)}{E_\rho(C)}\right|_{V_B=0}, \tag{4.123}$$

$$Y_{21} = \left.\frac{I_C}{V_B}\right|_{V_C=0} = \left.\frac{-bH_\phi(C)}{E_\rho(B)}\right|_{V_C=0}, \tag{4.124}$$

$$Y_{22} = \left.\frac{I_C}{V_C}\right|_{V_B=0} = \left.\frac{-bH_\phi(C)}{E_\rho(C)}\right|_{V_B=0}. \tag{4.125}$$

From evaluation of (4.122)–(4.125), we have $Y_{11} = Y_{22}$ and $Y_{12} = Y_{21}$. Using the derived expressions for the elements in the $\bar{\bar{Y}}$ matrix, the g parameters of the π network for the transmission-line section between $z = -L$ and $z = z_p$ can be written as

$$g_1 = g_3 = Y_{11} + Y_{12} = \frac{bk_z}{\omega\mu_0}\left\{\coth[jk_z(z_p + L)] - \frac{1}{\sinh[jk_z(z_p + L)]}\right\}, \tag{4.126}$$

$$g_2 = -Y_{12} = \frac{bk_z}{\omega\mu_0}\frac{1}{\sinh[jk_z(z_p + L)]}. \tag{4.127}$$

Similarly, the g parameters for the transmission-line section between $z = L$ and $z = z_p$ are derived as

$$g_1' = g_3' = \frac{-bk_z}{\omega\mu_0}\left\{\coth[jk_z(z_p - L)] - \frac{1}{\sinh[jk_z(z_p - L)]}\right\}, \tag{4.128}$$

$$g_2' = \frac{bk_z}{\omega\mu_0}\frac{-1}{\sinh[jk_z(z_p - L)]}. \tag{4.129}$$

B. Wall Admittance The wall admittances at the radiating edges ($z = -L$ and $z = L$) and the nonradiating edges ($\phi = \phi_1$ and $\phi = \phi_1 + 2\phi_0$) have also been derived [13]. The self-admittance y_s is defined as

$$y_s(B) = -\frac{b\langle B, B\rangle}{\iint_{S_B} E_\rho^2(B)\, dS}, \tag{4.130}$$

where $\langle B,B\rangle$ is the self-reactance for the equivalent magnetic current source located at B and is given by

$$\langle B, B\rangle = -\iint_{S_B} H_\phi(B) E_\rho(B)\, dS\,. \tag{4.131}$$

Similarly, the mutual admittance between the two sources at A and B is obtained from

$$y_m(A, B) = -\frac{b\langle A, B\rangle}{\iint_{S_B} E_\rho(A) E_\rho(B)\, dS}, \tag{4.132}$$

with

$$\langle A, B\rangle = -\iint_{S_B} H_\phi(A, B) E_\rho(B)\, dS\,, \tag{4.133}$$

where $H_\phi(A, B)$ is the magnetic field at B due to the source at A, $H_\phi(B)$ is the magnetic field at B due to the source at B, and S_B is the aperture area at B. It is also noted that since $\langle A, B\rangle = \langle B, A\rangle$ from the reciprocity theorem, we have $y_m(A, B) = y_m(B, A)$.

From the definitions above, the self-admittances $y_s(L)$ and $y_s(-L)$ are written as

$$y_s(L) = y_s(-L)$$
$$= \frac{2b^2 h}{\phi_0 \pi^2} \sum_{p=0}^{\infty} \frac{\sin^2 p\phi_0}{\sigma_p p} \int_0^\infty \left[\frac{j\omega\epsilon_0 H_p^{(2)\prime}(\xi)}{\xi p H_p^{(2)}(\xi)} + \frac{p u^2 H_p^{(2)}(\xi)}{j\omega\mu_0 \xi^3 H_p^{(2)\prime}(\xi)} \right] du\,, \tag{4.134}$$

with

$$\sigma_p = \begin{cases} 2\,, & p = 0\,, \\ 1\,, & p > 0\,, \end{cases} \tag{4.135}$$

$$\xi = b\sqrt{k_0^2 - u^2}\,. \tag{4.136}$$

The mutual admittance $y_m(L, -L)$ is derived as

$$y_m(L, -L) = \frac{2b^2h}{\phi_0\pi^2} \sum_{p=0}^{\infty} \frac{\sin^2 p\phi_0}{\sigma_p p} \times \int_0^\infty \left[\frac{j\omega\epsilon_0 H_p^{(2)\prime}(\xi)}{\xi p H_p^{(2)}(\xi)} + \frac{pu^2 H_p^{(2)}(\xi)}{j\omega\mu_0 \xi^3 H_p^{(2)\prime}(\xi)} \right] \cos(2uL)\, du\,. \tag{4.137}$$

Similarly, for the wall admittance at the nonradiating edges, the self-admittances $y_s^n(\phi_1)$, $y_s^n(\phi_1 + 2\phi_0)$ and the mutual admittance $y_m^n(\phi_1, \phi_1 + 2\phi_0)$ are derived as

$$y_s^n(\phi_1) = y_s^n(\phi_1 + 2\phi_0) = \frac{-j2h}{\pi^2 b^2 \omega\mu_0 L} \sum_{p=0}^{\infty} \times \int_0^\infty \frac{u^2\xi(1 + \cos 2uL)}{(\pi/2L)^2 - u^2} \frac{H_p^{(2)}(\xi)}{\sigma_p H_p^{(2)\prime}(\xi)}\, du\,, \tag{4.138}$$

$$y_m^n(\phi_1, \phi_1 + 2\phi_0) = \frac{-j2h}{\pi^2 b^2 \omega\mu_0 L} \sum_{p=0}^{\infty} \cos(2p\phi_0) \times \int_0^\infty \frac{u^2\xi(1 + \cos 2uL)}{(\pi/2L)^2 - u^2} \frac{H_p^{(2)}(\xi)}{\sigma_p H_p^{(2)\prime}(\xi)}\, du\,. \tag{4.139}$$

From (4.138) and (4.139), the transverse wall admittance y_t can be evaluated; that is,

$$y_t(\phi_1) = y_t(\phi_1 + 2\phi_0) = y_s^n(\phi_1) - \frac{E_\rho(\phi_1 + 2\phi_0)}{E_\rho(\phi_1)}\, y_m^n(\phi_1)\,. \tag{4.140}$$

The real part (g_t) of y_t in (4.140) is then used for calculating the propagation constant k_z in (4.110).

C. Input Impedance From the equivalent lumped-circuit elements and the wall admittances derived above, the input impedance Z_0 seen by the current I_0 is given by

$$Z_0 = \frac{E_\rho(C)}{I_0} = \{[(Z_A + R_A)\|(Z_B + R_B)] + R_C\}\|Z_C\,, \tag{4.141}$$

with

$$Z_A = [y_s(L) - y_m(L, -L) + g_3']^{-1}\,, \tag{4.142}$$

$$Z_B = [y_s(-L) - y_m(L, -L) + g_3]^{-1}\,, \tag{4.143}$$

$$Z_C = (g_1 + g_1')^{-1}\,, \tag{4.144}$$

$$R_A = \frac{g_2}{\Delta}\,, \qquad R_B = \frac{g_2'}{\Delta}\,, \qquad R_C = \frac{y_m(L, -L)}{\Delta}\,, \tag{4.145}$$

$$\Delta = g_2 g_2' + g_2 y_m(L, -L) + g_2' y_m(L, -L)\,. \tag{4.146}$$

Based on the expression above and knowing that $I_t = -I_0 2\phi_0$ (4.119), the input impedance of the antenna at the TM_{01} mode seen by the total feed current I_t is written as

$$Z_{in} = -\frac{hE_\rho(C)}{I_t} = \frac{h}{2\phi_0}\{[(Z_A + R_A)\|(Z_B + R_B)] + R_C\}\|Z_C\,. \tag{4.147}$$

D. GTLM Solutions From the formulation above, GTLM solutions of the input impedance for a cylindrical rectangular microstrip patch antenna at the TM_{01} mode are obtained. Typical results are presented in Figures 4.22 to 4.24. From the results in Figure 4.22 it is interesting to note that the GTLM solution is in good agreement with the measured data, although some approximations are made for GTLM analysis. The results obtained in Figure 4.23 also show that the resonant frequency is slightly affected by the cylinder-radius variation, similar to the observations of full-wave analysis. The resonant input resistance is also observed to be increased

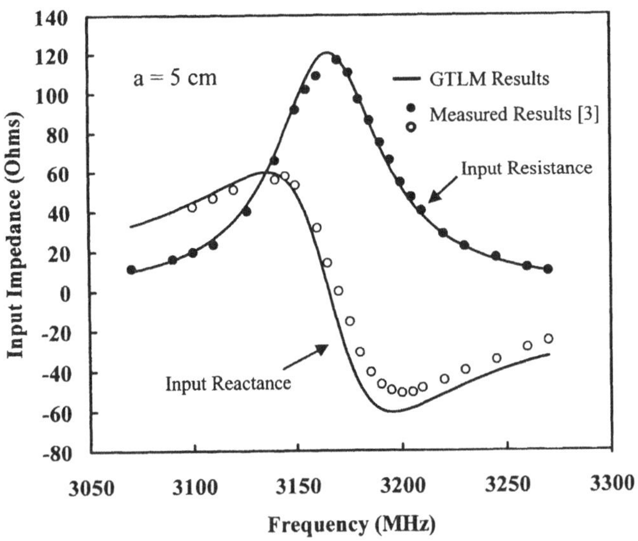

FIGURE 4.22 Input impedance versus frequency for the patch excited at the TM_{01} mode; $a = 5$ cm, $\epsilon_1 = 2.32$, $h = 0.795$ mm, $2L = 3$ cm, $2b\phi_0 = 4$ cm, $\phi_p = 90°$, $z_p = 1.0$ cm. (From Ref. [13], © 1994 John Wiley & Sons, Inc.)

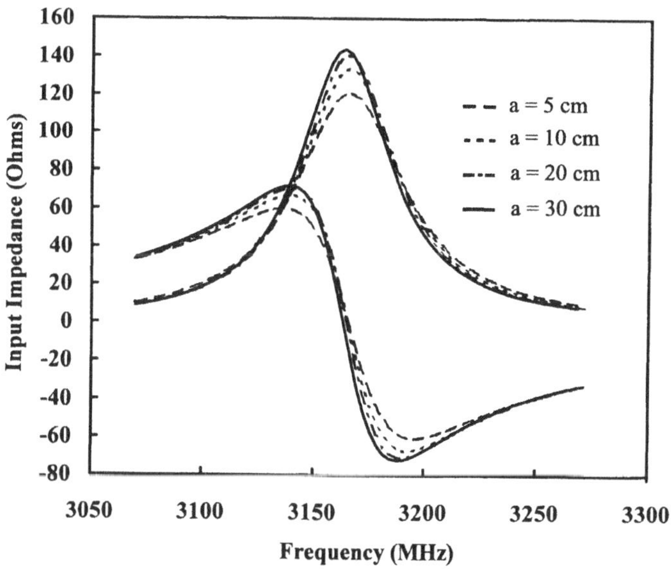

FIGURE 4.23 Input impedance versus frequency for various cylinder radii at the TM_{01} mode; antenna parameters are as in Figure 4.22. (From Ref. [13], © 1994 John Wiley & Sons, Inc.)

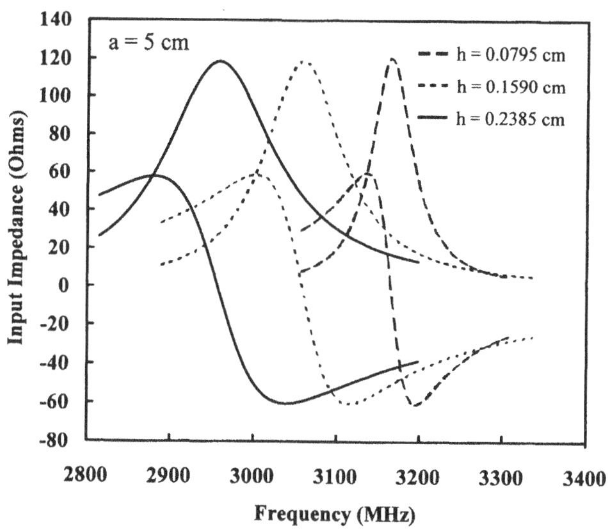

FIGURE 4.24 Input impedance versus frequency for various substrate thicknesses at the TM_{01} mode; antenna parameters are as in Figure 4.22. (From Ref. [13], © 1994 John Wiley & Sons, Inc.)

with increasing cylinder radius, which is similar to the characteristics predicted by the full-wave approach and cavity-model analysis. Figure 4.24 shows the input impedance results for various substrate thicknesses. It is seen that the input impedance bandwidth increases and the resonant frequency decreases with increasing substrate thickness. However, the resonant input resistance is found to be relatively insensitive to substrate-thickness variation. Figure 4.25 shows the resonant input resistance as a function of feed position for different cylinder radii. The resonant input resistance is seen to decrease with decreasing cylinder radius. Also, the resonant input resistance shows a maximum value when the feed position is at the edge and is zero at the patch center.

4.4.2 Circular Patch

The geometry shown in Figure 4.16 is treated. By considering the circular patch to be excited at the TM_m ($\equiv \Sigma_n TM_{mn}$) mode and solving the wave equation (4.74), the electric and magnetic fields inside the grounded substrate under the patch are written as [14]

$$E_p(l,\beta) = \begin{cases} E_1 J_m(kl)\cos[m(\beta-\beta_p)]\,, & 0 < l < l_p\,, \\ E_2[J_m(kl) + E_3 H_m^{(2)}(kl)]\cos[m(\beta-\beta_p)]\,, & l_p < l \le r_d\,, \end{cases} \tag{4.148}$$

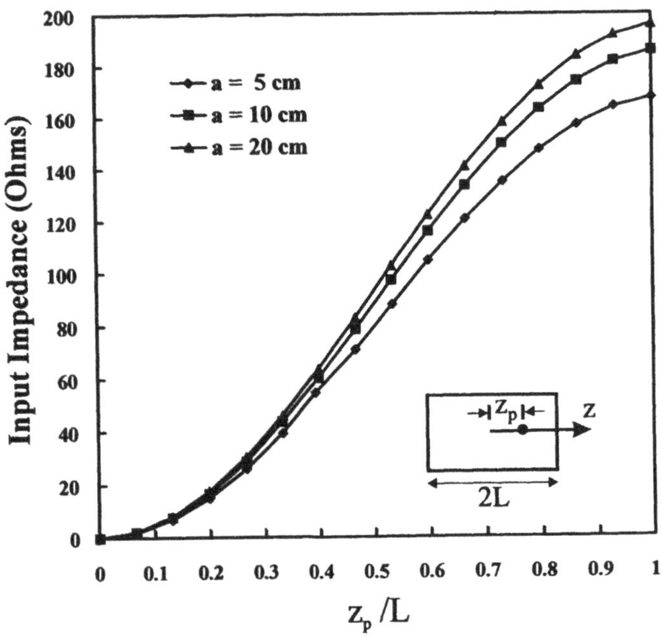

FIGURE 4.25 Resonant input resistance for $a = 5$, 10, and 20 cm against feed position; antenna parameters are as in Figure 4.22.

$$H_\beta(l,\beta)=\begin{cases}\dfrac{kE_1}{j\omega\mu_0}J'_m(kl)\cos[m(\beta-\beta_p)]\,, & 0<l<l_p\,,\\ \dfrac{k}{j\omega\mu_0}[E_2(J'_m(kl)+E_3H_m^{(2)\prime}(kl)]\cos[m(\beta-\beta_p)]\,, & l_p<l\le r_d\,,\end{cases}$$

(4.149)

where $k=\omega\sqrt{\mu_0\epsilon_0\epsilon_1}$, and E_1, E_2, and E_3 are coefficients to be determined. Then, based on GTLM theory, the circular patch is modeled as a transmission line taken in the direction of $\hat{l}$ and loaded with a wall admittance at the radiation aperture around the patch edge at $l=r_d$. The equivalent circuit for this patch antenna is shown in Figure 4.26, where g_1, g_2, and g_3 again are the circuit elements of an equivalent π network for the section of transmission line between $l=l_p$ and $l=r_d$; y_p is the admittance at $l=l_p^-$ (i.e., just off the feed position); y_w is the wall admittance around the patch edge; and I_m is the feed current corresponding to the TM_m mode. By defining the modal voltage as E_ρ and the modal current as $\pm lH_\beta$ and following the formulation in Section 4.4.1, we can derive

$$y_p=\frac{lH_\beta}{E_\rho}\bigg|_{l=l_p^-}=\frac{kl_p}{j\omega\mu_0}\frac{J'_m(kl_p)}{J_m(kl_p)}\,, \tag{4.150}$$

and

$$g_1=\frac{jkl_p}{\omega\mu_0}\left[\frac{\Delta_1(l_p,r_d)}{\Delta_2(l_p,r_d)}+\frac{\Delta_1(l_p,l_p)}{\Delta_2(r_d,l_p)}\right], \tag{4.151}$$

$$g_2=-\frac{jkl_p}{\omega\mu_0}\frac{\Delta_1(l_p,l_p)}{\Delta_2(r_d,l_p)}\,, \tag{4.152}$$

$$g_3=\frac{jkl_p}{\omega\mu_0}\frac{1}{\Delta_2(r_d,l_p)}[\Delta_1(l_p,l_p)-\Delta_1(r_d,l_p)]\,, \tag{4.153}$$

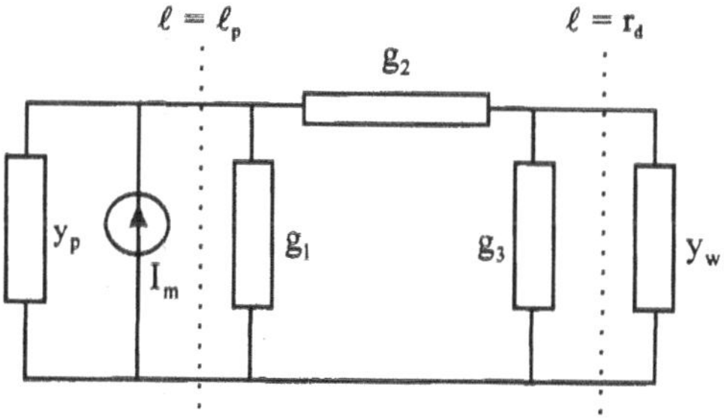

FIGURE 4.26 Equivalent circuit of a probe-fed cylindrical circular microstrip antenna, as shown in Figure 4.16, for the GTLM analysis.

with

$$\Delta_1(x, y) = J'_m(kx)H^{(2)}_m(ky) - J_m(ky)H^{(2)\prime}_m(kx)\,, \tag{4.154}$$

$$\Delta_2(x, y) = J_m(kx)H^{(2)}_m(ky) - J_m(ky)H^{(2)}_m(kx)\,. \tag{4.155}$$

For the wall admittance y_w, we have [14]

$$y_w = \frac{r_d^2 h}{4\pi^3 b}\frac{1}{j\omega\mu_0}\sum_{p=-\infty}^{\infty}\int_{-\infty}^{\infty}\left\{\frac{H_p^{(2)}(\xi)}{H_p^{(2)\prime}(\xi)}\left[\frac{\xi}{b}I_c^2(p,u) + \frac{2pu}{\xi}I_s(p,u)I_c(p,u) + \frac{bp^2u^2}{\xi^3}I_s^2(p,u)\right] + \frac{j\omega\epsilon_0\epsilon_1 b}{\xi}\frac{H_p^{(2)\prime}(\xi)}{H_p^{(2)}(\xi)}I_s^2(p,u)\right\}du\,, \tag{4.156}$$

where ξ is again in (4.136), and $I_s(p, u)$ and $I_c(p, u)$ are two integrals, given in (4.80) and (4.81), respectively. It should also be noted that b_w (the imaginary part of y_w) is very slow in converging, and therefore the self-susceptance b_w can be evaluated using the extension formula for the fringing field; that is,

$$b_w = \frac{-r_d H_\beta(r_d)}{jE_\rho(r_d)} = \frac{kr_d}{\omega\mu_0}\frac{J'_m(kr_d)H^{(2)\prime}_m(kr_{de}) - J'_m(kr_{de})H^{(2)\prime}_m(kr_d)}{J_m(kr_d)H^{(2)\prime}_m(kr_{de}) - J'_m(kr_{de})H^{(2)}_m(kr_d)}\,, \tag{4.157}$$

where r_{de} is the effective disk radius [15] given by (4.77). Also, from the condition of the magnetic field discontinuity at $l = l_p$, we have

$$I_m = \rho[H_\beta(l = l_p^+) - H_\beta(l = l_p^-)]\,. \tag{4.158}$$

Extracting the mth term in the Fourier expansion of the total feed current, I_t, supplied by the source; that is,

$$I_t\delta(\beta - \beta_p) = \sum_{k,\, k\neq m} A_k\cos[k(\beta - \beta_p)] + A_m\cos[m(\beta - \beta_p)]\,, \tag{4.159}$$

gives

$$I_m = \begin{cases} \dfrac{-I_t}{2\pi}\,, & m = 0\,, \\[2ex] \dfrac{-I_t}{\pi}\,, & m > 0\,. \end{cases} \tag{4.160}$$

With y_p, g_1, g_2, g_3, and y_w determined, the input impedance of the cylindrical circular patch antenna at the TM_m mode can readily be obtained from

$$Z_{\text{in}} = -\frac{hE_\rho(l = l_p)}{I_t} = \frac{-hI_m}{I_t}\frac{E_\rho(l = l_p)}{I_m}\,, \tag{4.161}$$

with

$$\frac{E_p(l = l_p)}{I_m} = \left[y_p + g_1 + \frac{g_2(g_3 + y_w)}{g_2 + g_3 + y_w} \right]^{-1} . \tag{4.162}$$

Thus for TM_{11}-mode excitation (i.e., $m = 1$), we have

$$Z_{\text{in}} = \frac{h}{\pi} \left[y_p + g_1 + \frac{g_2(g_3 + y_w)}{g_2 + g_3 + y_w} \right]^{-1} . \tag{4.163}$$

From (4.163), GTLM results for the input impedance of a cylindrical circular microstrip antenna with various cylinder radii are calculated and shown in Figures 4.27 and 4.28. In Figure 4.27 the feed positions are at $l_p = 1.82$ cm, $\beta_p = 0$ and 90°, and Figure 4.27 shows the results for the feed positions moving toward the patch edge at $l_p = 2.4$ cm, $\beta_p = 0$ and 90°. For both cases it is observed that the resonant input resistance increases with decreasing cylinder radius when $\beta_p = 0°$ (feed position in the x–y plane), whereas the resonant input resistance decreases with decreasing cylinder radius when $\beta_p = 90°$ (feed position in the x–z plane). Also, the input impedance levels for various cylinder radii are in good agreement with the data measured, and the difference between the GTLM results and the measured data is less than 6 MHz, or 0.4% for the parameters studied here.

Figure 4.29 shows the resonant input resistance for various feed positions. It is seen that the resonant input resistance decreases when the feed position moves toward the patch center and has a maximum value at the patch edge. The resonant input resistance has a maximum value when the feed position is in the $\beta_p = 0°$ plane (x–y plane) and has a minimum value in the $\beta_p = 90°$ plane (x–z plane). This behavior resembles the results obtained using cavity-model solutions [9]. Figure 4.30 presents more calculated and measured results of the resonant input resistance as a function of β_p. Again, characteristics similar to those in Figure 4.29 are observed. This behavior can be used for impedance matching of the cylindrical circular patch antenna by adjusting β_p.

4.4.3 Annular-Ring Patch

Given the geometry in Figure 4.17, we consider an annular-ring patch excited at the TM_m ($\equiv \Sigma_n TM_{mn}$) mode for the construction of an equivalent circuit for GTLM analysis. To begin with, the wave equation (4.87) is solved, and we have an expression for an E_ρ field under an annular-ring patch written as

$$E_\rho = [E_1 J_m(kl) + E_2 H_m^{(2)}(kl)] \cos m(\beta - \beta_p) , \qquad r_1 \le l \le r_2 , \tag{4.164}$$

where E_1 and E_2 are, again, coefficients to be determined. Also, the magnetic field under the patch is given by

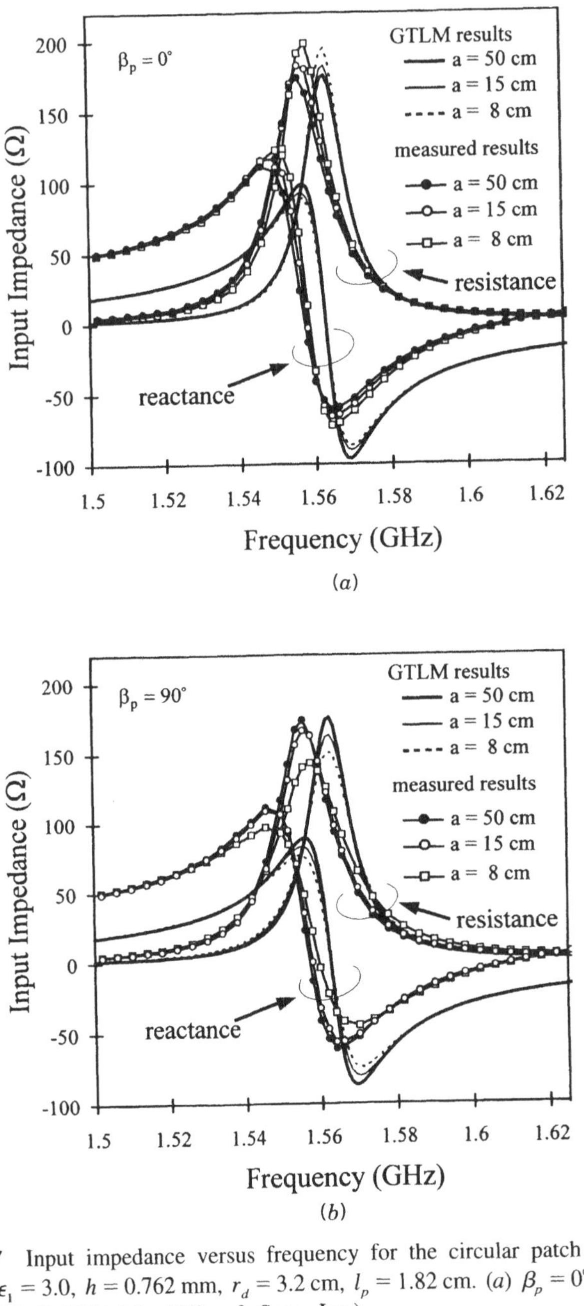

FIGURE 4.27 Input impedance versus frequency for the circular patch excited at the TM_{11} mode; $\epsilon_1 = 3.0$, $h = 0.762$ mm, $r_d = 3.2$ cm, $l_p = 1.82$ cm. (*a*) $\beta_p = 0°$; (*b*) $\beta_p = 90°$. (From Ref. [42], © 1996 John Wiley & Sons, Inc.)

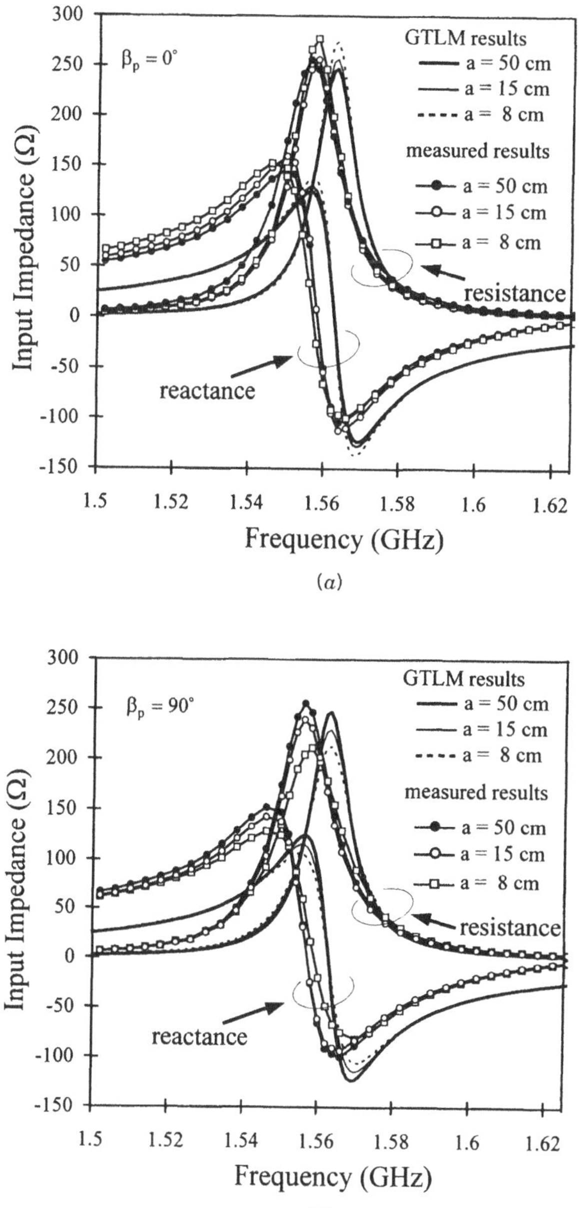

FIGURE 4.28 Input impedance versus frequency for the circular patch excited at the TM_{11} mode; $\epsilon_1 = 3.0$, $h = 0.762$ mm, $r_d = 3.2$ cm, $l_p = 2.4$ cm. (*a*) $\beta_p = 0°$; (*b*) $\beta_p = 90°$.

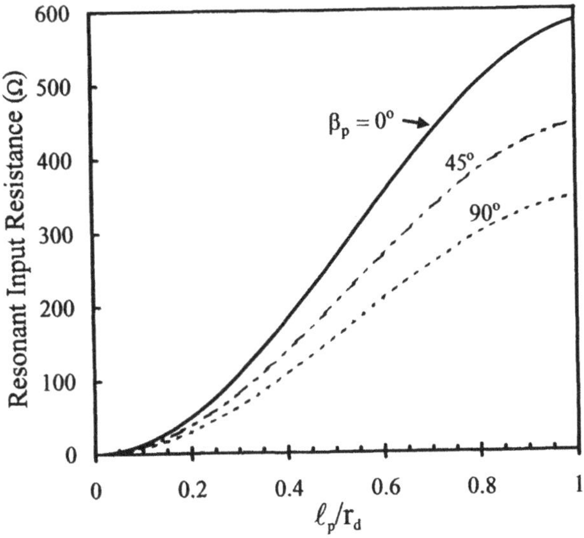

FIGURE 4.29 Resonant input resistance for various feed positions at the TM_{11} mode; $a = 5$ cm, $\epsilon_1 = 2.47$, $h = 0.8$ mm, $r_d = 1.88$ cm, $\beta_p = 0°, 45°, 90°$. (From Ref. [14], © 1995 John Wiley & Sons, Inc.)

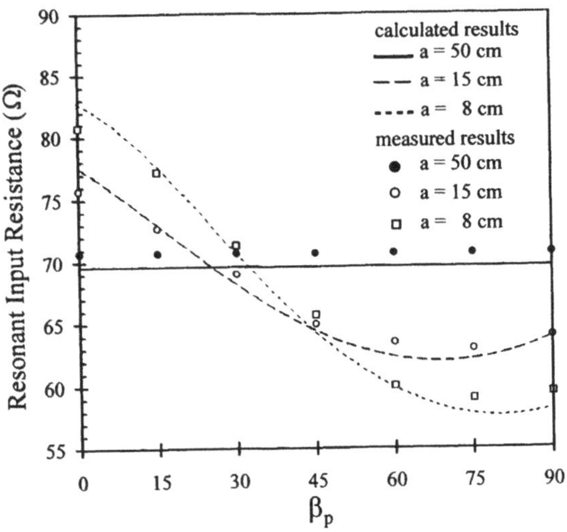

FIGURE 4.30 Resonant input resistance as a function of β_p with $l_p = 1.05$ cm; other parameters are as in Figure 4.27. (From Ref. [42], © 1996 John Wiley & Sons, Inc.)

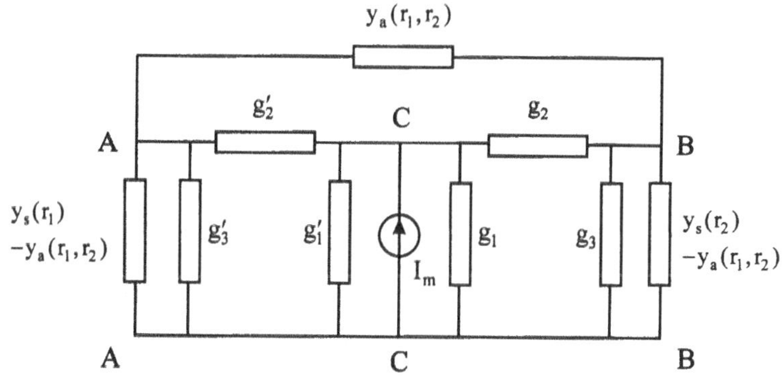

FIGURE 4.31 Equivalent circuit of a probe-fed cylindrical annular-ring microstrip antenna, as shown in Figure 4.17, for GTLM analysis.

$$H_\beta = \frac{-k}{j\omega\mu_0}[E_1 J'_m(kl) + E_2 H_m^{(2)\prime}(kl)]\cos m(\beta - \beta_p)\,, \qquad r_1 \le l \le r_2\,. \quad (4.165)$$

By following the GTLM formulation in Sections 4.4.1 and 4.4.2, the equivalent circuit shown in Figure 4.31 can be constructed, where two sections of the transmission line in the direction of $\hat{l}$ separated by the feed at (l_p, β_p) are replaced by equivalent π networks and terminated with wall admittances at the radiating apertures around the edges at $l = r_1$ and r_2. The mutual admittance, described by $y_a(r_1, r_2)$, between the two radiating apertures is also considered in the equivalent circuit.

The g parameters derived for the transmission-line section between $l = l_p$ and $l = r_2$ are written as [16]

$$g_1 = \frac{j}{\omega\mu_0 \Delta_2(l_p, r_2)}\left[kl_p \Delta_1(l_p, r_2) + \frac{2}{\pi}\right], \quad (4.166)$$

$$g_2 = \frac{-2j}{\omega\pi\mu_0 \Delta_2(l_p, r_2)}, \quad (4.167)$$

$$g_3 = \frac{-j}{\omega\mu_0 \Delta_2(r_2, l_p)}\left[kr_2 \Delta_1(r_2, l_p) + \frac{2}{\pi}\right], \quad (4.168)$$

where the functions Δ_1 and Δ_2 are as given in (4.154) and (4.155), respectively. For the g' parameters of the transmission-line section between $l = l_p$ and $l = r_1$, we have

$$g'_1 = \frac{-j}{\omega\mu_0 \Delta_2(l_p, r_1)}\left[kl_p \Delta_1(l_p, r_1) + \frac{2}{\pi}\right], \quad (4.169)$$

$$g_2' = \frac{-2j}{\omega \pi \mu_0 \Delta_2(r_1, l_p)}, \tag{4.170}$$

$$g_3' = \frac{-j}{\omega \mu_0 \Delta_2(r_1, l_p)} \left[kr_1 \Delta_1(r_1, l_p) + \frac{2}{\pi} \right]. \tag{4.171}$$

The mutual admittance $y_a(r_1, r_2)$ is given as

$$y_a(r_1, r_2) = \frac{r_1 r_2 h}{j4\pi^3 a\omega\mu_0} \sum_{p=-\infty}^{\infty} \int_{-\infty}^{\infty} \left\{ \frac{H_p^{(2)}(\xi)}{H_p^{(2)\prime}(\xi)} \left[\frac{\xi}{a} I_c(p, u, r_1) I_c(p, u, r_2) \right. \right.$$

$$+ \frac{pu}{\xi} I_s(p, u, r_1) I_c(p, u, r_2) + \frac{pu}{\xi} I_c(p, u, r_1) I_s(p, u, r_1)$$

$$\left. + \frac{ap^2u^2}{\xi^3} I_s(p, u, r_1) I_s(p, u, r_2) \right]$$

$$\left. - \frac{k_0^2 a}{\xi} \frac{H_p^{(2)\prime}(\xi)}{H_p^{(2)}(\xi)} I_s(p, u, r_1) I_s(p, u, r_2) \right\} du. \tag{4.172}$$

In (4.172), ξ is as given in (4.136), and $I_c(p, u, l)$ and $I_s(p, u, l)$ in (4.172) are two integrals written as

$$I_c(p, u, l) = \pi(-j)^{m+1} [J_{m+1}(\chi) \cos \alpha_A - J_{m-a}(\chi) \cos \alpha_B], \tag{4.173}$$

$$I_s(p, u, l) = \pi(-j)^{m+1} [J_{m+1}(\chi) \sin \alpha_A + J_{m-1}(\chi) \sin \alpha_B], \tag{4.174}$$

with

$$\alpha_A = -m\beta_p + (m+1) \tan^{-1} \frac{au}{p}, \tag{4.175}$$

$$\alpha_B = -m\beta_p + (m-1) \tan^{-1} \frac{au}{p}, \tag{4.176}$$

$$\chi = l \sqrt{\left(\frac{p}{a}\right)^2 + u^2}. \tag{4.177}$$

Also, the self-admittance $y_s(r_i)$, $i = 1, 2$, can be computed from $y_a(r_i, r_i)$ in (4.172).

With the circuit elements in the equivalent circuit of Figure 4.31 determined, the input impedance of the antenna seen by the total feed current I_t at the TM_m mode can be written as

$$Z_m = \frac{-hE_p(l = l_p)}{I_t} = \frac{h}{\pi\sigma_m} (\{[(Z'_A + R'_A)\|(Z'_B + R'_B)] + R'_C\}\|Z'_C) , \quad (4.178)$$

with

$$\sigma_m = \begin{cases} 2 , & m = 0 , \\ 1 , & m > 0 , \end{cases} \quad (4.179)$$

$$Z'_A = [y_s(r_1) - y_a(r_1, r_2) + g'_3]^{-1} , \quad (4.180)$$

$$Z'_B = [y_s(r_2) - y_a(r_1, r_2) + g_3]^{-1} , \quad (4.181)$$

$$Z'_C = (g_1 + g'_1)^{-1} , \quad (4.182)$$

$$R'_A = \frac{g_2}{g_2 y_a(r_1, r_2) + g'_2 y_a(r_1, r_2) + g_2 g'_2} , \quad (4.183)$$

$$R'_B = \frac{g'_2}{g_2} R_A , \quad (4.184)$$

$$R'_C = \frac{y_a(r_1, r_2)}{g_2} R_A . \quad (4.185)$$

From (4.178), the numerical results of a cylindrical annular-ring patch antenna excited at the TM_{12} mode are studied. The calculated and measured input impedances are presented in Figure 4.32. Because the resonant frequencies of various modes are strongly dependent on r_2/r_1 of the annular-ring patch, the value of r_2/r_1 can thus be used to control the number of resonant modes to be excited. For the case of $r_2/r_1 = 2$, we consider a total of six modes; that is, TM_{02}, TM_{12}, TM_{22}, TM_{32}, TM_{41}, and TM_{51} are included in the theoretical calculations (i.e., $Z_{in} = Z_0 + Z_1 + Z_2 + Z_3 + Z_4 + Z_5$), because their resonant frequencies are near those of the TM_{12} mode. With the feed positions at $l_p = 1.8$ cm and 1.6 cm with $\beta_p = 90°$ shown in Figure 4.32, results show that the resonant input resistance increases with increasing cylinder radius, while the resonant frequency decreases with increasing cylinder radius. This behavior is also verified by experiment. Figure 4.33 shows the resonant input resistance for different feed positions at $a = 8$ cm. It is seen that the resonant input resistance increases when the feed position moves close to the inner or outer edges.

4.5 SLOT-COUPLED CASE: FULL-WAVE SOLUTION

In this section, the slot-coupled cylindrical rectangular microstrip antenna shown in Figure 4.34 is analyzed. In such a configuration the coupling slot in the ground cylinder is much smaller than its resonant size, so that most radiation occurs from

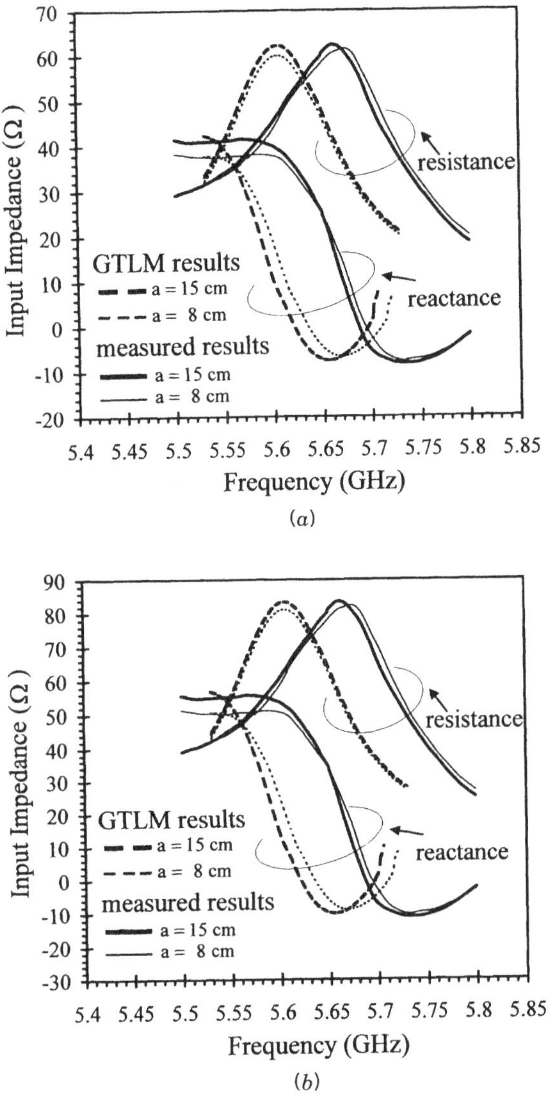

FIGURE 4.32 Input impedance versus frequency for the annular-ring patch excited at the TM_{12} mode; $\epsilon_1 = 3.0$, $h = 0.762$ mm, $r_1 = 1.5$ cm, $r_2 = 3.0$ cm, $\beta_p = 90°$. (*a*) $l_p = 1.8$ cm; (*b*) $l_p = 1.6$ cm. (From Ref. [16], © 1997 John Wiley & Sons, Inc., reprinted with permission.)

the resonant patch element rather than the coupling slot. On the other hand, when there is no microstrip patch, the printed slot can also be an efficient radiator. Since the analysis of a printed slot as a radiator is related to that for a microstrip patch antenna with the printed slot as an energy coupler, a printed slot used as a radiator is studied first.

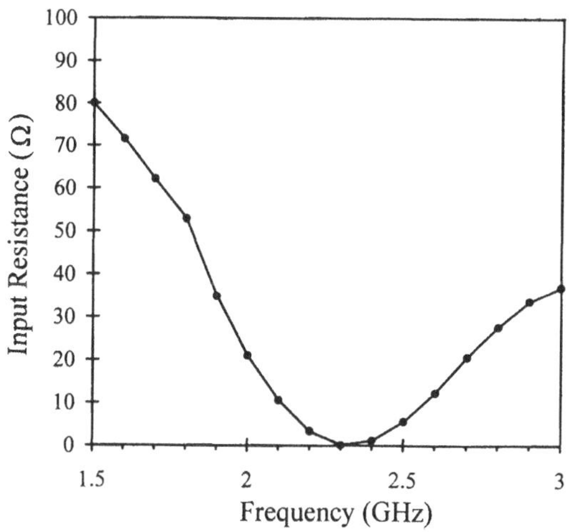

FIGURE 4.33 Input impedance versus feed position with $a = 8$ cm; other parameters are as in Figure 4.32. (From Ref. [16], © 1997 John Wiley & Sons, Inc., reprinted with permission.)

4.5.1 Printed Slot as a Radiator

The geometry of a microstrip-line-fed cylindrical printed slot is shown in Figure 4.35. The full-wave solution considered here is obtained by using a theoretical approach based on a combination of reciprocity analysis [17] and a moment-method calculation incorporating the Green's function formulation for the cylindrical structure. The exact Green's functions are derived to evaluate the necessary field components generated from electric and magnetic currents in the presence of a grounded cylindrical substrate. From the reciprocity analysis, expressions for the amplitudes of reflected and transmitted waves on the microstrip line and an equivalent circuit representing the slot discontinuity are obtained. Then, by representing the unknown current in the feed line in terms of a traveling-wave mode and expanding the unknown electric field in the slot using a set of piecewise sinusoidal (PWS) basis functions, an electric-field integral equation can be derived to solve for the above-mentioned unknown feed-line current and slot electric-field amplitudes. With these unknown amplitudes solved, the characteristics of a cylindrical printed slot antenna can be determined. This technique is versatile and can be used for the analysis and design of microstrip-line-fed printed antennas and circuits. This approach is a good alternative to simplifying the brute-force full-wave approach described in Section 2.5 for solving the resonance problem of a slot-coupled microstrip structure.

A. Reciprocity Analysis Given the geometry in Figure 4.35, the radiating slot has a length of L_s ($=2a\phi_s$) and a width of W_s and is printed on a cylindrical ground cylinder of radius a. The microstrip line is assumed to be infinitely long

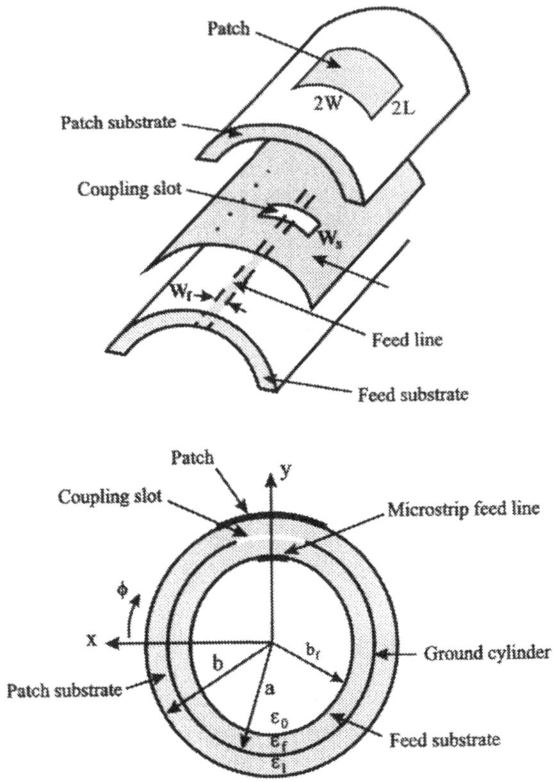

FIGURE 4.34 Geometry of a slot-coupled cylindrical rectangular microstrip antenna.

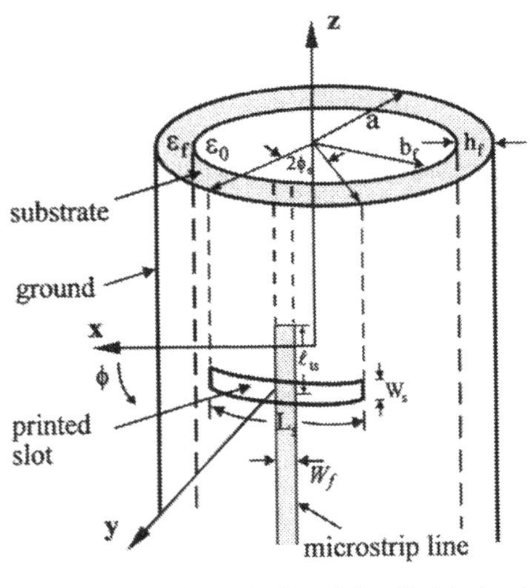

FIGURE 4.35 Geometry of a microstrip-line-fed cylindrical printed slot antenna.

with a width of W_f $(=2b_f\phi_f)$ and is printed on a substrate of thickness h_f $(=a-b_f)$ and relative permittivity ϵ_f. The regions of $\rho < b_f$ and $\rho > a$ are assumed to be air. By assuming that the radiating slot is narrow (i.e., $L_s >> W_s$), the electric field in the slot is first taken approximately as a PWS mode with unknown amplitude V_0; that is,

$$\mathbf{E}_s = \hat{z}V_0 E_z^s(\phi')\,, \qquad |z| < \frac{W_s}{2}\,, \quad |\phi'| < \phi_s\,, \qquad \phi' = \phi - \frac{\pi}{2}\,, \tag{4.186}$$

with

$$E_z^s = \frac{\sin k_e(a\phi_s - a|\phi'|)}{W_s \sin k_e a\phi_s}\,, \tag{4.187}$$

where $k_e = \sqrt{\epsilon_e}\,k_0$; $\epsilon_e = (\epsilon_f + 1)/2$ is chosen as the average of the two dielectric constants of the media adjacent to the slot. Then, by further assuming that the microstrip feed line propagates a quasi-TEM mode, we can have the transverse modal fields given by

$$\begin{aligned}\mathbf{E}^{\pm}(\rho, \phi, z) &= \mathbf{E}_t(\rho, \phi)e^{\pm j\beta z}\\ &= [\hat{\rho}E_\rho(\rho, \phi) + \hat{\phi}E_\phi(\rho, \phi)]e^{\pm j\beta z}\,,\end{aligned} \tag{4.188}$$

$$\begin{aligned}\mathbf{H}^{\pm}(\rho, \phi, z) &= \mathbf{H}_t(\rho, \phi)e^{\pm j\beta z}\\ &= [\hat{\rho}H_\rho(\rho, \phi) + \hat{\phi}H_\phi(\rho, \phi)]e^{\pm j\beta z}\,.\end{aligned} \tag{4.189}$$

The modal fields above are also assumed to be normalized, so that

$$\int_0^a \int_0^{2\pi} \mathbf{E}_t \times \mathbf{H}_t \cdot \hat{z}\rho\, d\rho\, d\phi = 1\,. \tag{4.190}$$

Since it is known that the printed slot will result in a voltage discontinuity on the feed line, we can have the microstrip-line fields in different regions written as follows: In region 1 ($z<0$, $0 \le \rho \le a$, $0 \le \phi \le 2\pi$),

$$\begin{aligned}\mathbf{E}_1(\rho, \phi, z) &= \mathbf{E}^+(\rho, \phi, z) + R\mathbf{E}^-(\rho, \phi, z)\\ &= \mathbf{E}_t(\rho, \phi)(e^{+j\beta z} + \mathrm{Re}^{-j\beta z})\,,\end{aligned} \tag{4.191}$$

$$\begin{aligned}\mathbf{H}_1(\rho, \phi, z) &= \mathbf{H}^+(\rho, \phi, z) - R\mathbf{H}^-(\rho, \phi, z)\\ &= \mathbf{H}_t(\rho, \phi)(e^{+j\beta z} - \mathrm{Re}^{-j\beta z})\,,\end{aligned} \tag{4.192}$$

and in region 2 ($z>0$, $0 \le \rho \le a$, $0 \le \phi \le 2\pi$),

$$\begin{aligned}\mathbf{E}_2(\rho, \phi, z) &= T\mathbf{E}^+(\rho, \phi, z) \\ &= T\mathbf{E}_t(\rho, \phi)e^{+j\beta z},\end{aligned} \tag{4.193}$$

$$\begin{aligned}\mathbf{H}_2(\rho, \phi, z) &= T\mathbf{H}^+(\rho, \phi, z) \\ &= T\mathbf{H}_t(\rho, \phi)e^{+j\beta z},\end{aligned} \tag{4.194}$$

where R and T are the voltage reflection and transmission coefficients due to the slot discontinuity on the microstrip feed line, respectively.

Then, to determine the unknown coefficients of R and T, the reciprocity theorem is employed. Applying the reciprocity theorem to the total fields $\mathbf{E}$ and $\mathbf{H}$, and the positive traveling-wave fields $\mathbf{E}^+$ and $\mathbf{H}^+$, we have

$$\oint_S \mathbf{E} \times \mathbf{H}^+ \cdot d\mathbf{S} = \oint_S \mathbf{E}^+ \times \mathbf{H} \cdot d\mathbf{S}, \tag{4.195}$$

where $S = S_a + S_0 + S_g$ is a closed surface, S_a the aperture surface (slot region), S_0 the effective cross section of the microstrip line enclosed by the ground cylinder at $z = \pm\infty$ (i.e., $0 \le \rho \le a$, $0 \le \phi \le 2\pi$, $z = \pm\infty$), and S_g the cylindrical ground surface, excluding S_a (i.e., $\rho = a$, $0 \le \phi \le 2\pi$, $-\infty < z < \infty$ excluding S_a).

It is first noted that on the surface of S_g, the tangential electric field must vanish; that is, $\hat{\rho} \times \mathbf{E} = \hat{\rho} \times \mathbf{E}^+ = 0$. The right-hand side of (4.195) can then be rewritten as

$$\begin{aligned}\oint_S \mathbf{E}^+ \times \mathbf{H} \cdot d\mathbf{S} &= \int_{S_a} \mathbf{E}^+ \times \mathbf{H} \cdot \hat{n}\, dS + \int_{S_0} \mathbf{E}^+ \times \mathbf{H} \cdot \hat{n}\, dS \\ &= \int_{S_a} (\hat{n} \times \mathbf{E}^+) \cdot \mathbf{H}\, dS + \int_{S_0} \mathbf{E}^+ \times \mathbf{H}_2 \cdot \hat{z}\, dS + \int_{S_0} \mathbf{E}^+ \times \mathbf{H}_1 \cdot (-\hat{z})\, dS \\ &= \int_{S_0} \mathbf{E}^+ \times (\mathbf{H}_2 - \mathbf{H}_1) \cdot \hat{z}\, dS,\end{aligned} \tag{4.196}$$

In the expression above, the vector identity of $(\mathbf{A} \times \mathbf{B}) \cdot \hat{n} = (\hat{n} \times \mathbf{A}) \cdot \mathbf{B}$ is applied and thus the term $\int_{S_a} (\hat{n} \times \mathbf{E}^+) \cdot \mathbf{H}\, dS$ vanishes because $\mathbf{E}^+$ and $\mathbf{H}$ have no $\hat{z}$-directed components and $\hat{n}$ is a unit vector pointing outward on the specific surface. Substituting (4.192) and (4.194) into (4.196) gives

$$\oint_S \mathbf{E}^+ \times \mathbf{H} \cdot d\mathbf{S} = (T-1) \int_{S_0} \mathbf{E}_t \times \mathbf{H}_t e^{j2\beta z} \cdot \hat{z}\, dS + R. \tag{4.197}$$

Similarly, the left-hand side of (4.195) can be derived as

$$\oint_S \mathbf{E} \times \mathbf{H}^+ \cdot d\mathbf{S} = \int_{S_a} \mathbf{E} \times \mathbf{H}^+ \cdot \hat{n}\, dS + \int_{S_0} \mathbf{E} \times \mathbf{H}^+ \cdot \hat{n}\, dS\,, \tag{4.198}$$

with

$$\begin{aligned}\oint_{S_a} \mathbf{E} \times \mathbf{H}^+ \cdot \hat{n}\, dS &= \int_{S_a} (\hat{\rho} \times \mathbf{E}) \cdot \mathbf{H}^+\, dS \\ &= \int_{S_a} \hat{\rho} \times (\mathbf{E}_1 + \mathbf{E}_2 + \mathbf{E}_s) \cdot \mathbf{H}^+\, dS \\ &= \int_{S_a} (\hat{\rho} \times \mathbf{E}_s) \cdot \mathbf{H}^+\, dS \\ &= -\int_{S_a} V_0 E_z^s H_\phi\, dS\,,\end{aligned} \tag{4.199}$$

and

$$\begin{aligned}\oint_{S_0} \mathbf{E} \times \mathbf{H}^+ \cdot \hat{n}\, dS &= \int_{S_0} \mathbf{E}_1 \times \mathbf{H}^+ \cdot (-\hat{z})\, dS + \int_{S_0} \mathbf{E}_2 \times \mathbf{H}^+ \cdot \hat{z}\, dS \\ &= (T-1) \int_{S_0} \mathbf{E}_t \times \mathbf{H}_t e^{j2\beta z} \cdot \hat{z}\, dS - R\,.\end{aligned} \tag{4.200}$$

In (4.199), the phase term $e^{+j\beta z}$ is neglected because the phase shift across the narrow slot at $z = 0$ is very small. Then, substitution of (4.197), (4.199), and (4.200) into (4.195) yields

$$R = -\frac{V_0}{2} \int_{S_0} E_z^s H_\phi\, dS = \frac{V_0}{2} \Delta v\,, \tag{4.201}$$

where

$$\Delta v = -\int_{S_0} E_z^s H_\phi\, dS\,. \tag{4.202}$$

The expression of (4.202) shows the reaction between the slot field and the microstrip-line field, representing the voltage discontinuity in the microstrip line across the slot. By following a similar derivation procedure described above to another reciprocity theorem of the form

$$\oint_S \mathbf{E} \times \mathbf{H}^- \cdot d\mathbf{S} = \oint_S \mathbf{E}^- \times \mathbf{H} \cdot d\mathbf{S}\,, \tag{4.203}$$

we can have the expression of T written as

$$T = 1 + \frac{V_0}{2} \int_{S_a} E_z^s H_\phi \, dS = 1 - \frac{V_0}{2} \Delta v \, . \tag{4.204}$$

The results of (4.201) and (4.204) give

$$T = 1 - R \, , \tag{4.205}$$

which implies that the slot discontinuity appears as a simple series impedance Z_s to the microstrip line, as shown in Figure 4.36. Based on this equivalent-circuit representation, the problem with the presence of tuning stub and other external circuitry can also be solved using transmission-line theory.

B. Equivalent Series Impedance Once the equivalent series impedance in Figure 4.36 is solved, the input impedance of the cylindrical printed slot antenna can readily be obtained. To solve for Z_s, the equivalence principle is applied, and the slot can be closed off and then replaced by an equivalent magnetic surface current $\mathbf{M}_s$ ($=\mathbf{E}_s \times \hat{\rho}$) just outside the ground cylinder and $-\mathbf{M}_s$ just inside the ground cylinder. Next, the boundary condition that the tangential component of the magnetic field must be continuous across the slot is imposed; that is,

$$H_\phi^e(a^+, \phi) = H_\phi^i(a^-, \phi) + H_\phi^f(a^-, \phi) \, , \tag{4.206}$$

where H_ϕ^e and H_ϕ^i are, respectively, the magnetic fields just outside and inside the

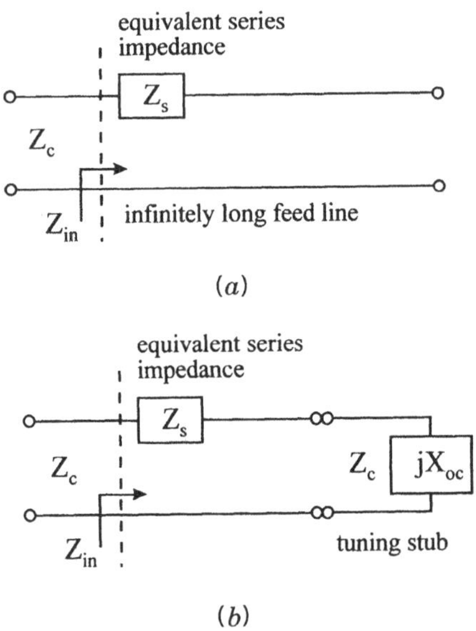

FIGURE 4.36 Equivalent circuit of a cylindrical printed slot antenna: (*a*) with an infinitely long microstrip feed line; (*b*) with an open-ended microstrip feed line.

slot due to the equivalent magnetic current, and H^f_ϕ is the magnetic field just inside the slot due to the electric current $\mathbf{J}_f$ in the microstrip line.

Next, by defining a Green's function $G^{HM}_{\phi\phi}$ [18] to account for the $\hat{\phi}$-directed magnetic field across the slot due to a unit $\hat{\phi}$-directed magnetic current, we have

$$H^e_\phi - H^i_\phi = \int_{S_a} G^{HM}_{\phi\phi}(\phi, z; \phi_0, z_0) M_s(\phi_0, z_0)\, dS_0 \,. \tag{4.207}$$

Also, at $\rho = a^-$, $z = 0^+$, we have

$$H^f_\phi = TH_\phi(a^-, \phi) = (1 - R)H_\phi(a^-, \phi)\,. \tag{4.208}$$

By substituting (4.207) and (4.208) into (4.206) and after some straightforward manipulation, we can derive

$$V_0 Y^s - \Delta v(1 - R)\,, \tag{4.209}$$

where Y^s is the slot admittance seen by the microstrip line and is defined as

$$Y^s = \int_{S_a} \int_{S_a} E^s_z(\phi) G^{HM}_{\phi\phi}(\phi, z; \phi_0, z_0) E^s_z(\phi_0)\, dS_0\, dS \,. \tag{4.210}$$

From (4.201), (4.205), and (4.209), we can derive

$$V_0 = \frac{2\Delta v}{\Delta v^2 + 2Y^s}\,, \tag{4.211}$$

$$R = \frac{\Delta v^2}{\Delta v^2 + 2Y^s}\,, \tag{4.212}$$

$$T = \frac{2Y^s}{\Delta v^2 + 2Y^s}\,. \tag{4.213}$$

Also, from the equivalent circuit shown in Figure 4.36*a*, we have

$$R = \frac{Z_{\text{in}} - Z_c}{Z_{\text{in}} + Z_c} = \frac{Z_s}{Z_s + 2Z_c}\,. \tag{4.214}$$

Then, from (4.212) and (4.214), the equivalent series impedance Z_s can be derived as

$$Z_s = Z_c \frac{2R}{1 - R} = Z_c \frac{\Delta v^2}{Y^s}\,, \tag{4.215}$$

where Z_c is the characteristic impedance of the cylindrical microstrip line, whose expression and characteristic are given in Chapter 8. It should also be noted that for the case in Figure 4.36*a*, the series impedance Z_s obtained is also the input impedance of the printed slot antenna seen at the slot position. However, for practical applications, the microstrip feed line is usually with an open-circuited tuning stub, and the equivalent circuit is given in Figure 4.36*b*. The tuning stub is used for reactance tuning and its length is usually selected to be about one-quarter guided wavelength for the wave propagating in the microstrip line. In this case the input impedance of the printed slot antenna seen at the slot position is given by

$$Z_{\text{in}} = Z_s - jZ_c \cot \beta l_{ts}$$
$$= Z_c \frac{2R}{1-R} - jZ_c \cot \beta l_{ts} , \tag{4.216}$$

where β is the effective propagation constant of the cylindrical microstrip line (see Chapter 8) and l_{ts} is the tuning-stub length.

C. Moment-Method Computation Since in practical cases the radiating slot is usually operated near the resonant frequency, a one-mode approximation to the slot electric field of the form (4.186) may not be a sufficient approximation. We thus present in the following the moment-method solution for the unknown slot electric field. In this case the slot electric field is expanded using a set of N PWS modes; that is,

$$\mathbf{E}_s = \hat{z} \sum_{n=1}^{N} V_n \frac{1}{W_s} f_{PWS}(\phi' - \phi_n) , \qquad |z| < \frac{W_s}{2} , \quad |\phi'| < \phi_s , \quad \phi' = \phi - \frac{\pi}{2} , \tag{4.217}$$

with

$$f_{\text{PWS}}(\phi') = \begin{cases} \dfrac{\sin k_e(a\phi_h - a|\phi'|)}{\sin k_e a\phi_h} , & |\phi'| < \phi_h , \\ 0 , & |\phi'| > \phi_h , \end{cases} \tag{4.218}$$

$$\phi_n = \left(\frac{2n}{N+1} - 1 \right) \phi_s , \qquad \phi_h = \frac{2\phi_s}{N+1} , \tag{4.219}$$

where V_n is the unknown expansion coefficient, $f_{\text{PWS}}(\phi')$ is the PWS basis function for the slot electric field, ϕ_n is the center point of the nth expansion mode, and $a\phi_h$ is the half-length of a PWS mode. Next, by following the reciprocity analysis described for the one-mode expansion case, the reflection coefficient R is derived as a matrix form and (4.201) is rewritten as

$$R = \frac{1}{2} [V_n]^{\text{T}} [\Delta v] , \tag{4.220}$$

with

$$[V_n] = ([Y^s] + \frac{1}{2}[\Delta v][\Delta v]^{\mathrm{T}})^{-1}[\Delta v], \tag{4.221}$$

where $[Y^s]$ is an admittance matrix, whose elements are written as

$$Y^s_{mp} = \frac{1}{2\pi}\sum_{n=-\infty}^{\infty}\int_{-\infty}^{\infty} dk_z \operatorname{sinc}^2\left(\frac{k_z W_s}{2}\right)\tilde{G}^{HM}_{\phi\phi}(n, k_z)F^2_{PWS}(n)\cos[n(\phi_m - \phi_p)], \tag{4.222}$$

and $[\Delta v]$ denotes the voltage discontinuity vector across the slot, whose expression in the spectral domain is derived as

$$\Delta v_m = \frac{1}{\sqrt{Z_c}}\sum_{n=-\infty}^{\infty} J_0(n\phi_f)\operatorname{sinc}^2\left(\frac{\beta W_s}{2}\right)\tilde{G}^{HJ}_{\phi z}(n, \beta)F_{\mathrm{PWS}}(n)\cos n\phi_m . \tag{4.223}$$

In (4.223), ϕ_f is the half-angle subtended by the microstrip line (i.e., $\phi_f = W_f/2b_f$), $\tilde{G}^{HJ}_{\phi z}$ is the spectral-domain $\hat{\phi}$-directed magnetic field at $\rho = a$ due to a $\hat{z}$-directed unit-amplitude electric current at $\rho = b_f$ [18]. Also, $F_{\mathrm{PWS}}(n)$ is the Fourier transform of the PWS function $f_{\mathrm{PWS}}(\phi')$ and is given by

$$F_{\mathrm{PWS}}(n) = \frac{2k_e(\cos n\phi_h - \cos k_e a\phi_h)}{[k_e^2 - (n/a)^2]\sin k_e a\phi_h} . \tag{4.224}$$

From the formulation above, the reflection coefficient R can be evaluated from (4.220), and in turn, the slot impedance Z_s and the input impedance Z_{in} can readily be obtained from (4.215) and (4.216), respectively.

For calculation of the far-zone radiated fields, the axial components of the electric and magnetic fields due to the equivalent spectral-domain magnetic current $\tilde{M}_s$ at the closed slot are first derived as

$$\begin{bmatrix} E_z(\rho, \phi, z) \\ H_z(\rho, \phi, z) \end{bmatrix} = \frac{1}{2\pi}\sum_{n=-\infty}^{\infty} e^{jn\phi}\int_{-\infty}^{\infty} dk_z\, H_n^{(1)}(\xi_\rho)e^{jk_z z}\begin{bmatrix} B_n^e(k_z) \\ B_n^h(k_z) \end{bmatrix}, \tag{4.225}$$

where

$$B_n^e(k_z) = \frac{\tilde{M}_s}{H_n^{(1)}(\xi_\rho)}, \qquad B_n^h(k_z) = \frac{-jnk_z\tilde{M}_s}{\omega\mu_0\xi_\rho a H_n^{(1)}(\xi_\rho)}, \qquad \xi_\rho = \rho\sqrt{k_0^2 - k_z^2}, \tag{4.226}$$

$$\tilde{M}_s = \sum_{p=1}^{N} V_p e^{-jn\phi_p}\operatorname{sinc}\left(\frac{k_z W_s}{2}\right)F_{\mathrm{PWS}}(n) . \tag{4.227}$$

The far-zone radiated fields in spherical coordinates can be given approximately by

$$\begin{bmatrix} E_\theta(r,\theta,\phi) \\ E_\phi(r,\theta,\phi) \end{bmatrix} = \frac{1}{\sin\theta}\begin{bmatrix} -1 & 0 \\ 0 & \eta_0 \end{bmatrix}\begin{bmatrix} E_z(\rho,\phi,z) \\ H_z(\rho,\phi,z) \end{bmatrix}. \tag{4.228}$$

However, it is noted that for a large distance from the cylinder such that $r >> 1$, the Hankel function $H_n^{(1)}(\xi_\rho)$ and the exponential function $e^{jk_z z}$ vary rapidly. The integral in (4.225) is thus evaluated using the method of steepest descent [19]; that is, the following approximation is used:

$$\int_{-\infty}^{\infty} f(k_z) H_n^{(1)}(\xi_\rho) e^{jk_z z}\, dk_z \xrightarrow{r\to\infty} 2\frac{e^{jk_0 r}}{r}(-j)^{n+1} f(k_0\cos\theta). \tag{4.229}$$

From the approximation above, the final forms of the far-zone radiated fields can be written as

$$E_\theta(r,\theta,\phi) = A\sum_{n=0}^{\infty} e_n \frac{(-j)^{n+1}F_{\text{PWS}}(n)}{H_n^{(1)}(k_0 a\sin\theta)} \sum_{p=1}^{N} V_p \cos[n(\phi'-\phi_p)], \tag{4.230}$$

$$E_\phi(r,\theta,\phi) = \frac{A}{k_0 a\sin\theta}\sum_{n=0}^{\infty} e_n \frac{n(-j)^{n+1}F_{\text{PWS}}(n)}{H_n^{(1)\prime}(k_0 a\sin\theta)} \sum_{p=1}^{N} V_p \sin[n(\phi'-\phi_p)], \tag{4.231}$$

where

$$A = \frac{e^{-jk_0 r}\cos\theta}{\pi r\sin\theta}\,\text{sinc}\left(\frac{k_z W_s}{2}\right), \qquad e_n = \begin{cases} 1, & n=0, \\ 2, & n\neq 0. \end{cases} \tag{4.232}$$

D. Results Figure 4.37 presents typical results for the calculated slot impedance Z_s. Three [$N=3$ in (4.217)] PWS basis functions for the unknown slot electric field are used in the computation, which shows good convergent solutions. The impedance level is seen to increase with increasing cylinder radius, and good agreement is obtained between the slot impedance for the case with a larger radius and the data measured for the planar case. The results of resonant frequency (determined from the zero crossing of the reactance curve) versus slot length are presented in Figure 4.38. The results for the planar case are calculated using a full-wave approach [17]. The resonant frequency is seen to be strongly affected by the slot-length variation. Figure 4.39 presents measured normalized input impedance seen at the slot position for various cylinder radii. Behavior similar to that predicted by the calculated results is observed. Figure 4.40 shows the radiation patterns for various cylinder radii. It is seen that due to the presence of a cylindrical ground plane, the bidirectional radiation of a planar slot antenna is

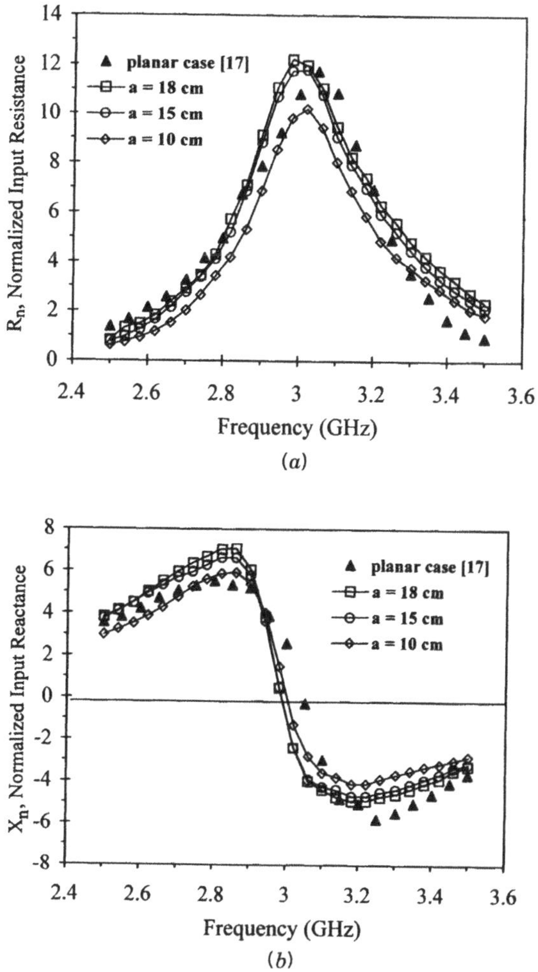

FIGURE 4.37 Normalized calculated input impedance, $R_n + jX_n$ ($= Z_s/50\ \Omega$), seen at the slot position versus frequency; $L_s = 40.2$ mm, $W_s = 0.7$ mm, $h_f = 1.6$ mm, $\epsilon_f = 2.2$, $W_f =$ 5 mm. (*a*) Input resistance; (*b*) input reactance. (From Ref. [18], © 1995 John Wiley & Sons, Inc.)

eliminated. However, back radiation exists in the lower hemisphere and is greater with a smaller cylinder radius.

4.5.2 Rectangular Patch with a Coupling Slot

Figure 4.34 shows the geometry under consideration. To account for the presence of the microstrip patch as seen by the coupling slot, the analysis described in Section 4.5.1 is modified. It is first noted that since the coupling slot in the ground cylinder is much smaller than its resonant size and is thus electrically small, a

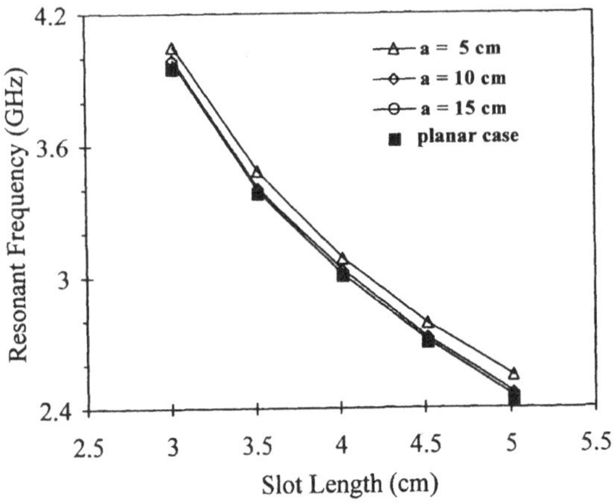

FIGURE 4.38 Variation of the resonant frequency with the slot length; parameters are as in Figure 4.37. (From Ref. [18], © 1995 John Wiley & Sons, Inc.)

single PWS mode of (4.186) is adequate to represent the unknown slot electric field. For the unknown excited patch surface current, a set of entire-domain sinusoidal basis functions of (2.61)–(2.62) is used. By imposing the boundary condition that the tangential electric field on the patch must vanish, a matrix equation of (2.85) is obtained, which gives the unknown coefficients of the expansion basis functions for the excited patch surface current. Also, with the unknown patch surface current obtained, the admittance Y^p seen looking into the microstrip patch at the slot position due to the patch current distribution can

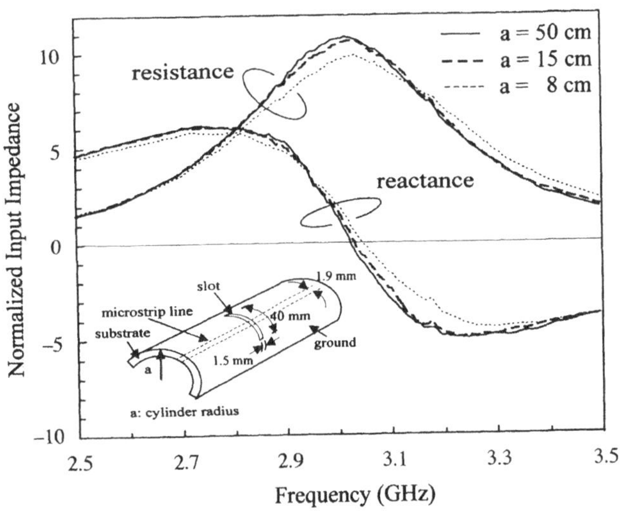

FIGURE 4.39 Measured normalized input impedance seen at the slot position for various cylinder radii; $L_s = 40$ mm, $W_s = 1.5$ mm, $h_f = 0.762$ mm, $\epsilon_f = 3.0$, $W_f = 1.9$ mm.

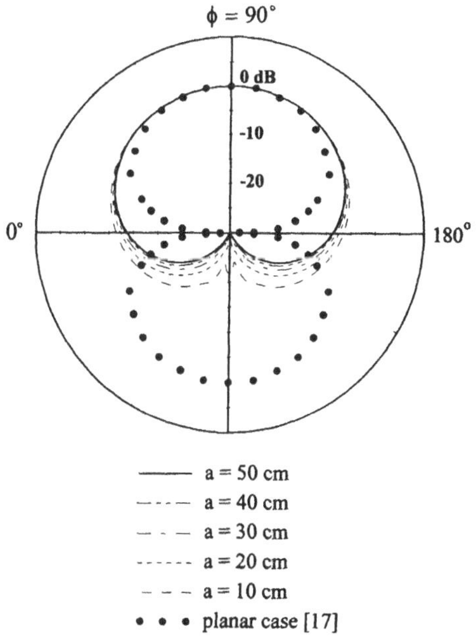

(a)

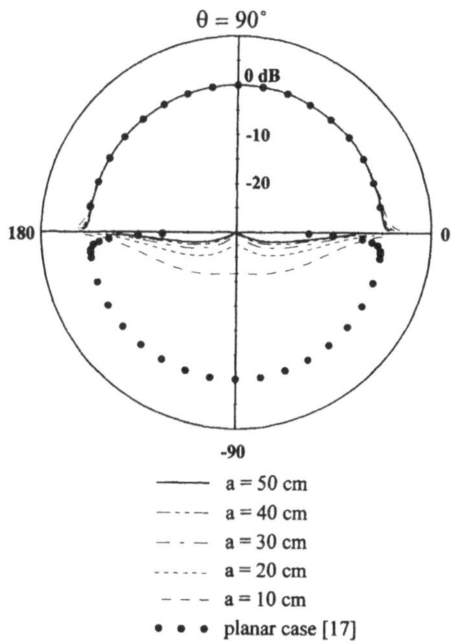

(b)

FIGURE 4.40 Radiation patterns for a cylindrical printed slot: (a) pattern versus ϕ in the $\theta = 90°$ plane; (b) pattern versus θ in the $\phi = 90°$ plane. Antenna parameters are as in Figure 4.37. (From Ref. [18], © 1995 John Wiley & Sons, Inc.)

readily be calculated from (2.93). Then, by modifying the slot admittance Y^s in (4.215) to be $Y^s + Y^p$ and knowing that Δv, the voltage discontinuity across the slot in (4.202) or in (4.223), is independent of the microstrip patch loading, the reflection coefficient R is obtained; that is,

$$R = \frac{\Delta v^2}{\Delta v^2 + 2(Y^s + Y^p)} . \qquad (4.233)$$

With the value of R determined, the equivalent series impedance Z_s can be evaluated from (4.215) and an equivalent circuit similar to that shown in Figure 4.36 can be constructed. Considering the tuning stub, the input impedance of the patch antenna seen at the slot is thus given by

$$Z_{\text{in}} = Z_c \frac{2R}{1-R} - jZ_c \cot \beta l_{ts}$$

$$= Z_c \frac{\Delta v^2}{Y^s + Y^p} - jZ_c \cot \beta l_{ts} . \qquad (4.234)$$

Typical results calculated for the input impedance of a slot-coupled cylindrical rectangular patch antenna from (4.234) with $a = 15$ cm are presented in Figure 4.41. The data measured for a planar structure are plotted for comparison. The results calculated are seen to approach the data measured for a cylinder radius of 15 cm. For a much smaller cylinder radius (about 1 cm), input impedances have been reported by Tam et al. [20]. The resonant input resistances are found to decrease with decreasing cylinder radius, similar to the observation for the probe-fed case. Figure 4.42 shows the results of input impedance for various slot lengths. It is seen that the resonant input resistance and resonant frequency are both very sensitive to slot-length variation. This behavior provides more degrees of freedom for a slot-coupled patch antenna in resonant-frequency fine tuning and impedance matching.

4.6 SLOT-COUPLED CASE: CAVITY-MODEL SOLUTION

In this section, a theoretical approach based on a combination of cavity-model theory and reciprocity analysis is used in the study of cylindrical microstrip antennas. The equivalent circuit shown in Figure 4.43 is first constructed by calculating the slot self-admittance Y_{slot} simply from a short-circuited slot line and obtaining the patch admittance Y_{patch} seen at the slot position using the cavity-model method. Since from reciprocity analysis, microstrip patch loading on the slot can be treated as an equivalent series load as seen by the microstrip feed line, we introduce an impedance transformer, derived using conformal transformation, in the equivalent circuit to transform the patch admittance and slot self-admittance into an equivalent series load in the microstrip feed line. The input impedance of

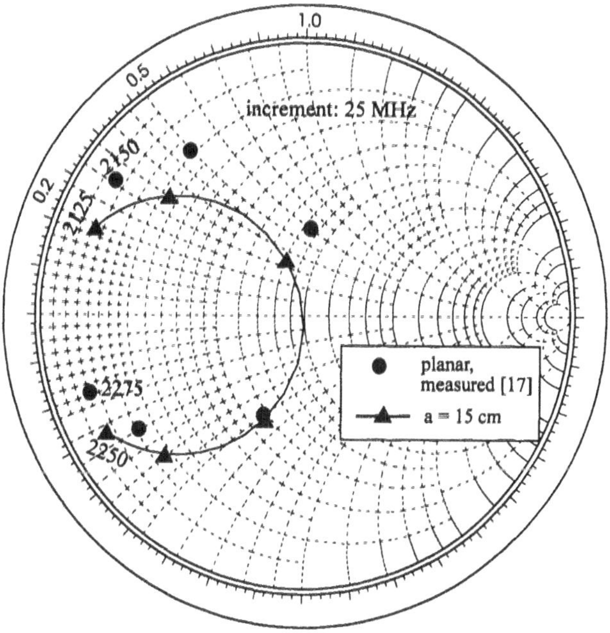

FIGURE 4.41 Input impedance of a slot-coupled cylindrical rectangular patch antenna; $a = 15\,\text{cm}$, $\epsilon_1 = \epsilon_f = 2.54$. $h = h_f = 1.6\,\text{mm}$, $2L = 40\,\text{mm}$, $2W = 30\,\text{mm}$, $L_s = 11.2\,\text{mm}$, $W_s = 1.55\,\text{mm}$, $W_f = 4.42\,\text{mm}$, $l_{ts} = 20\,\text{mm}$.

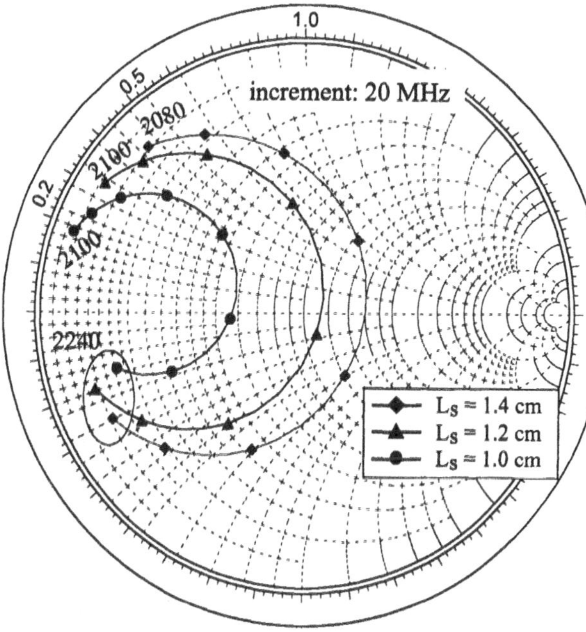

FIGURE 4.42 Input impedance for various slot lengths with $a = 15\,\text{cm}$, $W_s = 1.1\,\text{mm}$, $l_{ts} = 22\,\text{mm}$. Other parameters are as given in Figure 4.41.

the antenna can then readily be calculated using simple transmission-line theory. Details of the formulation for slot-coupled rectangular and circular microstrip antennas are described below.

4.6.1 Rectangular Patch

As referred to the equivalent circuit in Figure 4.43, the patch admittance, the slot self-admittance, and the transformation ratio need to be determined for evaluation of the input impedance of the antenna. Details of the formulation are given below.

A. Patch Admittance and Slot Self-Admittance Given in the geometry shown in Figure 4.34, the coupling slot is assumed to be narrow and is centered below the patch at $(\rho, \phi, z) = (a, \pi/2, 0)$. To begin with, the microstrip patch is considered as a cavity bounded by four perfect magnetic walls around the cavity and two electric walls on the top and bottom of the cavity. This assumption is valid for a thin-substrate condition. This cavity is then assumed to be excited at the TM_{01} mode by an equivalent magnetic current source located uniformly in the volume above the slot, and the magnetic current density of this source can be written as

$$\mathbf{J}_m = \hat{\phi}\frac{2E_s}{h}, \tag{4.235}$$

with

$$E_s = \frac{V_0}{W_s}\frac{\sin k_e(a\phi_s - a|\phi - \pi/2|)}{\sin k_e a\phi_s}, \qquad k_e = k_0\sqrt{\frac{\epsilon_1 + \epsilon_f}{2}}, \tag{4.236}$$

where E_s is the slot electric field, similar to the one PWS-mode approximation in (4.186), and V_0 is again the voltage at the slot center. Then, by solving Maxwell's equations with the J_m excitation given by

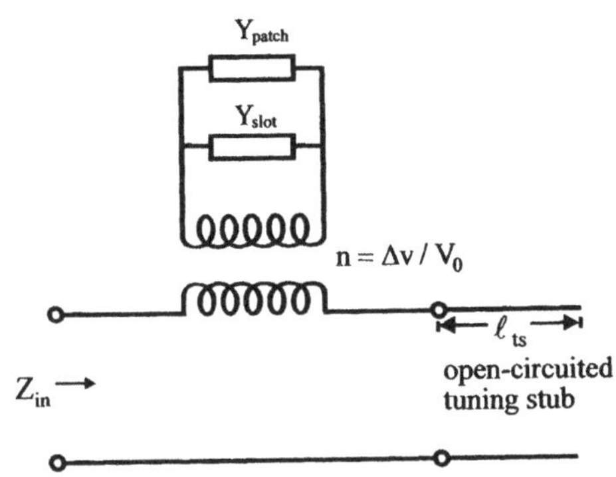

FIGURE 4.43 Equivalent circuit of a slot-coupled cylindrical microstrip antenna for cavity-model analysis; n is the transformation ratio.

$$\nabla \times \mathbf{E} = -j\omega\mu_0 \mathbf{H} - \mathbf{J}_m , \tag{4.237}$$

$$\nabla \times \mathbf{H} = j\omega\epsilon_0\epsilon_1 \mathbf{E} , \tag{4.238}$$

simple expressions for the electric and magnetic fields inside the cavity at the TM_{01} mode are derived as

$$\mathbf{E} = \hat{\rho} E_0 \cos \frac{\pi z}{2L} , \tag{4.239}$$

$$\mathbf{H} = \hat{\phi} H_0 \sin \frac{\pi z}{2L} , \tag{4.240}$$

with

$$E_0 = \frac{-V_0 \pi}{hk_e WL^2} \frac{\operatorname{sinc}(\pi W_s / 4L)}{k_{\text{eff}}^2 - (\pi/2L)^2} \frac{1 - \cos k_e a\phi_s}{\sin k_e a\phi_s} , \tag{4.241}$$

$$H_0 = \frac{-j2L\omega\epsilon_0\epsilon_1 E_0}{\pi} . \tag{4.242}$$

In (4.241), k_{eff} is the effective wavenumber given in (4.40). From these fields, an equivalent magnetic current source around the cavity can be evaluated, which is allowed to radiate into space. Then the radiation loss P_{rad} can be derived as [21]

$$P_{\text{rad}} = \frac{1}{240\pi} \int_0^{\pi} \int_0^{2\pi} (E_\theta E_\theta^* + E_\phi E_\phi^*) r^2 \sin\theta \, d\phi \, d\theta , \tag{4.243}$$

with

$$E_\theta = \frac{A_0}{\sin\theta} (1 + e^{-j2k_0 L \cos\theta}) \sum_{p=-\infty}^{\infty} \frac{j^{p+1} e^{jp(\phi - \pi/2 + \phi_0)} I_1(\phi_0, p)}{H_p^{(2)}(k_0 a \sin\theta)} , \tag{4.244}$$

$$E_\phi = \frac{-jA_0 \cos\theta}{ak_0 \sin^2\theta} (1 + e^{-j2k_0 L \cos\theta}) \sum_{p=-\infty}^{\infty} \frac{pj^{p+1} e^{jp(\phi - \pi/2 + \phi_0)} I_1(\phi_0, p)}{H_p^{(2)\prime}(k_0 a \sin\theta)}$$
$$- \frac{jA_0 I_2(2L, -k_0 \cos\theta)}{a} \sum_{p=-\infty}^{\infty} \frac{j^{p+1} e^{jp(\phi - \pi/2 + \phi_0)} (1 - e^{-j2p\phi_0})}{H_p^{(2)\prime}(k_0 a \sin\theta)} , \tag{4.245}$$

where

$$A_0 = \frac{V_0 e^{-jk_0 r}}{2\pi r k_e W L^2} \frac{\operatorname{sinc}(\pi W_s/4L)}{k_{\text{eff}}^2 - (\pi/2L)^2} \frac{1 - \cos k_e a\phi_s}{\sin k_e a\phi_s}, \tag{4.246}$$

$$I_1(\phi_0, p) = \int_0^{2\phi_0} e^{jp\phi}\, d\phi, \tag{4.247}$$

$$I_2(L, u) = \int_{-L}^{0} e^{-juz} \cos\frac{\pi z}{L}\, dz. \tag{4.248}$$

By further evaluating the conductor (copper) loss P_{cu} of the microstrip patch in (4.36), the dielectric loss P_d in the substrate within the cavity in (4.37), the stored electric energy W_e inside the cavity in (4.38*a*), and the stored magnetic energy W_m inside the cavity in (4.38*b*), we have

$$P_{\text{cu}} = \frac{(2a+h)WL}{a}\sqrt{\frac{\omega\mu_0}{2\sigma_c}}\left(\frac{2L\omega\epsilon_0\epsilon_1 E_0}{\pi}\right)^2, \tag{4.249}$$

$$P_d = \omega\epsilon_0\epsilon_1 hWLE_0^2\left(1 + \frac{h}{2a}\right)\tan\delta, \tag{4.250}$$

$$W_e = \epsilon_0\epsilon_1 hWLE_0^2\left(1 + \frac{h}{2a}\right), \tag{4.251}$$

$$W_m = \frac{\mu_0 h(2a+h)WL^3(\omega\epsilon_0\epsilon_1 E_0)^2}{\pi^2 a}. \tag{4.252}$$

From (4.243)–(4.252), the admittance of the rectangular microstrip patch seen at the slot position can readily be calculated from

$$Y_{\text{patch}} = \frac{2[P_{\text{rad}} + P_{\text{cu}} + P_d + 2j\omega(W_e - W_m)]}{|V_0|^2}, \tag{4.253}$$

where $V_0 = E_0 W_s$.

As for the evaluation of the slot self-admittance, the coupling slot is considered approximately as two short-circuited slot lines of length $L_s/2$ connected in parallel. In this case we have

$$Y_{\text{slot}} = \frac{-2j\cot(k_e L_s/2)}{Z_{cs}}$$

$$= -j\frac{\sqrt{(\epsilon_1 + \epsilon_f)/2}\,\ln(16L/\pi W_s)}{295.85}\cot k_e a\phi_s, \tag{4.254}$$

where Z_{cs} is the characteristic impedance of the slot line [22].

B. Transformation Ratio The transformation ratio is defined as the ratio of the voltage discontinuity across the slot (Δv) to the voltage at the slot center (V_0); that is,

$$n = \frac{\Delta v}{V_0} . \tag{4.255}$$

To obtain the voltage discontinuity in the microstrip feed line due to the slot cut, we use three successive conformal transformations to deal with the electric-field distribution in the cross section of the cylindrical microstrip line [23–25]. These successive transformations are written as

$$\zeta = re^{j\phi} , \tag{4.256}$$

$$Z = x + jy = \frac{\pi}{2} - \phi + j \ln \frac{r}{b_f} , \tag{4.257}$$

$$U = -\cosh^2 \frac{\pi Z}{2 \ln(a/b_f)} , \tag{4.258}$$

$$W' = \frac{\mathrm{sn}^{-1}[(-U)^{1/2}, \alpha_t] - \mathrm{sn}^{-1}(\alpha_t^{-1}, \alpha_t)}{\mathrm{sn}^{-1}(1, \alpha_t) - \mathrm{sn}^{-1}(\alpha_t^{-1}, \alpha_t)} , \tag{4.259}$$

where

$$\alpha_t = \left[\cosh \frac{\pi W_f}{4b_f \ln(a/b_f)} \right]^{-1} , \tag{4.260}$$

$$\mathrm{sn}^{-1}(x, k) = \int_0^x \frac{d\lambda}{[(1 - \lambda^2)(1 - k^2\lambda^2)]^{1/2}} . \tag{4.261}$$

In (4.261), $\mathrm{sn}^{-1}(x, k)$ is an inverse elliptic function. The expression of (4.256) describes the cylindrical structure of the original problem, while (4.257) maps the original cylindrical microstrip line into the Z plane (a planar microstrip line with a finite ground plane), which is then mapped through the U plane (an upper half-plane), by the transformation of (4.258), to the final W' plane (a rectangular region of plane-parallel capacitor) by the transformation of (4.259). In the final W' plane, the electric-field distribution inside the plane-parallel capacitor can easily be obtained, which in turn gives the electric-field distribution in the original cylindrical microstrip-line structure to be written as

$$E = -\frac{e^{-jk_s z}}{2 \ln(a/b_f)\sqrt{F(\alpha_t)}} \left[\frac{j}{\zeta(1 + \alpha_t^2 U)^{1/2}} \right]^* , \tag{4.262}$$

with

$$F(\alpha_t) = \frac{1}{[\ln(a/b_f)]^2} \int_{3\pi/2}^{-\pi/2} \int_{b_f}^{a} \left[\frac{je^{-j2\phi}}{r(1+\alpha_t^2 U)^{1/2}} \right]^* dr\, d\phi , \qquad (4.263)$$

$$k_s = \frac{2\pi}{\lambda_0} \sqrt{\frac{\epsilon_1 + \epsilon_f}{2}} , \qquad (4.264)$$

where the asterisk denotes a complex conjugate and k_s is the effective propagation constant in the slot. Next, the normalized magnetic field h^l for a quasi-TEM mode propagating in the microstrip line is evaluated by assuming that

$$\int\int_S \mathbf{E} \times \mathbf{H} \cdot \hat{z}\, dS = 1 , \qquad (4.265)$$

and we have

$$h^l = \frac{-je^{jk_s z}}{2\ln(a/b_f)\sqrt{F(\alpha_t)}} \left(\frac{e^{-j\phi}}{r\sqrt{1+\alpha_t^2 U}} \right)^* . \qquad (4.266)$$

With h^l determined, the voltage discontinuity Δv can be calculated from

$$\Delta v = \int_{\text{slot}} E_s \cdot h^l\, dS , \qquad (4.267)$$

and we have

$$\Delta v = \frac{-jV_0}{2\ln(a/b_f)\sqrt{F(\alpha_t)}} \int_{\pi/2+\phi_s}^{\pi/2-\phi_s} \frac{\sin k_e(a\phi_s - a|\pi/2-\phi|)}{\sin k_e a\phi_s}$$

$$\times \frac{e^{-j\phi}\, d\phi}{\{1+\alpha_t^2 \sinh^2[\pi(\pi/2-\phi)/2\ln(a/b_f)]\}^{1/2}} . \qquad (4.268)$$

From (4.268), the transformation ratio $n = \Delta v / V_0$ can be obtained.

C. Input Impedance and Results With Y_{patch}, Y_{slot}, and n obtained, the input impedance of the antenna seen at the slot position can be calculated from

$$Z_{\text{in}} = \frac{n^2}{Y_{\text{patch}} + Y_{\text{slot}}} - jZ_c \cot k_l l_{ts} , \qquad (4.269)$$

where Z_c and k_l ($=k_0\sqrt{\epsilon_{\text{eff}}}$) are, respectively, the characteristic impedance and propagation constant of the microstrip line. The expression of Z_c is given approximately by [23]

$$Z_c = \begin{cases} \dfrac{188.34}{\sqrt{\epsilon_f}}\left\{\dfrac{B}{2} + 0.441 + 0.082\left(\dfrac{\epsilon_f - 1}{\epsilon_f^2}\right) + \dfrac{\epsilon_f + 1}{2\pi\epsilon_f}\left[1.451 + \ln\left(\dfrac{B}{2} + 0.94\right)\right]\right\}^{-1}, & B > 1, \\ \dfrac{84.78}{\sqrt{\epsilon_f + 1}}\left[\ln\dfrac{8}{B} + \dfrac{B^2}{32} - \dfrac{1}{2}\left(\dfrac{\epsilon_f - 1}{\epsilon_f + 1}\right)\left(\ln\dfrac{\pi}{2} + \dfrac{1}{\epsilon_f}\ln\dfrac{4}{\pi}\right)\right], & B < 1, \end{cases} \tag{4.270}$$

where $B = W_f/[b_f \ln(a/b_f)]$. The effective relative permittivity ϵ_{eff} can also be evaluated approximately from [26]

$$\epsilon_{\text{eff}} = 0.5(\epsilon_1 + 1) + 0.5(\epsilon_1 - 1)\left(1 + \frac{10h_f}{W_f}\right)^{-\alpha_1\alpha_2}, \tag{4.271}$$

with

$$\alpha_1 = 1 + \frac{1}{49}\ln\frac{(W_f/h_f)^4 + (W_f/52h_f)^2}{(W_f/h_f)^4 + 0.432} + \frac{1}{18.7}\ln\left[1 + \left(\frac{W_f}{18.1h_f}\right)^3\right], \tag{4.272}$$

$$\alpha_2 = 0.564\left(\frac{\epsilon_f - 0.9}{\epsilon_f + 3}\right)^{0.053}. \tag{4.273}$$

From the formulation above, the input impedance of a slot-coupled cylindrical rectangular patch antenna can be calculated. Figure 4.44 presents typical results calculated and measured for various cylinder radii. It is seen that the impedance

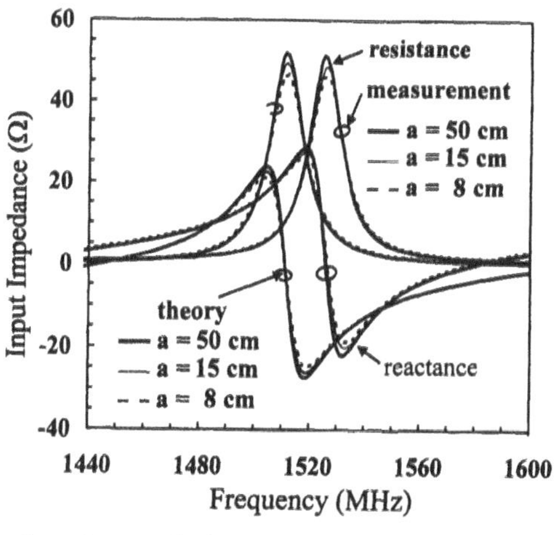

FIGURE 4.44 Input impedance calculated and measured versus frequency; $\epsilon_1 = \epsilon_f = 3.0$, $h = h_f = 0.762$ mm, $\tan\delta = 0.004$, $2L = 55.3$ mm, $2W = 30$ mm, $L_s = 9.8$ mm, $W_s = 1.2$ mm, $W_f = 1.9$ mm, $l_{ts} = 28.4$ mm.

levels of the results calculated are in good agreement with the data measured. The deviation between the calculated and measured resonant frequencies is only about 14 MHz, or 1% for the case studied here. There is also a small shifting of the measured resonant frequencies due to the curvature variation, similar to that for the probe-fed case.

4.6.2 Circular Patch

The geometry of a slot-coupled cylindrical circular microstrip antenna is shown in Figure 4.45. The circular patch is centered at $(\rho, \phi, z) = (b, 0°, 0)$ and has a radius of r_d. The coupling slot in the ground cylindrical surface is also assumed to be narrow and is centered below the circular patch. Other parameters have similar meanings, as described in Figure 4.34. Similar to a rectangular patch, the slot-coupled cylindrical circular patch antenna can be described by the equivalent circuit shown in Figure 4.43. It is first noted that since the transformation ratio is independent of the microstrip patch loading, the expression derived in (4.268) for a rectangular patch can be applied in the present case with a circular patch. As for

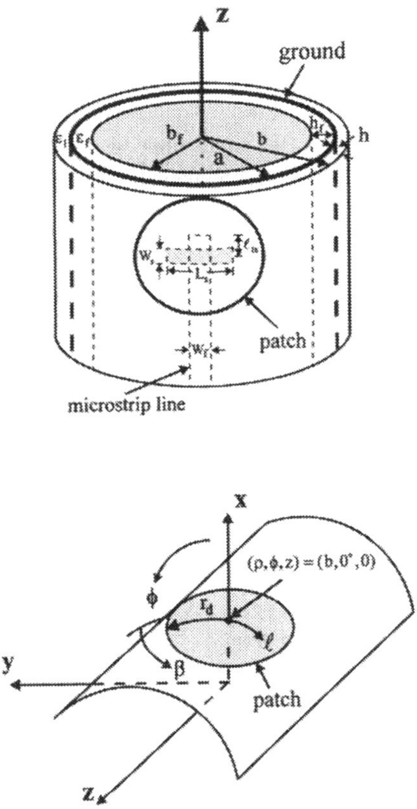

FIGURE 4.45 Geometry of a slot-coupled cylindrical circular microstrip antenna.

the patch admittance and slot self-admittance, the formulation is described below.

To facilitate the problem, the coordinates $(\hat{\rho}, \hat{l}, \hat{\beta})$ as well as the cylindrical coordinates $(\hat{\rho}, \hat{\phi}, \hat{z})$ are used. The relationships of these two coordinates are given in (4.73). By following cavity-model analysis, the substrate region under the circular microstrip patch is again considered as a cylindrical cavity bounded by a magnetic wall around the cavity and two electric walls on the top and bottom of the cavity. This cavity is then assumed to be excited by an equivalent magnetic current source located uniformly in the volume above the slot. This equivalent magnetic current source has the same expression as given in (4.235). Then, by considering the TM_{11}-mode excitation and solving Maxwell's equations (4.237)–(4.238), the electric and magnetic fields inside the cavity can be derived as

$$\mathbf{E} = \hat{\rho} E_0 J_1(k_{11} l) \sin \beta , \tag{4.274}$$

$$\mathbf{H} = \hat{\beta} H_0 J_1'(k_{11} l) \sin \beta - \hat{l} \frac{H_0}{k_{11} l} J_1(k_{11} l) \cos \beta , \tag{4.275}$$

with

$$E_0 = \frac{2V_0 k_{11}^2}{\pi h k_e (k_{\text{eff}}^2 - k_{11}^2)} \frac{1-\cos(k_e L_s/2)}{\sin(k_e L_s/2)} \left[\int_0^{r_d} J_1(k_{11} l)\, dl \right]^{-1} , \tag{4.276}$$

$$H_0 = -j \frac{\omega \epsilon_0 \epsilon_1}{k_{11}} E_0 , \tag{4.277}$$

where k_{11} satisfies $J_1'(k_{11} r_d) = 0$ and k_{eff} is given in (4.40). With the **E** and **H** fields obtained, an equivalent magnetic current source around the cavity can be evaluated, which in turn gives the far-zone radiated electric fields expressed as [27]

$$E_\theta = \frac{j h r_d E_0 J_1(k_{11} r_d) e^{-jk_0 r}}{2\pi^2 r a \sin\theta} \left[\frac{I_s(0, k_0 \cos\theta)}{H_0^{(2)}(a k_0 \sin\theta)} + 2 \sum_{p=1}^{\infty} \frac{j^{p+1} I_s(p, k_0 \cos\theta) \cos p\phi}{H_p^{(2)}(a k_0 \sin\theta)} \right], \tag{4.278}$$

$$E_\phi = \frac{h r_d E_0 J_1(k_{11} r_d) e^{-jk_0 r}}{\pi^2 r a} \sum_{p=1}^{\infty} \left\{ \frac{j^{p+1} \sin p\phi}{H_p^{(2)\prime}(a k_0 \sin\theta)} \right.$$
$$\left. \times \left[I_c(p, k_0 \cos\theta) + \frac{p I_s(p, k_0 \cos\theta) \cos\theta}{a k_0 \sin^2\theta} \right] \right\}, \tag{4.279}$$

with

$$I_s(p,u) = \int_0^{2\pi} \sin^2\beta \, e^{-j(pr_d/a)\cos\beta - jur_d \sin\beta} \, d\beta \,, \tag{4.280}$$

$$I_c(p,u) = \int_0^{2\pi} \sin\beta \cos\beta \, e^{-j(pr_d/a)\cos\beta - jur_d \sin\beta} \, d\beta \,. \tag{4.281}$$

The foregoing expressions of $I_s(p, u)$ and $I_c(p, u)$ can also be obtained from the integrals in (4.80) and (4.81) with $m = 1$, $\beta_p = 90°$. By substituting (4.278)–(4.281) into (4.243), the radiation loss P_{rad} can be determined. Also, the conductor loss, dielectric loss, and stored electric and magnetic energy inside the cavity are, respectively, derived as

$$P_{\text{cu}} = \frac{\pi(\omega\epsilon_0\epsilon_1 r_d E_0)^2}{2k_{11}^2} \sqrt{\frac{\omega\mu_0}{2\sigma_c}} \, [1 - (k_{11}r_d)^{-2}] J_1^2(k_{11}r_d) \,, \tag{4.282}$$

$$P_d = \frac{\pi h\omega\epsilon_0\epsilon_1 r_d^2 E_0^2 \tan\delta}{4} \, [1 - (k_{11}r_d)^{-2}] J_1^2(k_{11}r_d) \,, \tag{4.283}$$

$$W_e = \frac{\pi h\omega\epsilon_0\epsilon_1 r_d^2 E_0^2}{8} \, [1 - (k_{11}r_d)^{-2}] J_1^2(k_{11}r_d) \,, \tag{4.284}$$

$$W_m = \frac{\pi h\mu_0(\omega\epsilon_0\epsilon_1 r_d E_0)^2}{8k_{11}^2} \, [1 - (k_{11}r_d)^{-2}] J_1^2(k_{11}r_d) \,. \tag{4.285}$$

With the radiation and dielectric losses and stored electric and magnetic energy obtained above, the admittance of the circular patch, Y_{patch}, seen at the slot position can then be calculated from (4.253). As for the slot self-admittance, the following expression can be used:

$$Y_{\text{slot}} = -j \frac{\sqrt{(\epsilon_1 + \epsilon_f)/2} \, \ln(16r_d/\pi W_s)}{295.85} \cot k_e a\phi_s \,, \tag{4.286}$$

which is similar to that in (4.254) with the disk radius r_d in place of the half-length L of the rectangular patch. Then, by using the results in Section 4.6.1 for the

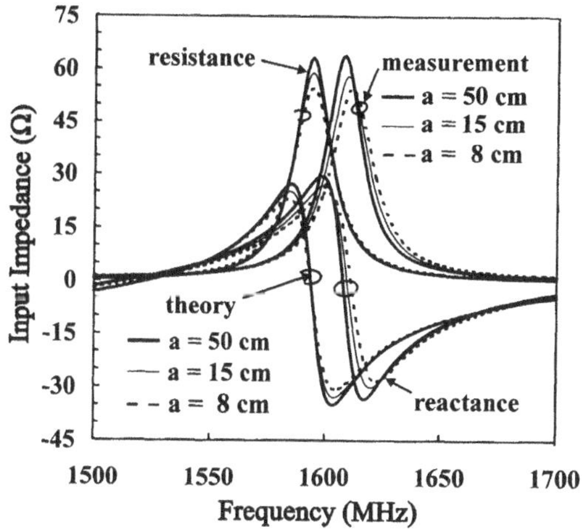

FIGURE 4.46 Input impedance calculated and measured versus frequency; $\epsilon_1 = \epsilon_f = 3.0$, $h = h_f = 0.762$ mm, $\tan\delta = 0.004$, $r_d = 30.2$ mm, $L_s = 13.8$ mm, $W_s = 1.4$ mm, $W_f = 1.9$ mm, $l_{ts} = 24.8$ mm. (From Ref. [27], © 1996 John Wiley & Sons, Inc.)

characteristic impedance and propagation constant of the microstrip feed line, the input impedance of a slot-coupled cylindrical circular patch antenna can be evaluated from (4.269). Figure 4.46 presents typical results calculated for the patch antenna with various cylinder radii. The measured data are shown for comparison. It is seen that the results calculated for the resonant input resistance agree well with the corresponding data measured. For the resonant frequency, only 16 MHz, a 1% deviation between the data calculated and measured, is observed, similar to the results in Section 4.6.1 for a rectangular patch. However, there is no variation in the resonant frequency calculated due to the curvature variation, which is in contrast to the results measured. The resonant frequency measured is shifted slightly to a higher frequency when the cylinder radius decreases, which is similar to the results for a probe-fed cylindrical circular patch antenna with the feed in the axial direction of the ground cylinder; that is, $\beta_p = 90°$.

The resonant input resistance calculated as a function of slot length for various cylinder radii is also presented in Figure 4.47. The frequencies marked in the figure are resonant frequencies at which the input resistances are evaluated. The resonant input resistance is seen to increase with increasing cylinder radius and slot length. It is also noted that for the parameters studied here, the resonant input resistance approaches a constant value with a cylinder radius greater than about 30 cm. That is, in this case the curvature effects on the input impedance of the antenna can be ignored. Figure 4.48 shows the results calculated for another set of antenna parameters. The cavity-model solutions obtained for the resonant frequencies are also not affected by curvature variation. On the other hand, the resonant input resistance for a large cylinder radius is seen to agree well with the data measured for the planar case [28]. The variation in resonant input resistance with slot length and cylinder radius is similar to that shown in Figure 4.47.

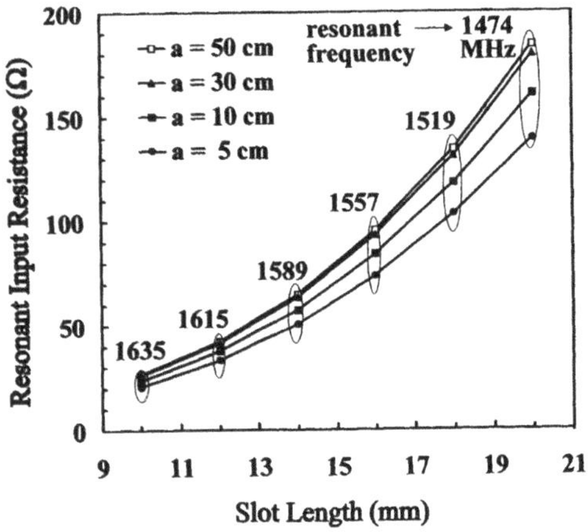

FIGURE 4.47 Resonant input resistance calculated as a function of slot length for various cylinder radii; the antenna parameters are as in Figure 4.46. (From Ref. [27], © 1996 John Wiley & Sons, Inc.)

4.7 SLOT-COUPLED CASE: GTLM SOLUTION

In Section 4.4, various equivalent circuits for the microstrip antennas with different patch shapes have been developed using the GTLM theory. These equivalent circuits can also be applied in the study of slot-coupled patch antennas. Examples for the slot-coupled cylindrical rectangular and circular patch antennas are described below.

4.7.1 Rectangular Patch

By referring to the geometry in Figure 4.34 and considering that the patch is excited at the TM_{01} mode with the maximum radiation taking place from the edges at $z = \pm L$, an equivalent circuit can be developed by combining the equivalent circuits in Figure 4.21 for a rectangular microstrip patch and in Figure 4.43 for a slot-coupled feeding structure. The equivalent circuit is shown in Figure 4.49, where two different impedance transformations are used [29]. The first transformation ratio, n_1 $(=L_s/2W)$, is the fraction of patch current flowing through the slot to the total patch current; and the second transformation ratio, n_2, is calculated from $\Delta v/V_0$, similar to (4.268). For the self admittance y_s at $z = \pm L$, the mutual admittance y_m between two radiating edges, and the circuit elements $(g_1, g_2, g_3,$ and $g_1', g_2', g_3')$ of the two π networks (which replace the sections of transmission line between the slot and the radiating edge), the expressions are as derived in Section 4.4.1. From these equivalent lumped-circuit elements, the patch admittance Y_{patch} can be expressed as

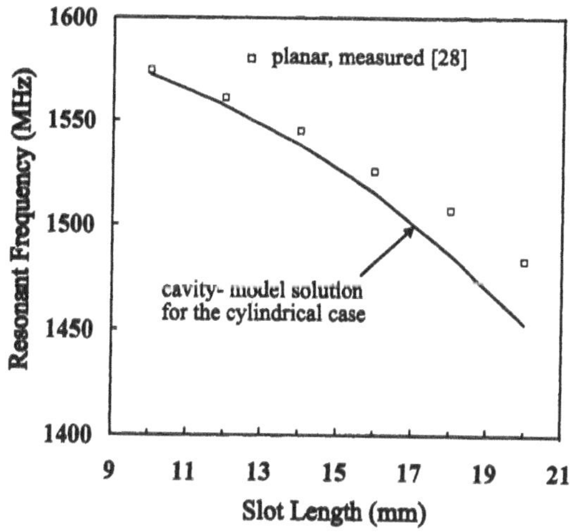

(a)

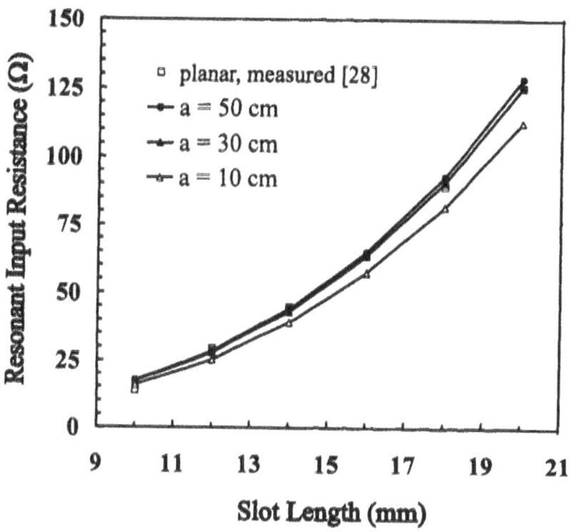

(b)

FIGURE 4.48 (a) Resonant frequency and (b) resonant input resistance calculated as a function of slot length for various cylinder radii; $\epsilon_1 = \epsilon_f = 2.55$, $h = h_f = 0.8$ mm, $r_d =$ 33.81 mm, $W_s = 1.5$ mm, $W_f = 2.22$ mm, $l_{ts} = 30$ mm. (From Ref. [27], © 1996 John Wiley & Sons, Inc.)

$$Y_{\text{patch}} = [(y_3 \| y_4 \| y_5) + (g_1 \| y_2)] \| y_1 \, , \tag{4.287}$$

with

$$y_1 = \frac{y_7 g_1' + (y_6 \| y_8) \| y_7 + (y_6 \| y_8) \| g_1']}{y_7} \, , \tag{4.288}$$

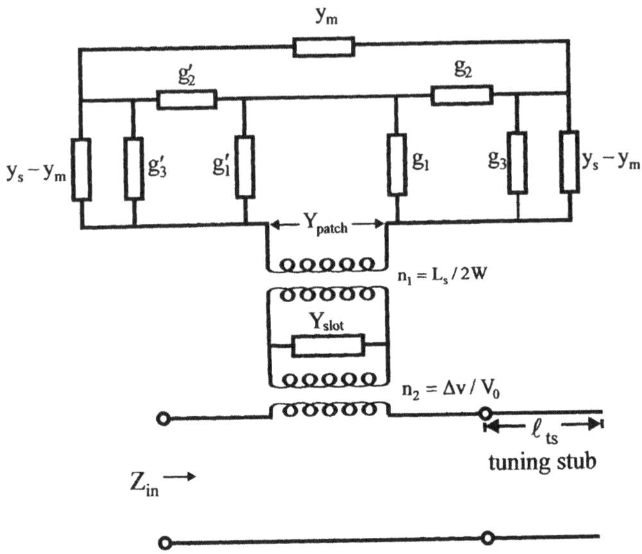

FIGURE 4.49 Equivalent circuit of a slot-coupled cylindrical rectangular microstrip antenna at the TM_{01} mode for GTLM analysis.

$$y_2 = \frac{y_7 g_1' + (y_6 \| y_8) \| y_7 + (y_6 \| y_8) \| g_1'}{y_6 \| y_8}, \tag{4.289}$$

$$y_3 = \frac{y_7 g_1' + (y_6 \| y_8) \| y_7 + (y_6 \| y_8) \| g_1'}{g_1'}, \tag{4.290}$$

$$y_4 = y_s - y_m + g_3, \tag{4.291}$$

$$y_5 = \frac{y_m g_2' + y_m g_2 + g_2 g_2'}{g_2'}, \tag{4.292}$$

$$y_6 = \frac{y_m g_2' + y_m g_2 + g_2 g_2'}{g_2}, \tag{4.293}$$

$$y_7 = \frac{y_m g_2' + y_m g_2 + g_2 g_2')}{y_m}, \tag{4.294}$$

$$y_8 = y_s - y_m + g_3'. \tag{4.295}$$

With the patch admittance obtained above and the slot self-admittance in (4.254), the input impedance of the antenna can be calculated from

$$Z_{in} = \frac{n_2^2}{n_1^2 Y_{patch} + Y_{slot}} - jZ_c \cot k_l l_{ts} . \tag{4.296}$$

Typical calculated input impedance results obtained from (4.296) are presented in Figure 4.50. Again, it is seen that the resonant input resistance increases with increasing slot length and cylinder radius, and good agreement between the resonant input resistance for the case with a larger cylinder radius and the planar data [30] is also observed. As for the resonant frequency results, the GTLM solutions are about the same as those obtained from cavity-model analysis. Therefore, the equivalent circuit derived using GTLM theory can be a useful tool for the design of slot-coupled cylindrical microstrip antennas.

4.7.2 Circular Patch

Based on the geometry in Figure 4.45, by considering the circular patch to be excited at the TM_{11} mode and applying the GTLM formulation in Section 4.4.2, an equivalent circuit for a slot-coupled cylindrical circular patch antenna is as constructed in Figure 4.51. Unlike the circuit in Figure 4.49 for a rectangular patch, we use only one impedance transformer in the present equivalent circuit.

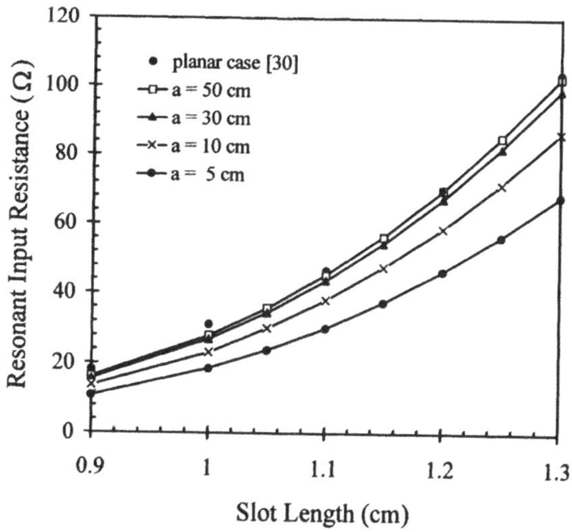

FIGURE 4.50 Resonant input resistance versus slot length; $\epsilon_1 = \epsilon_f = 2.54$, $h = h_f =$ 1.6 mm, $2L = 40$ mm, $2W = 30$ mm, $W_s = 1.1$ mm, $W_f = 4.42$ mm, $l_{ts} = 20$ mm.

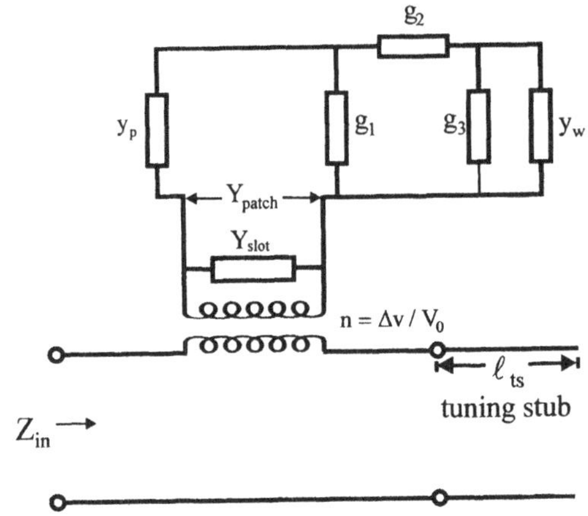

FIGURE 4.51 Equivalent circuit of a slot-coupled cylindrical circular microstrip antenna at the TM_{11} mode for GTLM analysis.

This transformer uses the same transformation ratio n $(=\Delta v/V_0)$ as in (4.268) and slot self-admittance as in (4.286). The patch admittance is derived as

$$Y_{\text{patch}} = \frac{y_p[g_1(g_2 + g_3 + y_w) + g_2(g_3 + y_w)]}{(y_p + g_1)(g_2 + g_3 + y_w) + g_2(g_3 + y_w)}, \tag{4.297}$$

where the circuit elements are as interpreted and derived in Section 4.4.2. By applying the patch admittance in (4.297) and the slot self-admittance in (4.286) to (4.269), the GTLM solution of the input impedance of a slot-coupled cylindrical circular patch antenna can be evaluated. The results calculated for the parameters given in Figure 4.48 are shown in Figure 4.52. By comparing the results in Figures 4.52 and 4.48, good agreement is observed between GTLM solutions and cavity-model results.

4.8 MICROSTRIP-LINE-FED CASE

In addition to the feeding mechanism using probe feed and slot coupling described in Sections 4.2 to 4.7, using a microstrip feed line to excite a cylindrical microstrip antenna directly has also been studied [31,32]. In [31], a full-wave approach incorporating the use of dyadic Green's functions and moment-method calculation is applied to solve the input impedance problem of a microstrip-line-fed cylindrical

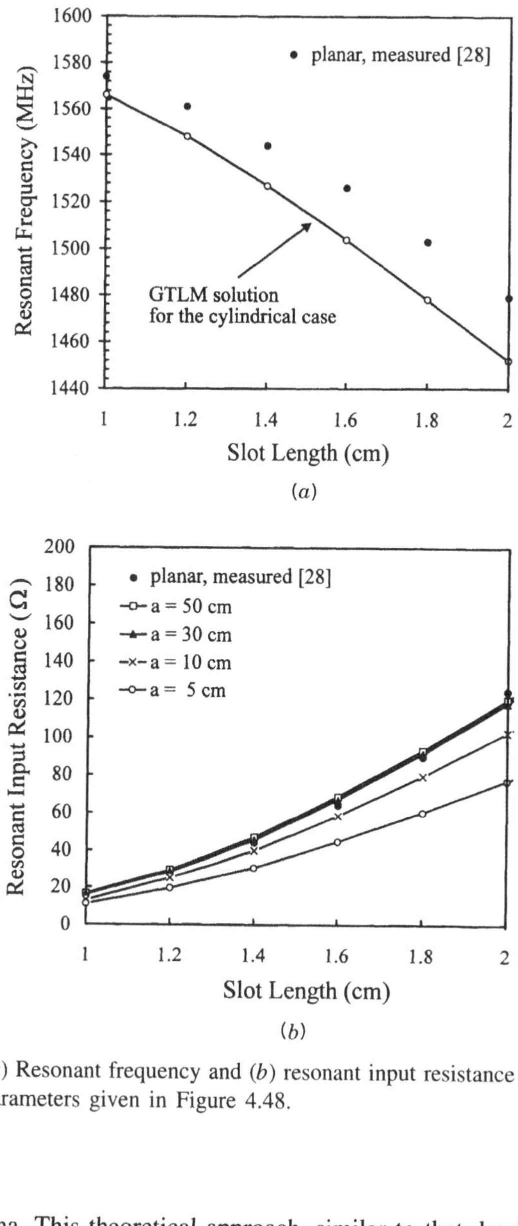

FIGURE 4.52 (*a*) Resonant frequency and (*b*) resonant input resistance versus slot length for the antenna parameters given in Figure 4.48.

microstrip antenna. This theoretical approach, similar to that described in Section 4.2, is relatively complicated and computationally inefficient. A simpler method for using GTLM theory has also been reported [32], providing efficient computation with good accuracy for practical designs. The GTLM results obtained from solving a microstrip-line-fed cylindrical microstrip antenna are described below.

By considering the geometry shown in Figure 4.53, a microstrip antenna is fed directly by a microstrip line printed along the axial direction of a ground cylinder.

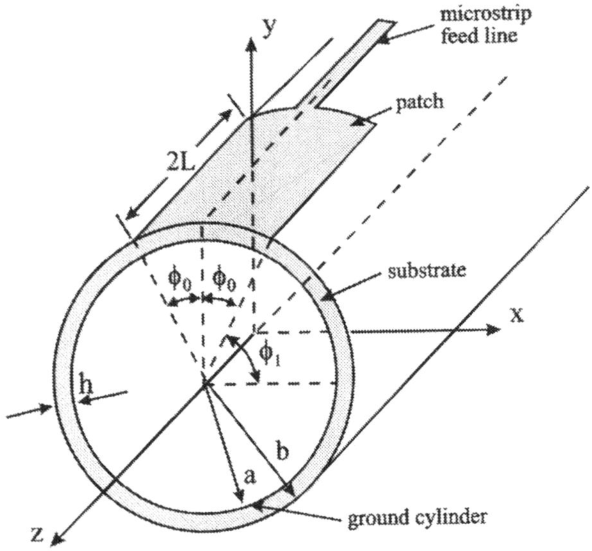

FIGURE 4.53 Geometry of a microstrip-line-fed cylindrical rectangular microstrip antenna.

The width of the microstrip line is W_f and the antenna is considered at TM_{01}-mode excitation. By following the GTLM formulation in Section 4.4, the equivalent circuit shown in Figure 4.54 can be derived. To model the parasitic effects of the feed line on the microstrip antenna, the self-admittance of the radiating edge facing the feed line is reduced by a factor of

$$r = 1 - \frac{W_f}{2b\phi_0}. \tag{4.298}$$

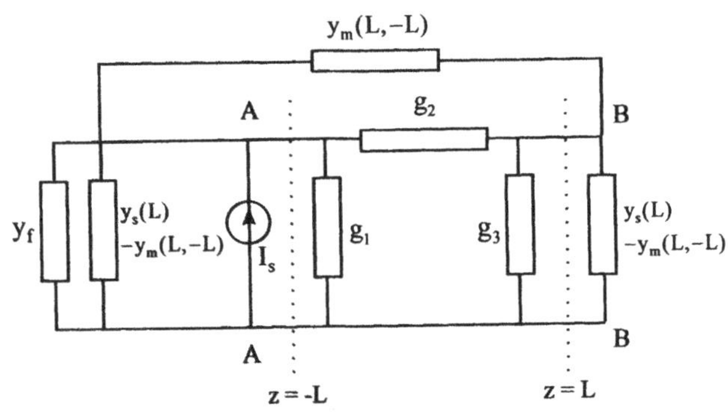

FIGURE 4.54 Equivalent circuit of a microstrip-line-fed cylindrical rectangular patch antenna at the TM_{01} mode for GTLM analysis.

This reduction takes into account partial coverage of the radiating edge by the feed line. The reduction in self-admittance at $z = -L$ can then be considered as an addition of a parallel admittance, given as

$$y_f = (r - 1)y_s(-L) . \qquad (4.299)$$

It can also be found that the equivalent circuit in Figure 4.54 corresponds to the case of $z_p = -L$ in Figure 4.21. The circuit elements of the π network are written as

$$g_1 = g_3 = -\frac{bk_z}{\omega\mu_0}\left[\coth(-j2Lk_z) + \frac{1}{\sinh(j2Lk_z)}\right], \qquad (4.300)$$

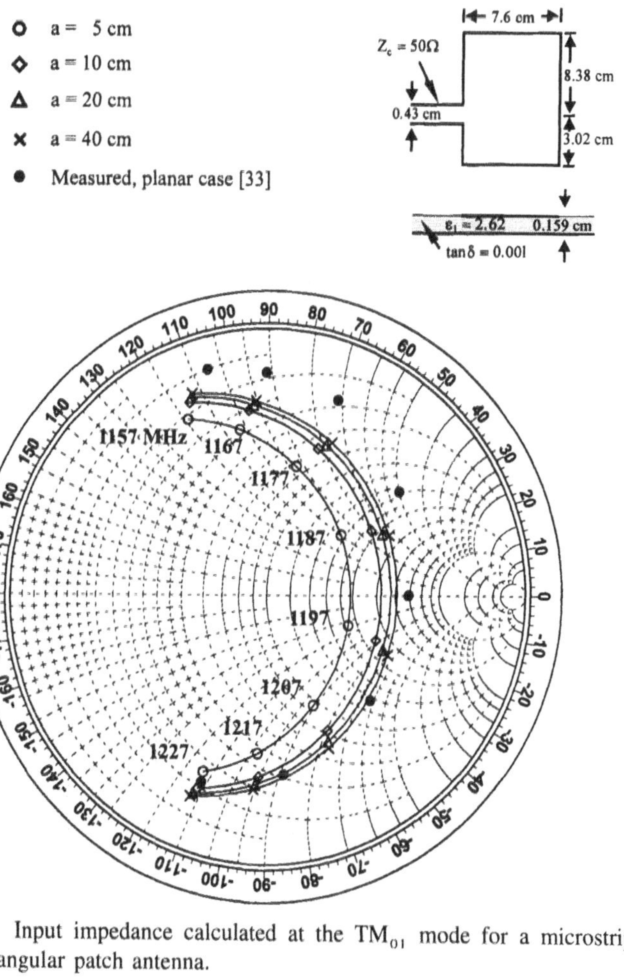

FIGURE 4.55 Input impedance calculated at the TM_{01} mode for a microstrip-line-fed cylindrical rectangular patch antenna.

$$g_2 = \frac{bk_z}{\omega\mu_0 \sinh(j2Lk_z)} . \tag{4.301}$$

The expressions above can be obtained by setting $z_p = -L$ in (4.128) and (4.129). For the self-admittance y_s and the mutual admittance y_m, the expressions are the same as in (4.134) and (4.137), respectively. With the equivalent-circuit elements obtained, the input impedance of the patch antenna seen at $z = -L$ by the microstrip feed line is written as

$$Z_{\text{in}} = \frac{h}{2\phi_0 Y}, \tag{4.302}$$

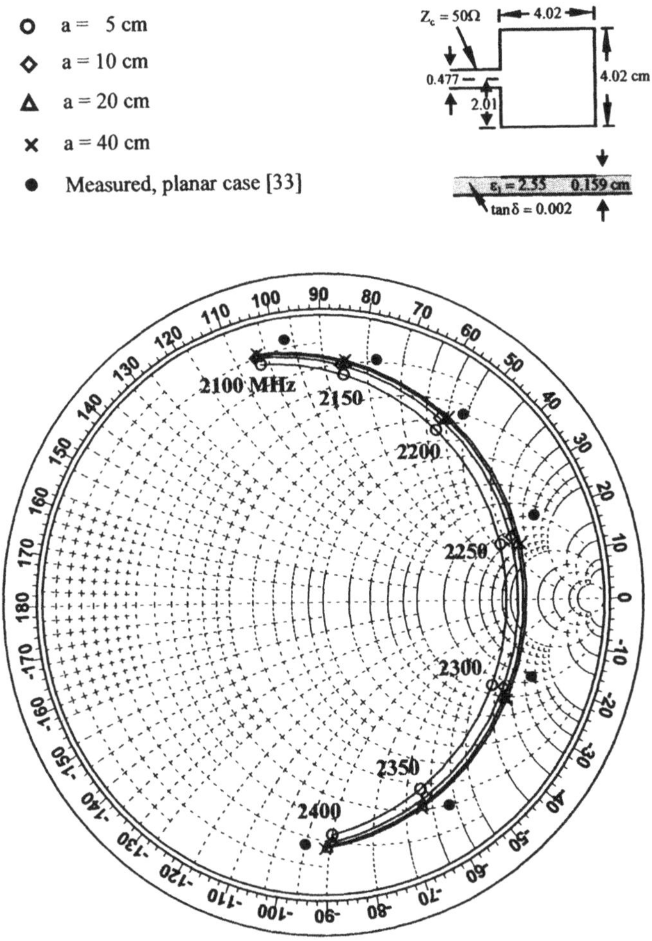

FIGURE 4.56 Input impedance calculated at the TM_{01} mode for a microstrip-line-fed cylindrical square patch antenna.

with

$$Y = y_f + g_3 + y_s(-L) - y_m(L, -L) + \frac{[g_1 + y_s(L) - y_m(L, -L)][y_m(L, -L) + g_2]}{g_1 + g_2 + y_s(L)}. \tag{4.303}$$

The GTLM solution of the input impedance for a microstrip-line-fed rectangular patch antenna is obtained by evaluating (4.302). Figures 4.55 and 4.56 present typical results. The rectangular patch shown in Figure 4.55 has an aspect ratio of 1.5 (i.e., $b\phi_0/L = 1.5$), and the patch in Figure 4.56 is square ($b\phi_0/L = 1.0$). For both cases the results obtained for $a > 40$ cm show very small variations with those for $a = 40$ cm, and therefore are not shown. It can also be seen that when the cylinder radius is increased, the input impedance curve approaches the data measured in the planar case [33]. This provides credence for the conviction that GTLM solutions are reliable.

4.9 CYLINDRICAL WRAPAROUND PATCH ANTENNA

The configuration of a cylindrical wraparound microstrip patch antenna with a probe feed is shown in Figure 4.57. The wraparound patch can also be excited by a parallel feed network [34,35]. In such a feed design [34], the wraparound patch is excited by a corporate feed at a number of points around the patch edge, and the spacing between feed points is designed to be less than one wavelength in the dielectric substrate. This arrangement can result in a nearly uniform electric field distribution inside the wraparound patch antenna and makes possible excitation of only the TM_{01} mode. Based on cavity-model analysis [35], in this case the far-zone radiated fields are given approximately by

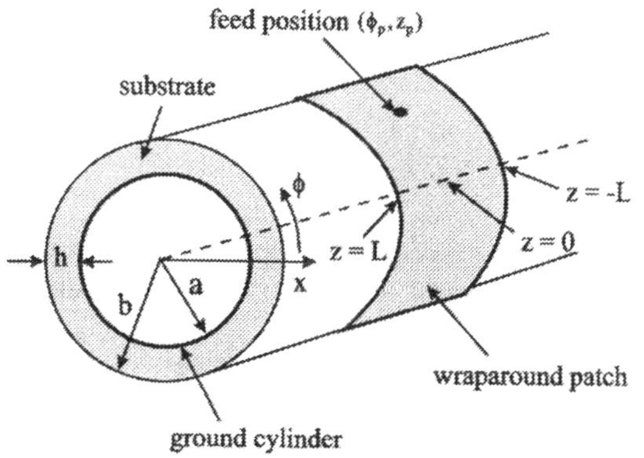

FIGURE 4.57 Geometry of a cylindrical wraparound microstrip antenna.

$$E_\theta \simeq \frac{j2E_0 h e^{-jk_0 r}}{\pi r \sin\theta} \frac{\cos(k_0 L \cos\theta)}{H_0^{(2)}(bk_0 \sin\theta)}, \tag{4.304}$$

$$E_\phi \simeq 0, \tag{4.305}$$

where E_0 is a constant and L is the half-length of the wraparound patch in the axial direction. From the results it is seen that the radiation pattern can be considered as the pattern of a half-wave dipole multiplied by a factor given by

$$f(\theta) = \left| \frac{1}{\sin\theta} \frac{\cos(k_0 L \cos\theta)}{H_0^{(2)}(k_0 b \sin\theta)} \right| . \tag{4.306}$$

For the E-plane case (the plane containing the z axis), the pattern is found to be strongly affected by the ratio b/λ_0. When b/λ_0 is less than 1.0 (i.e., the cylinder radius is less than approximately one operating wavelength), a smooth E-plane pattern is expected. On the other hand, when b/λ_0 is much larger than 1.0, ripples usually occur in the E-plane pattern. As for the H-plane pattern [the pattern in the roll ($\theta = 90°$) plane], we have nearly omnidirectional radiation. This characteristic makes the wraparound patch antenna an attractive candidate for applications in spinning missiles that require omnidirectional coverage.

For the input impedance of the wraparound patch antenna, the case with probe-feed excitation at the TM_{mn} mode has been solved [36]. By considering the geometry in Figure 4.57 and applying the cavity-model approximation, the electric field under the wraparound patch is given for the source-free case by

$$E_\rho = \sum_{m,n} E_0 \cos m\phi \cos\left[\frac{n\pi}{2L}(z - L)\right]. \tag{4.307}$$

The resonant frequency and wavenumber for the TM_{mn} mode excitation are written, respectively, as

$$f_{mn} = \frac{c}{2\sqrt{\epsilon_1}} \left[\left(\frac{m}{\pi b}\right)^2 + \left(\frac{n}{2L}\right)^2 \right]^{1/2}, \tag{4.308}$$

$$k_{mn} = \left[\left(\frac{m}{b}\right)^2 + \left(\frac{n\pi}{2L}\right)^2 \right]^{1/2}. \tag{4.309}$$

When a probe feed at (ϕ_p, z_p), modeled as a unit-amplitude current ribbon of width w_p is considered, the electric field under the wraparound patch is rewritten as

$$E_\rho = j\omega\mu_0 \sum_{m,n} C_{mn} \cos m\phi \cos\left[\frac{n\pi}{2L}(z - L)\right], \tag{4.310}$$

with

$$C_{mn} = \frac{w_p}{k_{\text{eff}}^2 - k_{mn}^2} \frac{e_m e_n}{4\pi a L} \cos m\phi_p \cos\left[\frac{n\pi}{2L}(z_p - L)\right] j_0\left(\frac{m w_p}{2a}\right), \quad (4.311)$$

where e_m, e_n, $j_0(x)$, and k_{eff} are as defined in Section 4.3.1. The input impedance is then given by

$$Z_{\text{in}} = \frac{-ha}{w_p^2} \int_{\phi_p - w_p/2a}^{\phi_p + w_p/2a} E_\rho \, d\phi$$

$$= \sum_{m=0}^{\infty} \sum_{n=0}^{\infty} \frac{h e_m e_n}{8\pi^2 a L \epsilon_0 \epsilon_1} \cos^2 m\phi_p \cos^2\left[\frac{n\pi}{2L}(z_p - L)\right]$$

$$\times j_0^2\left(\frac{m w_p}{2a}\right) \frac{\delta_{\text{eff}} f^3 - jf(f^2 - f_{mn}^2)}{\delta_{\text{eff}}^2 f^4 - (f^2 - f_{mn}^2)^2}. \quad (4.312)$$

From an evaluation of (4.312), the input impedance at the TM_{01} mode is obtained. The results are similar to those for a rectangular patch; that is, the resonant input resistance has a zero value when the feed position is at the wraparound patch center ($z_p = 0$) and increases with feed position moving toward the patch edge ($z_p = -L$ or L).

4.10 CIRCULAR POLARIZATION CHARACTERISTICS

The circular polarization (CP) condition for a nearly square patch antenna printed on a thin cylindrical substrate has also been investigated [37]. The thin-substrate condition allows the use of cavity-model analysis. By referring to the geometry shown in Figure 4.58 and applying cavity-model theory as described in Section 4.3.1, expressions for the far-zone radiated fields of the TM_{01} (axial-direction excitation) and TM_{10} (ϕ-direction excitation) modes can be derived: For the TM_{01} mode,

$$E_{\theta 1} = A_1 \cos(k_0 L \cos\theta) \sum_{p=0}^{\infty} \frac{e_p j^p \sin p\phi_0}{p H_p^{(2)}(k_0 a \sin\theta)} \cos\left[p\left(\phi - \frac{\pi}{2}\right)\right], \quad (4.313)$$

$$E_{\phi 1} = \frac{A_1 \cos\theta \cos(k_0 L \cos\theta)}{k_0 a \sin\theta} \frac{(\pi/2k_0 L)^2 - 1}{(\pi/2k_0 L)^2 - \cos^2\theta}$$

$$\times \sum_{p=0}^{\infty} \frac{e_p j^p \sin p\phi_0}{H_p^{(2)\prime}(k_0 a \sin\theta)} \sin\left[p\left(\phi - \frac{\pi}{2}\right)\right], \quad (4.314)$$

and for the TM_{10} mode,

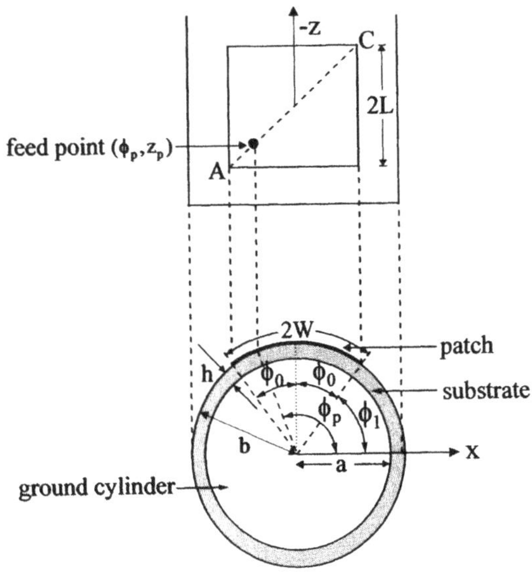

FIGURE 4.58 Geometry of a circularly polarized cylindrical nearly square microstrip antenna.

$$E_{\theta 2} = A_2 \sin(k_0 L \cos\theta) \sum_{p=0}^{\infty} \frac{e_p j^p \cos p\phi_0}{H_p^{(2)}(k_0 a \sin\theta)} \frac{p \sin[p(\phi - \pi/2)]}{(\pi/2\phi_0)^2 - p^2}, \tag{4.315}$$

$$E_{\phi 2} = \frac{A_2 \sin\theta \sin(k_0 L \cos\theta)}{k_0 a \cos\theta} \sum_{p=0}^{\infty} \frac{e_p j^p \cos p\phi_0 \cos[p(\phi - \pi/2)]}{H_p^{(2)\prime}(k_0 a \sin\theta)}$$
$$- \frac{A_2 \cos\theta \sin(k_0 L \cos\theta)}{k_0 a \cos\theta} \sum_{p=0}^{\infty} \frac{e_p j^p \cos p\phi_0 \cos[p(\phi - \pi/2)]}{H_p^{(2)\prime}(k_0 a \sin\theta)}$$
$$\times \frac{p^2}{(\pi/2\phi_0)^2 - p^2}, \tag{4.316}$$

where

$$A_1 = \frac{-2\omega\mu_0 h C_{01}}{\pi^2 r \sin\theta} e^{-jk_0(r + L\cos\theta)}, \tag{4.317}$$

$$A_2 = \frac{-2\omega\mu_0 h C_{10}}{\pi^2 r \sin\theta} e^{-jk_0(r + L\cos\theta)}. \tag{4.318}$$

In the expressions above, C_{01} and C_{10} can be obtained from (4.54) and e_p is defined in (4.60).

For CP excitation, the feed position is selected to be on the diagonal of the nearly square patch. In this case, two orthogonal modes of TM_{01} and TM_{10} with equal amplitudes and 90° phase difference can be excited, which results in CP radiation with the center frequency between the resonant frequencies of the two modes. We have the CP condition written as

$$\angle E_\theta - \angle E_\phi = \pm 90^\circ , \tag{4.319}$$

$$|E_\theta| = |E_\phi| , \tag{4.320}$$

with

$$E_\theta = E_{\theta 1} + E_{\theta 2} , \qquad E_\phi = E_{\phi 1} + E_{\phi 2} . \tag{4.321}$$

The condition above can be achieved by careful selection of the aspect ratio of the curved patch and the corresponding operating frequency. Another important factor to be considered is the impedance matching of the patch antenna through a single feed, which can be achieved by selecting a feed position along the diagonal of the patch.

From the formulation above, typical numerical results have been calculated. The effect of the aspect ratio on the CP radiation is studied first. The results are presented in Figure 4.59. In the calculation, the dimension of the straight edge of the patch is chosen to be fixed, while the dimension of the curved edge is varied to allow for variation of the aspect ratio. The results are shown for three different

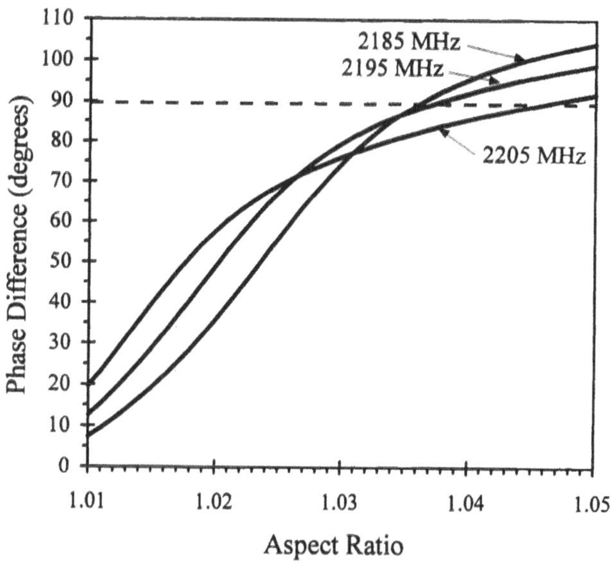

FIGURE 4.59 Dependence of the phase difference ($\angle E_\phi - \angle E_\theta$) on the aspect ratio ($= W/L$); $a = 10$ cm, $2L = 4.14$ cm, $\epsilon_1 = 2.64$, $h = 1.59$ mm, feed at point A. (From Ref. [37], © 1993 IEEE, reprinted with permission.)

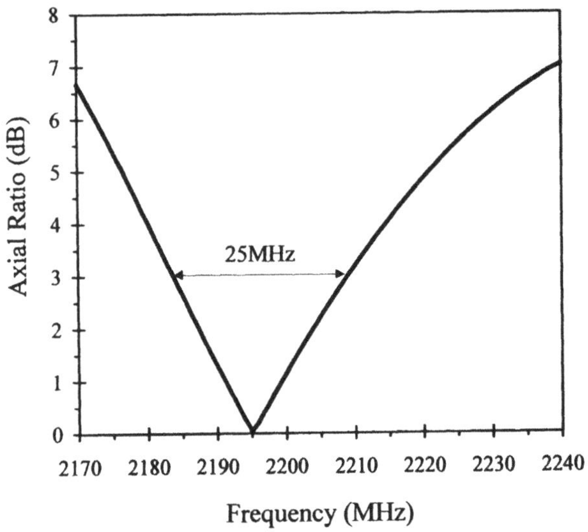

FIGURE 4.60 Axial ratio versus frequency; $a = 10$ cm, $2L = 4.14$ cm, $\epsilon_1 = 2.64$, $h =$ 1.59 mm, $W/L = 1.038$. (From Ref. [37], © 1993 IEEE, reprinted with permission.)

operating frequencies. It is seen that each frequency has a corresponding aspect ratio to satisfy the 90° phase difference of the radiated fields. It should be noted that there are still many frequencies with aspect ratios that satisfy the 90° phase difference requirement for CP radiation. Figure 4.60 presents the axial ratio versus frequency for choosing an aspect ratio (W/L) of 1.038. This aspect ratio results in a CP radiation of center frequency at 2195 MHz and 3-dB axial ratio bandwidth about 25 MHz (or 1.14%). The phase differences within the 3-dB bandwidth were also calculated and found to range from 92 to 82° and were asymmetric with respect the center frequency. When the cylinder radius increases and approaches the planar case, the phase difference within the bandwidth becomes more symmetric with respect to the center frequency.

The return loss versus the feed position on the diagonal of the patch has also been studied. It is found that the return loss is sensitive to the feed position and the optimal feed position is found to be at a position 0.33 times the diagonal length. It is also noted that by varying the feed position along the diagonal, the 3-dB bandwidth, the optimal aspect ratio, and the center frequency are found to remain unchanged. Also, within the 3-dB axial bandwidth, the return loss is quite stable and shows fairly good matching.

Dependence of the optimal aspect ratio on the cylinder radius has also been analyzed with results as shown in Figure 4.61. The optimal aspect ratio decreases with increasing cylinder radius and approaches a limiting value of 1.033 for a radius greater than about 35 cm. The corresponding center frequency for the optimal aspect ratio shown in Figure 4.61 is also found to increase with increasing cylinder radius. For $a > 40$ cm, the center frequency approaches a limiting value of

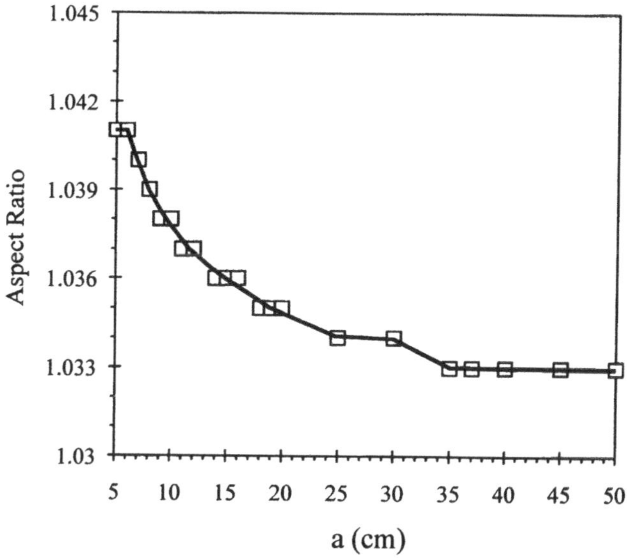

FIGURE 4.61 Dependence of the optimal axial ratio on the cylindrical radius; $2L = 4.14$ cm, $\epsilon_1 = 2.64$, $h = 1.59$ mm. (From Ref. [37], © 1993 IEEE, reprinted with permission.)

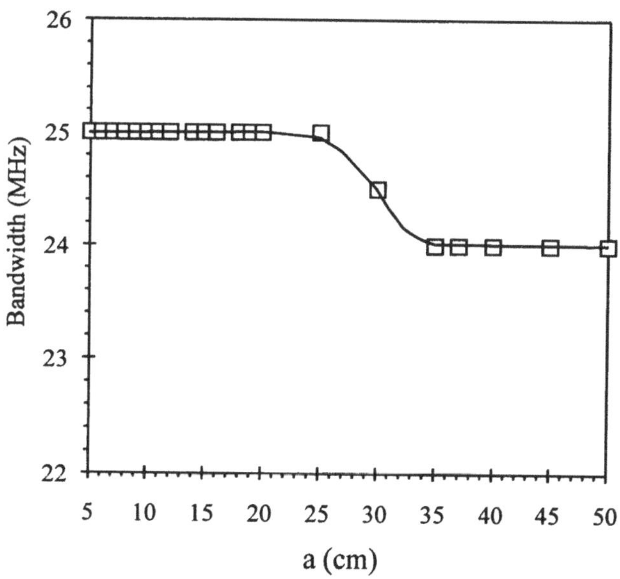

FIGURE 4.62 CP bandwidth, determined from a 3-dB axial ratio, versus cylindrical radius for the aspect ratio in Figure 4.61.

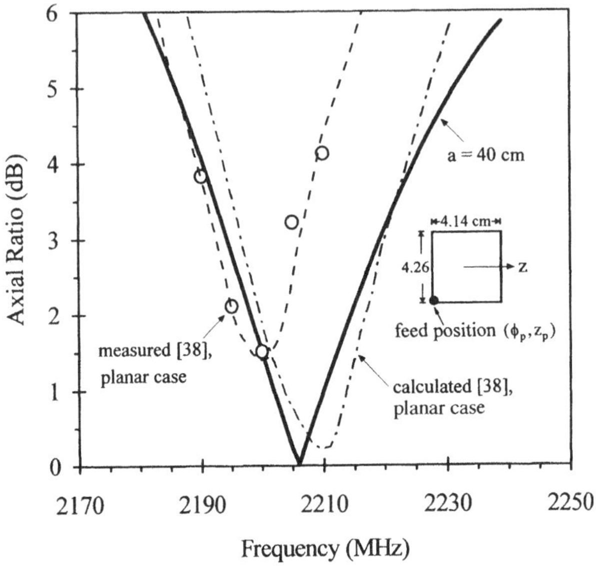

FIGURE 4.63 Comparison of a cylindrical CP patch antenna with a large cylinder radius and the planar CP patch antenna; $2L = 4.14$ cm, $2W = 4.26$ cm, $W/L = 1.029$, $h = 1.59$ mm, $\epsilon_1 = 2.62$.

about 2202 MHz. For the 3-dB axial ratio bandwidth versus the cylinder radius, the results are as shown in Figure 4.62. A very small change (about 1 MHz) in bandwidth is observed. A comparison of the cylindrical CP patch antenna with a large cylinder radius and the planar CP patch antenna [38] is also shown in Figure 4.63. For $a > 40$ cm, the results obtained are almost the same as the case of 40 cm shown in the figure and therefore are not shown. From these results it can be concluded that with the parameters studied here, the curvature effect on the CP characteristics of the present cylindrical patch antenna can be neglected for $a > 40$ cm (or about $3\lambda_0$).

4.11 CROSS-POLARIZATION CHARACTERISTICS

The cross-polarization characteristics are also important in microstrip antenna designs. Rectangular and triangular patch antennas mounted on a cylindrical body have been studied [39,40] and in both cases a strong dependence of the cross-polarization level on cylinder radius is observed. Details of the results are given below.

4.11.1 Rectangular Patch

By considering the geometry of a thin substrate shown in Figure 4.1, the far-zone radiated fields can be obtained using cavity-model analysis. The expressions are given in (4.57) and (4.58). Also, from the definition in [41], when the rectangular patch is excited at the TM_{10} mode, the copolarized and cross-polarized fields in a particular direction are, respectively, given by $E_{10}(\theta, \phi)$ and $\Sigma_{n=1}^{\infty} E_{0n}(\theta, \phi)$, where E_{mn} represents the magnitude of the elctric field corresponding to the TM_{mn} mode and can be calculated from (4.57) and (4.58). The cross-polarization level is defined here as the ratio of the maximum magnitude of the copolarized field $E_{10}(\theta, \phi)$ to the maximum magnitude of cross-polarized field $\Sigma_{n=1}^{\infty} E_{0n}(\theta, \phi)$ in a specific plane. The dependence of the cross-polarization level on the aspect ratio for various cylinder radii is first studied. The rectangular patch is excited at the TM_{10} mode (ϕ-direction excitation) and the results are presented in Figure 4.64. It is observed that for the case of $a = 50$ cm, the cross-polarization level has a peak value of about 22 dB at $L/W = 1.55$. When the cylinder radius decreases, the cross-polarization level also decreases and occurs at an aspect ratio below 1.55.

The dependence of the cross-polarization level on the resonant frequency and feed position for various cylinder radii has also been studied. The aspect ratio of the rectangular patch is chosen to be 1.5 and the TM_{10} mode is considered. Figure 4.65 shows the results for feeding the patch along the curved edge (the edge parallel to the excitation direction, known as the nonradiating edge), and Figure 4.66 presents the results for feeding the patch along the straight edge (the edge perpendicular to the excitation direction, known as the radiating edge). Results show that the cross-polarization level has a maximum value at the center of the straight edge and has a minimum value at the center of the curved edge. That is,

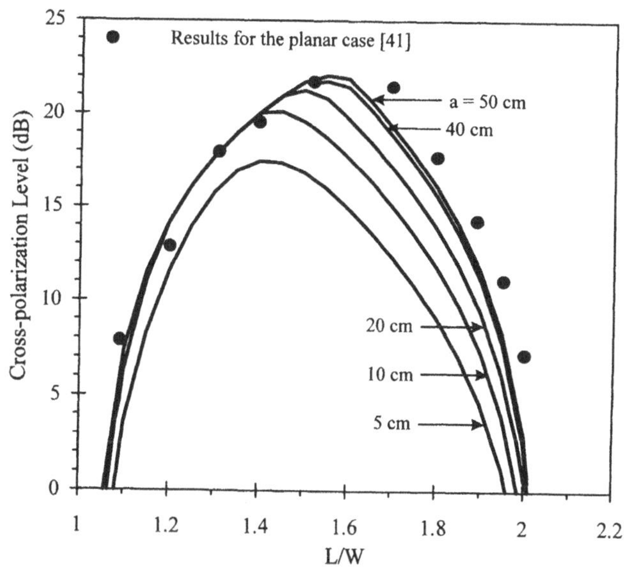

FIGURE 4.64 Dependence of the cross-polarization level, in the $\phi = 90°$ plane, on the aspect ratio (L/W) for different cylinder radii; $W = 2.62$ cm, $\epsilon_1 = 2.91$, $h = 1.48$ mm, $(\phi_p, z_p) = (0.7/b, L)$. (From Ref. [39], © 1993 John Wiley & Sons, Inc.)

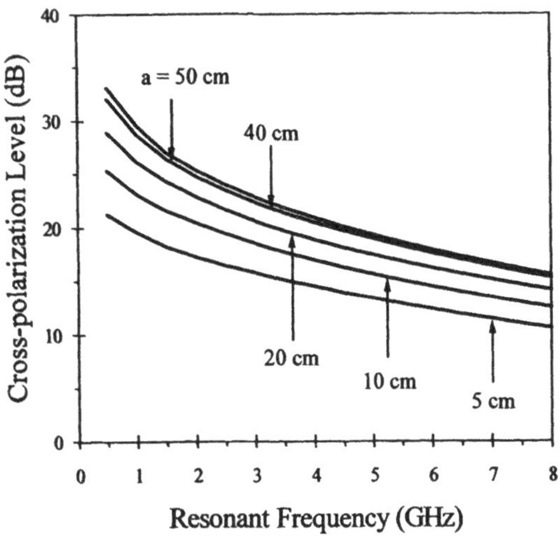

(*a*)

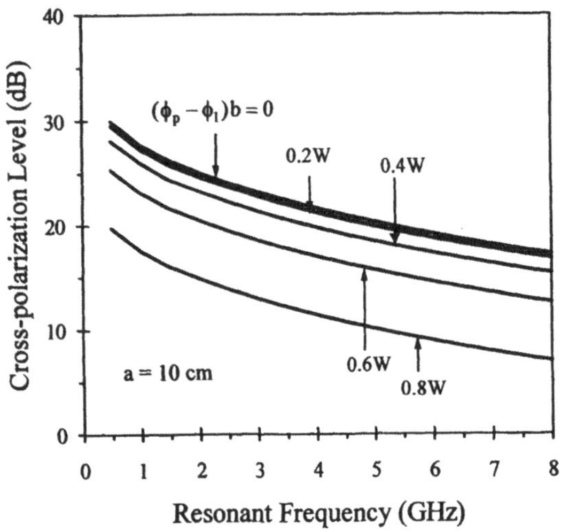

(*b*)

FIGURE 4.65 Dependence of the cross-polarization level in the $\phi = 90°$ plane on the resonant frequency for various (*a*) cylinder radii with $(\phi_p, z_p) = [(0.6W/b) + \phi_1, L]$ at the curved edge and (*b*) feed positions along the curved edge with $a = 10$ cm; $W = 2.62$ cm, $L = 1.5W$, $\epsilon_1 = 2.32$, $h = 1.48$ mm. (From Ref. [39], © 1993 John Wiley & Sons, Inc.)

the feed position at the center of the radiating edge can result in an optimal cross-polarization level. Also, the patch with lower resonant frequency has a better cross-polarization level.

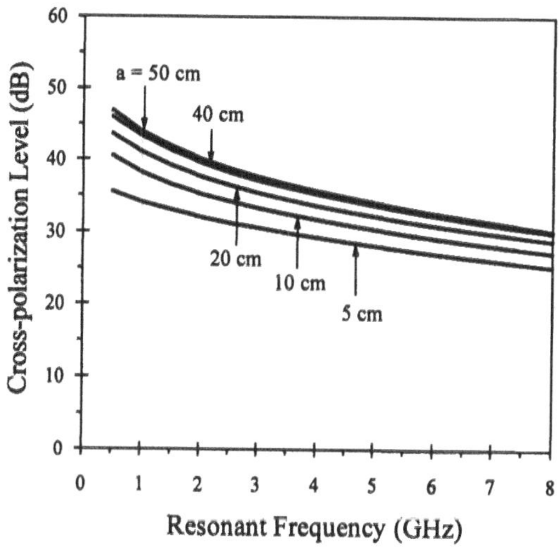

(a)

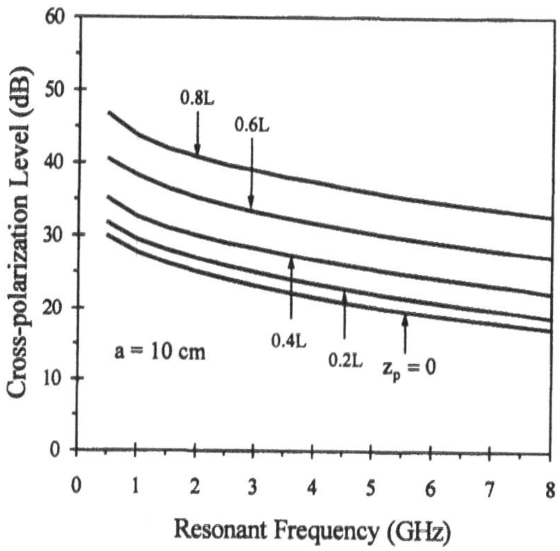

(b)

FIGURE 4.66 Dependence of the cross-polarization level in the $\phi = 90°$ plane on the resonant frequency for various (*a*) cylinder radii with $(\phi_p, z_p) = (\phi_1, 0.6L)$ at the straight edge and (*b*) feed positions along the straight edge with $a = 10$ cm; $W = 2.62$ cm, $L = 1.5W$, $\epsilon_1 = 2.32$, $h = 1.48$ mm. (From Ref. [39], © 1993 John Wiley & Sons, Inc.)

4.11.2 Triangular Patch

The cross-polarization characteristics of triangular microstrip antennas mounted on a cylindrical body have been studied using full-wave analysis [41]. Results of the far-zone copolarized and cross-polarized radiated fields at the fundamental mode (TM_{10}; axial-direction excitation) for various cylinder radii and flare angles have been analyzed. Significant dependence of the cross-polarization level on the cylinder radius and flare angle is observed. Here the cross-polarization level is also defined as the ratio of the maximum amplitude of the copolarized field to the maximum amplitude of the cross-polarized field in a specific plane. Considering the geometry shown in Figure 4.1, an isosceles triangular patch with a flare angle of α is treated. As derived in Section 4.2 using full-wave analysis, the far-zone radiated fields at the TM_{10} mode in spherical coordinates can be given approximately by

$$\begin{bmatrix} E_\theta(\theta, \phi) \\ E_\phi(\theta, \phi) \end{bmatrix} \simeq \frac{-je^{jk_0 r}}{\pi r \sin\theta} \begin{bmatrix} -1 & 0 \\ 0 & \eta_0 \end{bmatrix} \sum_{n=-\infty}^{\infty} \frac{(-j)^n e^{jn\phi}}{H_n^{(1)}(k_0 b \sin\theta)}$$

$$\times \begin{bmatrix} X_{11} & X_{12} \\ X_{21} & X_{22} \end{bmatrix}^{-1} \begin{bmatrix} \tilde{J}_{s\phi}(n, k_0 \cos\theta) \\ \tilde{J}_{sz}(n, k_0 \cos\theta) \end{bmatrix}, \qquad (4.322)$$

where $\tilde{J}_{s\phi}$ and $\tilde{J}_{sz}$ are, respectively, the $\hat{\phi}$- and $\hat{z}$-directed patch surface current densities in the spectral domain for the TM_{10} mode; expressions for X_{ij}'s can be found in (2.35)–(2.38). With E_θ and E_ϕ determined from (4.322), the copolarized and cross-polarized fields in the E plane ($\phi = 90°$ plane) can be evaluated from $|E_\theta||_{\phi=90°}$ and $|E_\phi||_{\phi=90°}$, respectively; in the H plane ($\theta = 90°$ plane), the copolarized and cross-polarized fields are determined from $|E_\theta||_{\theta=90°}$ and $|E_\phi||_{\theta=90°}$, respectively. Since it is also found that there are no cross-polarized fields in the E plane [i.e., $|E_\phi||_{\phi=90°} = 0$ in (4.322)], the present results show only the cross-polarization results in the H plane of the cylindrical triangular patch antenna.

Typical results of copolarized and cross-polarized patterns in the H plane for various cylinder radii with $\alpha = 50°$ are shown in Figure 4.67. The radiation patterns are obtained at a resonant frequency of the TM_{10} mode for each case. From the results it is seen that the cross-polarization level is larger for smaller cylinder radius. This suggests that a triangular patch antenna mounted on a cylindrical body of small radius has a better linear polarization characteristic, which is different from the results for a cylindrical rectangular patch antenna described in Section 4.11.1.

The dependence of the copolarized and cross-polarized patterns on the flare angle of the triangular patch has also been studied. Results are presented in Figure 4.68. It is noted that since the copolarized patterns calculated are almost the same for various flare angles, only one curve for the copolarized pattern is shown in the figure. From the results, significant dependence of the cross-polarization level on the flare angle is observed. The cross-polarization level increases with decreasing flare angle. More results for various cylinder radii and flare angles are given in

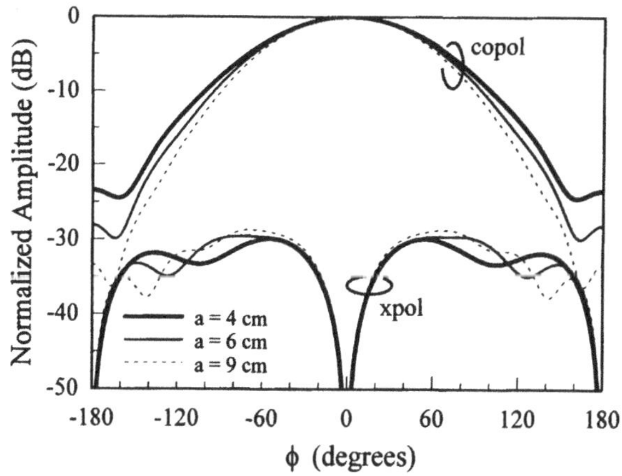

FIGURE 4.67 Copolarized and cross-polarized radiation patterns in the H plane (x–y plane) for various cylinder radii; $\epsilon_1 = 4.39$, $h = 1.6$ mm, $d_h = 43.3$ mm, $\phi_p = 90°$, $z_p =$ 7.15 mm, $\alpha = 50°$. (From Ref. [40], © 1998 IEE, reprinted with permission.)

Figure 4.69. Experiments for the planar triangular patch antenna are also conducted, and the data measured for the cross-polarization level are plotted for comparison. Good agreement between the theoretical results for the case of a larger cylinder radius and the data measured for a planar structure is observed. Effects of feed-position variation on the cross-polarization level have also been studied. Results show a small (about 0.3 dB) variation in cross-polarization level for feed position moving from the tip to the bottom of the triangular patch.

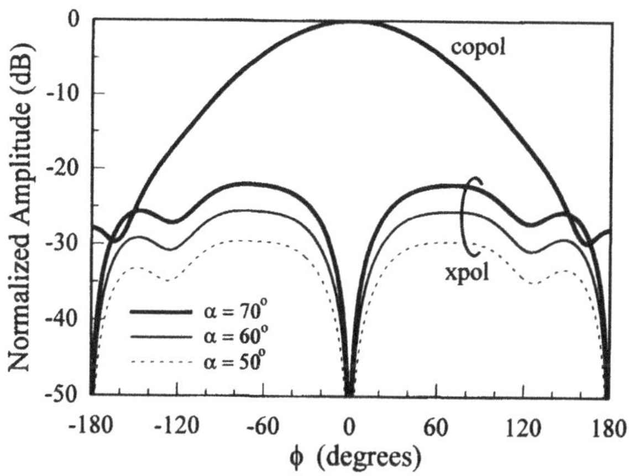

FIGURE 4.68 Copolarized and cross-polarized radiation patterns in the H plane for various flare angles with $a = 6$ cm; other parameters are as in Figure 4.60. (From Ref. [40], © 1998 IEE, reprinted with permission.)

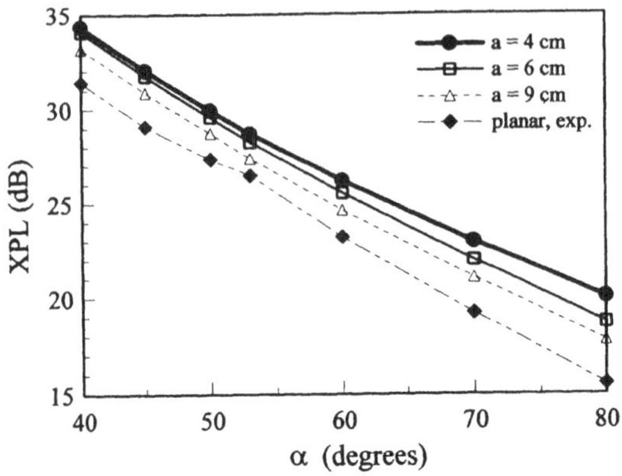

FIGURE 4.69 Cross-polarization level versus flare angle for various cylinder radii; the antenna parameters are as in Figure 4.60. (From Ref. [40], © 1998 IEE, reprinted with permission.)

REFERENCES

1. S. Y. Ke and K. L. Wong, "Input impedance of a probe-fed superstrate-loaded cylindrical-rectangular microstrip antenna," *Microwave Opt. Technol. Lett.*, vol. 7, pp. 232–236, Apr. 5, 1994.
2. K. L. Wong, S. M. Wang, and S. Y. Ke, "Measured input impedance and mutual coupling of rectangular microstrip antennas on a cylindrical surface," *Microwave Opt. Technol. Lett.*, vol. 11, pp. 49–50, Jan. 1996.
3. J. S. Dahele, R. J. Mitchell, K. M. Luk, and K. F. Lee, "Effect of curvature on characteristics of rectangular patch antenna," *Electron. Lett.*, vol. 23, pp. 748–749, July 2, 1987.
4. S. C. Pan and K. L. Wong, "Characteristics of a cylindrical triangular microstrip antenna," *Microwave Opt. Technol. Lett.*, vol. 15, pp. 49–52, May 1997.
5. R. F. Harrington, *Time-Harmonic Electromagnetic Fields*, McGraw-Hill, New York, 1961, Chap. 5.
6. K. M. Luk, K. F. Lee, and J. S. Dahele, "Analysis of the cylindrical-rectangular patch antenna," *IEEE Trans. Antennas Propagat.*, vol. 37, pp. 143–147, Feb. 1989.
7. S. Y. Ke, Radiation and Coupling of Probe-Fed Rectangular Microstrip Antennas on Cylindrical and Planar Surfaces, Ph.D. dissertation, Department of Electrical Engineering, National Sun Yat-Sen University, Kaohsiung, Taiwan, Feb. 1995.
8. K. L. Wong and S. T. Fang, "Cavity-model solutions of cylindrical triangular microstrip patch antennas," *Microwave Opt. Technol. Lett.*, vol. 15, pp. 377–380, Aug. 20, 1997.
9. K. M. Luk and K. F. Lee, "Characteristics of the cylindrical-circular patch antenna," *IEEE Trans. Antennas Propagat.*, vol. 38, pp. 1119–1123, July 1990.
10. H. D. Chen and K. L. Wong, "Input impedance and radiation pattern of a probe-fed cylindrical annular-ring microstrip antenna," *Microwave Opt. Technol. Lett.*, vol. 8, pp. 152–156, Feb. 20, 1995.

11. A. K. Bhattacharyya and R. Garg, "Input impedance of annular-ring microstrip antenna using circuit theory approach," *IEEE Trans. Antennas Propagat.*, vol. 33, pp. 369–374, Apr. 1985.

12. A. K. Bhattacharyya and R. Garg, "Generalized transmission line model for microstrip patches," *IEE Proc.*, pt. H, vol. 132, pp. 93–98, Apr. 1985.

13. K. L. Wong, Y. H. Liu, and C. Y. Huang, "Generalized transmission line model for cylindrical-rectangular microstrip antenna," *Microwave Opt. Technol. Lett.*, vol. 7, pp. 729–732, Nov. 1994.

14. K. L. Wong, C. Y. Huang, and Y. H. Liu, "Generalized transmission line model for cylindrical-circular microstrip antenna," *Microwave Opt. Technol. Lett.*, vol. 8, pp. 63–68, Feb. 5, 1995.

15. L. C. Shen, "Analysis of a circular-disc printed-circuit antenna," *IEE Proc.*, pt. H, vol. 126, pp. 1120–1122, Dec. 1979.

16. C. Y. Huang and W. S. Chen, "Input impedance of a probe-fed cylindrical annular-ring microstrip antenna," *Microwave Opt. Technol. Lett.*, vol. 16, pp. 41–44, Sept. 1997.

17. D. M. Pozar, "A reciprocity method of analysis for printed slot and slot-coupled microstrip antenna," *IEEE Trans. Antennas Propagat.*, vol. 34, pp. 1439–1446, Dec. 1986.

18. R. B. Tsai, K. L. Wong, and H. C. Su, "Analysis of a microstrip-line-fed radiating slot on a cylindrical surface," *Microwave Opt. Technol. Lett.*, vol. 8, pp. 193–196, Mar. 1995.

19. J. R. Wait, *Electromagnetic Radiation from Cylindrical Structures*, Peter Peregrinus, London, 1988.

20. W. Y. Tam, A. K. Y. Lai, and K. M. Luk, "Full-wave analysis of aperture-coupled cylindrical rectangular microstrip antenna," *Electron. Lett.*, vol. 30, pp. 1461–1462, Sept. 1, 1994.

21. K. L. Wong and J. S. Chen, "Cavity-model analysis of a slot-coupled cylindrical-rectangular microstrip antenna," *Microwave Opt. Technol. Lett.*, vol. 9, pp. 124–127, June 20, 1995.

22. S. B. Cohn, "Slot-line on a dielectric substrate," *IEEE Trans. Microwave Theory Tech.*, vol. 17, pp. 768–778, Oct. 1969.

23. L. R. Zeng and Y. Wang, "Accurate solutions of elliptical and cylindrical striplines and microstrip lines," *IEEE Trans. Microwave Theory Tech.*, vol. 34, pp. 259–264, Feb. 1986.

24. J. S. Rao and B. N. Das, "Impedance of off-centered stripline fed series slot," *IEEE Trans. Antennas Propagat.*, vol. 26, pp. 893–895, Nov. 1978.

25. M. Kumar and B. N. Das, "Coupled transmission lines," *IEEE Trans. Microwave Theory Tech.*, vol. 25, pp. 7–10, Jan. 1977.

26. E. O. Hammerstad and O. Jensen, "Accurate models for microstrip computer-aided design," *1980 IEEE MTT-S International Symposium Digest*, pp. 407–409.

27. J. S. Chen and K. L. Wong, "Input impedance of a slot-coupled cylindrical-circular microstrip patch antenna," *Microwave Opt. Technol. Lett.*, vol. 11, pp. 21–24, Jan. 1996.

28. C. Baumer, "Analysis of slot-coupled, circular microstrip patch antenna," *Electron. Lett.*, vol. 28, pp. 1454–1455, July 20, 1992.

29. C. Y. Huang and K. L. Wong, "Analysis of a slot-coupled cylindrical-rectangular microstrip antenna," *Microwave Opt. Technol. Lett.*, vol. 8, pp. 251–253, Apr. 5, 1995.

30. P. L. Sullivan and D. H. Schaubert, "Analysis of an aperture coupled microstrip antenna," *IEEE Trans. Antennas Propagat.*, vol. 34, pp. 977–984, Aug. 1986.

31. F. C. Silva, S. B. A. Fonseca, A. J. M. Soares, and A. J. Giarola, "Analysis of microstrip antennas on circular-cylindrical substrates with a dielectric overlay," *IEEE Trans. Antennas Propagat.*, vol. 39, pp. 1398–1403, Sept. 1991.

32. K. L. Wong, C. Y. Huang, and Y. H. Liu, "Analysis of a microstripline-fed cylindrical-rectangular microstrip antenna using generalized transmission line model," *Proc. Natl. Sci. Counc. ROC(A)*, vol. 19, pp. 452–456, Sept. 1995.

33. M. D. Deshpande and M. C. Bailey, "Input impedance of microstrip antennas," *IEEE Trans. Antennas Propagat.*, vol. 30, pp. 645–650, July 1982.

34. R. E. Munson, "Conformal microstrip antennas and microstrip phased arrays," *IEEE Trans. Antennas Propagat.*, vol. 22, pp. 74–78, Jan. 1974.

35. C. Yang and T. Z. Ruan, "Radiation characteristics of wraparound microstrip antenna on cylindrical body," *Electron. Lett.*, vol. 29, pp. 512–514, Mar. 18, 1993.

36. K. L. Wong and S. Y. Ke, "Characteristics of the cylindrical wraparound microstrip patch antenna," *Proc. Natl. Sci. Counc. ROC(A)*, vol. 17, pp. 438–442, Nov. 1993.

37. K. L. Wong and S. Y. Ke, "Cylindrical-rectangular microstrip antenna for circular polarization," *IEEE Trans. Antennas Propagat.*, vol. 41, pp. 246–249, Feb. 1993.

38. K. R. Carver and J. W. Mink, "Microstrip antenna technology," *IEEE Trans. Antennas Propagat.*, vol. 29, pp. 2–24, Jan. 1981.

39. S. Y. Ke and K. L. Wong, "Cross-polarization characteristics of rectangular microstrip patch antennas on a cylindrical surface," *Microwave Opt. Technol. Lett.*, vol. 6, pp. 911–914, Dec. 20, 1993.

40. S. T. Fang, S. C. Pan, and K. L. Wong, "Crosspolarisation characteristics of cylindrical triangular microstrip antennas," *Electron. Lett.*, vol. 34, pp. 6–7, Jan. 8, 1998.

41. M. L. Oberhart, Y. T. Lo, and R. Q. H. Lee, "New simple feed network for an array module of four microstrip elements," *Electron. Lett.*, vol. 23, pp. 436–437, Apr. 23, 1987.

42. C. Y. Huang and K. L. Wong, "Input impedance and mutual coupling of probe-fed cylindrical-circular microstrip patch antennas," *Microwave Opt. Technol. Lett.*, vol. 11, pp. 260–263, Apr. 5, 1996.

CHAPTER FIVE

Characteristics of Spherical and Conical Microstrip Antennas

5.1 INTRODUCTION

Microstrip antennas mounted on spherical or conical ground surfaces are discussed in this chapter. For microstrip antennas mounted on a spherical host, various characteristics of the microstrip antennas with circular and annular-ring patches are analyzed. Theoretical formulations using the full-wave approach [1–4], cavity-model analysis [5,6], and GTLM theory [7,8] are described. Cross-polarization characteristics [13,14] of spherical circular and annular-ring microstrip antennas are also analyzed. For microstrip antennas mounted on a conical ground plane, circular [9], annular-ring (wraparound) [10], and annular-ring-segment [11,12] microstrip patches have been studied, mainly using cavity-model analysis. Theoretical formulation and numerical results are presented and discussed in subsequent sections.

5.2 SPHERICAL MICROSTRIP ANTENNAS

Due to their symmetrical structures, circular and annular-ring patches are most suitable for mounting on a spherical body. Several theoretical approaches, such as full-wave analysis, the cavity-model method, and GTLM theory have been used to study spherical microstrip antennas. The related spherical microstrip structures for the source-free case were discussed in Chapter 3. In this chapter a case with probe-feed excitation is studied. A formulation of the input impedance and far-zone radiated fields is given. Numerical results for input impedance, especially the excited surface current distribution in the circular patch at the TM_{11} mode and in the annular-ring patch at the TM_{12} mode, are calculated and analyzed. With the surface current distribution calculated, the mode degeneracy problem in an annular-ring microstrip antenna with excitation of the TM_{12} mode is discussed. Radiation patterns of copolarized and cross-polarized radiated fields for both circular and annular-ring patches are also shown.

5.2.1 Full-Wave Solution

Figure 5.1 shows configurations of spherical circular and annular-ring microstrip antennas. The circular patch has a radius of r_d $(= b\theta_0)$, while the annular-ring patch has an inner radius of r_1 $(= b\theta_1)$ and an outer radius of r_2 $(= b\theta_2)$. The spherical substrate has a thickness of h and a relative permittivity of ϵ_1. The outer medium of $r > b$ is again free space with permittivity ϵ_0 and permeability μ_0. The microstrip patch is excited by a probe feed at (θ_p, ϕ_p) with unit-amplitude current density given by

$$\mathbf{J}^P(r') = \frac{\hat{r}\delta(\theta' - \theta_p)\delta(\phi' - \phi_p)}{r'^2 \sin\theta'}, \qquad a \le r' \le b. \tag{5.1}$$

Applying the boundary condition that the total tangential electric field must vanish on the patch we obtain,

$$\mathbf{E}^D(\theta, \phi) + \mathbf{E}^P(\theta, \phi) = 0, \tag{5.2}$$

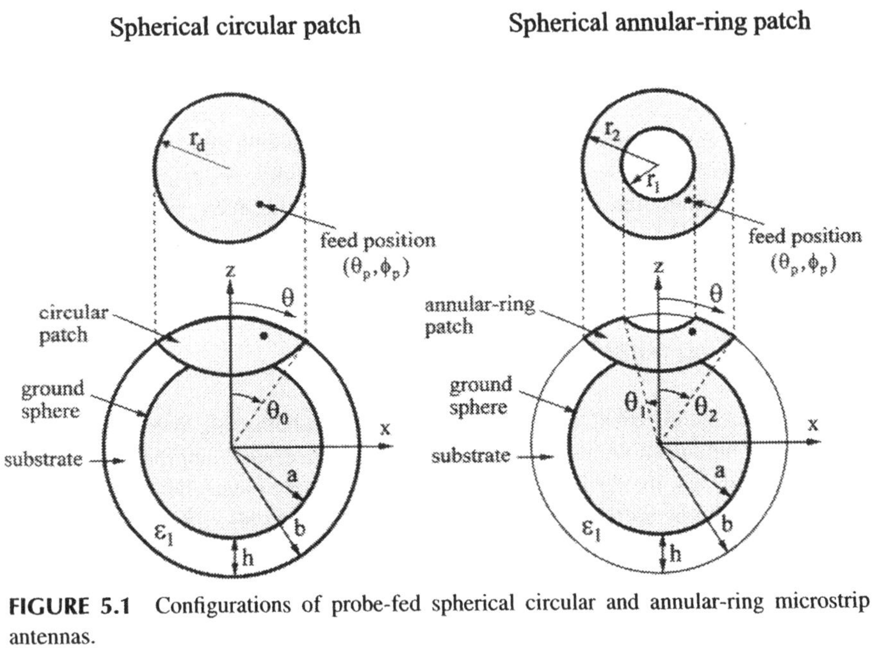

FIGURE 5.1 Configurations of probe-fed spherical circular and annular-ring microstrip antennas.

where $\mathbf{E}^D(\theta, \phi)$ is the electric field due to the patch current and $\mathbf{E}^P(\theta, \phi)$ is the electric field due to the probe current with the patch being absent. From the results derived in (3.79), in the spectral domain we have $\mathbf{E}^D(\theta, \phi)$ given by

$$\tilde{E}^D = \bar{\bar{\tilde{Y}}}^{-1}\tilde{J}, \tag{5.3}$$

with

$$\bar{\bar{\tilde{Y}}} = \begin{bmatrix} Y_{11} & 0 \\ 0 & Y_{22} \end{bmatrix}. \tag{5.4}$$

The elements Y_{11} and Y_{22} for the circular and annular-ring patches are derived, respectively, in Sections 3.2.3 and 3.3.1; $\mathbf{E}^D(\theta, \phi)$ and $\mathbf{J}$ are the vector Legendre transforms, defined in Chapter 3, of $\mathbf{E}^D(\theta, \phi)$ and the patch surface current $\mathbf{J}(\theta, \phi)$, respectively.

To find the field $E^P(\theta, \phi)$, we consider the spherical microstrip structure shown in Figure 5.1 without the microstrip patch excited by a line current source of (5.1). Since the probe feed is treated here as a line current pointing along the $\hat{r}$ direction, the Green's function ϕ_r, a scalar potential function, is needed in the formulation, which has been derived in (3.57). Considering the absence of the microstrip patch, the potential function ϕ_r^s in the substrate layer ($a < r < b$) can be written as

$$\phi_r^s = \cos m(\phi - \phi_p) \sum_{n=m}^{\infty} S_g(n, m) P_n^m(\cos\theta) P_n^m(\cos\theta_p)$$
$$\times [I_n(r) + a_n \hat{H}_n^{(1)}(kr) + b_n \hat{H}_n^{(2)}(kr)], \tag{5.5}$$

with

$$I_n(r) = r \int_a^b \frac{\mu_0 G_n(r, r')}{r'} dr', \tag{5.6}$$

where $S_g(n, m)$ is given in (3.58) and $G_n(r, r')$ is shown by (3.55); $I_n(r)$ accounts for the presence of the line current source of (5.1), and the term $a_n \hat{H}_n^{(1)}(kr) + b_n \hat{H}_n^{(2)}(kr)$ is for the presence of the grounded spherical substrate. For the air region of $r > b$, the potential function ϕ_r^a, which represents an outgoing wave, is written as

$$\phi_r^a = \cos m(\phi - \phi_p) \sum_{n=m}^{\infty} S_g(n, m) P_n^m(\cos\theta) P_n^m(\cos\theta_p) c_n \hat{H}_n^{(2)}(k_0 r). \tag{5.7}$$

By imposing the boundary conditions at $r = a$ and b, the unknown coefficients a_n, b_n, and c_n of the potential functions above are derived as

$$a_n = \frac{\Delta_a}{\Delta}, \qquad b_n = \frac{\Delta_b}{\Delta}, \qquad c_n = \frac{\Delta_c}{\Delta}, \tag{5.8}$$

where

$$\Delta_a = \begin{vmatrix} -I_n^{H'}(a)/k & \hat{H}_n^{(2)'}(ka) & 0 \\ -I_n^{J'}(b)/k & \hat{H}_n^{(2)'}(kb) & -\sqrt{\epsilon_1}\,\hat{H}_n^{(2)'}(k_0 b) \\ -I_n^J(b) & \hat{H}_n^{(2)}(kb) & -\hat{H}_n^{(2)}(k_0 b) \end{vmatrix}, \tag{5.9}$$

$$\Delta_b = \begin{vmatrix} \hat{H}_n^{(1)'}(ka) & -I_n^{H'}(a)/k & 0 \\ \hat{H}_n^{(1)'}(kb) & -I_n^{J'}(b)/k & -\sqrt{\epsilon_1}\,\hat{H}_n^{(2)'}(k_0 b) \\ \hat{H}_n^{(1)}(kb) & -I_n^J(b) & -\hat{H}_n^{(2)}(k_0 b) \end{vmatrix}, \tag{5.10}$$

$$\Delta_c = \begin{vmatrix} \hat{H}_n^{(1)'}(ka) & \hat{H}_n^{(2)'}(ka) & -I_n^{H'}(a)/k \\ \hat{H}_n^{(1)'}(kb) & \hat{H}_n^{(2)'}(kb) & -I_n^{J'}(b)/k \\ \hat{H}_n^{(1)}(kb) & \hat{H}_n^{(2)}(kb) & -I_n^J(b) \end{vmatrix}, \tag{5.11}$$

$$\Delta = \begin{vmatrix} \hat{H}_n^{(1)'}(ka) & \hat{H}_n^{(2)'}(ka) & 0 \\ \hat{H}_n^{(1)'}(kb) & \hat{H}_n^{(2)'}(kb) & -\sqrt{\epsilon_1}\hat{H}_n^{(2)'}(k_0 b) \\ \hat{H}_n^{(1)}(kb) & \hat{H}_n^{(2)}(kb) & -\hat{H}_n^{(2)}(k_0 b) \end{vmatrix}, \tag{5.12}$$

with

$$I_n^H(r) = \frac{\mu_0}{jk}\,\hat{J}_n(kr)\int_a^b \frac{\hat{H}_n^{(2)}(kr')}{r'^2}\,dr', \tag{5.13}$$

$$I_n^J(r) = \frac{\mu_0}{jk}\,\hat{H}_n^{(2)}(kr)\int_a^b \frac{\hat{J}_n(kr')}{r'^2}\,dr'. \tag{5.14}$$

Then, by substituting ϕ_r^a into (3.31) and performing a vector Legendre transform, the spectral amplitude of the electric field $\tilde{E}^P$ at $r = b$ surface can be obtained and written as

$$\tilde{E}^P\Big|_{r=b} = \begin{bmatrix} \dfrac{c_n k_0 e^{-jm\phi_p}}{j\omega\mu_0\epsilon_0 b} S_g(n, m) P_n^m(\cos\theta_p)\hat{H}_n^{(2)'}(k_0 b) \\ 0 \end{bmatrix}. \tag{5.15}$$

Substituting (5.3) and (5.15) into (5.2) and applying Galerkin's moment-method procedure, we obtain the following matrix equation:

$$[Z_{ij}][I_j] = [V_i], \tag{5.16}$$

with

$$Z_{ij} = \sum_{n=m}^{\infty}\left[\frac{2n(n+1)(n+m)!}{(2n+1)(n-m)!}\,\tilde{J}_i^*\bar{\bar{\tilde{Y}}}^{-1}\tilde{J}_j\right], \tag{5.17}$$

$$V_i = -\sum_{n=m}^{\infty} \left[\frac{2n(n+1)(n+m)!}{(2n+1)(n-m)!} \tilde{J}_i^* \tilde{E}^p \right]. \tag{5.18}$$

The asterisks in the expressions above denote the complex conjugate transpose, and I_j in (5.16) is the unknown coefficient of the jth expansion function J_j, with the total patch surface current density in spectral domain shown by

$$\tilde{J}(n) = \begin{bmatrix} \tilde{J}_U(n) \\ \tilde{J}_L(n) \end{bmatrix} = \sum_{j=1}^{N} I_j \tilde{J}_j(n). \tag{5.19}$$

It is also noted that for circular and annular-ring patches, cavity-model functions are used as expansion functions, whose expressions have been derived in (3.94)–(3.95) and (3.122)–(3.123), respectively. Once the I_j are determined, the surface current distributions in the microstrip patch can readily be obtained.

For calculation of the input impedance, we apply the formula

$$Z_{\text{in}} = -\int_v (\mathbf{E}_r^D + \mathbf{E}_r^P)\mathbf{J}^P \, dv, \tag{5.20}$$

where $\mathbf{E}_r^D$ and $\mathbf{E}_r^P$ are the $\hat{r}$ components of the electric field in the substrate layer due to the patch current and probe current, respectively; $\mathbf{J}^P$ the probe current given in (5.1); and v the volume over the probe. By applying the formulation in Chapter 3, the electric field $\mathbf{E}_r^D$ can be derived as

$$E_r^D = \frac{1}{j\omega\epsilon_0\epsilon_1} \sum_{m=0}^{\infty} \cos m\phi \sum_{n=m}^{\infty} \frac{n(n+1)}{r^2} [A_n \hat{H}_n^{(1)}(kr) + B_n \hat{H}_n^{(2)}(kr)] P_n^m(\cos\theta), \tag{5.21}$$

with

$$A_n = \frac{j\omega\epsilon_0\epsilon_1 b}{kY_{11}[\hat{H}_n^{(1)\prime}(kb) - \hat{H}_n^{(1)\prime}(ka)\hat{H}_n^{(2)\prime}(kb)/\hat{H}_n^{(2)\prime}(ka)]} \sum_{j=1}^{N} I_j J_{jU}(n), \tag{5.22}$$

$$B_n = -A_n \frac{\hat{H}_n^{(1)\prime}(ka)}{\hat{H}_n^{(2)\prime}(ka)}, \tag{5.23}$$

and the electric field $\mathbf{E}_r^P$ is written as

$$E_r^P = \frac{1}{j\omega\epsilon_0\epsilon_1} \sum_{m=0}^{\infty} \cos m(\phi - \phi_p) \sum_{n=m}^{\infty} \left\{ \frac{n(n+1)}{r^2} S_g(n,m) P_n^m(\cos\theta) P_n^m(\cos\theta_p) \right.$$
$$\left. \times [I_n(r) + a_n \hat{H}_n^{(1)}(kr) + b_n \hat{H}_n^{(2)}(kr)] \right\}, \tag{5.24}$$

where a_n and b_n are as given in (5.8).

For the far-zone radiated fields from the spherical microstrip antennas, we have [1]

$$E_\theta(\theta, \phi) = \frac{be^{-jk_0 r}}{r} \sum_{m=0}^{\infty} \cos m(\phi - \phi_p) \sum_{n=m}^{\infty} \times \left[\frac{-j^n \tilde{J}_U(n) P_n^{m\prime}(\cos\theta) \sin\theta}{Y_{11} \hat{H}_n^{(2)\prime}(k_0 b)} - \frac{m j^{n+1} \tilde{J}_L(n) P_n^m(\cos\theta)}{Y_{22} \hat{H}_n^{(2)}(k_0 b) \sin\theta} \right], \tag{5.25}$$

$$E_\phi(\theta, \phi) = \frac{be^{-jk_0 r}}{r} \sum_{m=0}^{\infty} \sin m(\phi - \phi_p) \sum_{n=m}^{\infty} \times \left[\frac{m j^n \tilde{J}_U(n) P_n^m(\cos\theta)}{Y_{11} \hat{H}_n^{(2)\prime}(k_0 b) \sin\theta} - \frac{j^{n+1} \tilde{J}_L(n) P_n^{m\prime}(\cos\theta) \sin\theta}{Y_{22} \hat{H}_n^{(2)}(k_0 b)} \right], \tag{5.26}$$

where $\tilde{J}_U$ and $\tilde{J}_L$ [see (5.19) or (3.97)] are the spectral amplitudes of the excited surface current on the microstrip patch. Once E_θ and E_ϕ are obtained, the radiation patterns of the copolarized and cross-polarized fields can be evaluated. Using the third definition of Ludwig [15], the copolarized field E_{copol} and cross-polarized field E_{xpol} are expressed as

$$E_{\text{copol}} = E_\theta \cos\phi - E_\phi \sin\phi\,, \tag{5.27}$$

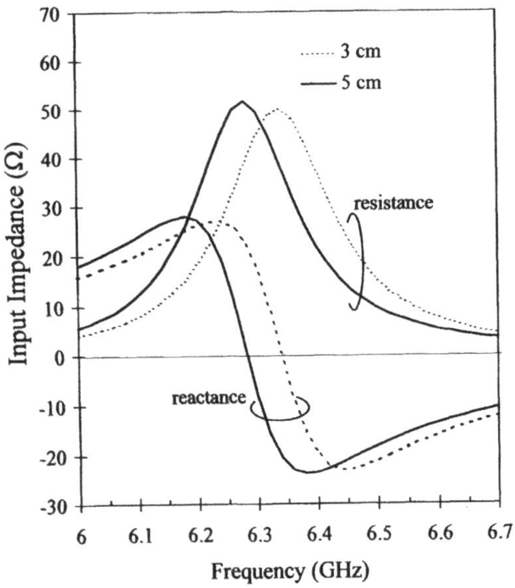

FIGURE 5.2 Input impedance versus frequency for a spherical circular microstrip antenna; $a = 5$ cm, 3 cm, $h = 1.0$ mm, $\epsilon_1 = 2.65$, $r_d = 8.3$ mm, $\theta_p/\theta_0 = 0.267$.

$$E_{\text{xpol}} = E_\theta \sin\phi + E_\phi \cos\phi \,. \tag{5.28}$$

From the expressions above, the cross-polarization level (XPL) [16,17], defined to be the maximum magnitude of E_{copol} to the maximum magnitude of E_{xpol} in a specified plane, can also be determined. The theoretical results of the input impedance evaluated from (5.20), the surface current distributions in the microstrip patch obtained from (5.19), and the radiation characteristics calculated from (5.25)–(5.28) for the circular and annular-ring microstrip patches are described below.

A. Circular Patch Numerical results for input impedance at the TM_{11} mode are shown first in Figure 5.2. In the calculation, the cavity-model basis functions [(3.94)–(3.95)] for the unknown patch surface current density are selected to include TM_{mn}, $m = 0$ to 5, $n = 1$ to 5, which show good convergence results [1]. The resonant input resistance occurs at lower frequencies for larger sphere radii, in general similar to the results observed for cylindrical microstrip antennas. The input impedance level is also seen to increase with increasing sphere radius.

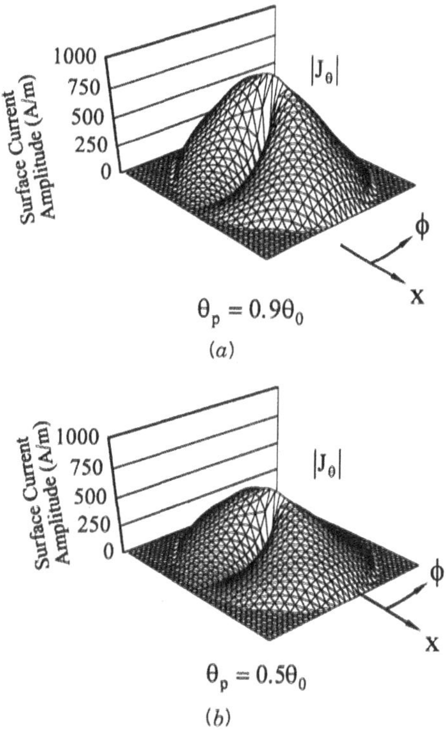

FIGURE 5.3 $\hat{\theta}$-Directed current distribution of the spherical circular microstrip antenna at 2.94 GHz (resonant frequency of the TM_{11} mode); $a = 3$ cm, $h = 1.59$ mm, $\epsilon_1 = 2.47$, $r_d = 18.8$ mm, $\phi_p = 0°$. (*a*) $\theta_p = 0.9\theta_0$; (*b*) $\theta_p = 0.5\theta_0$.

Figures 5.3 and 5.4 present, respectively, $\hat{\theta}$- and $\hat{\phi}$-directed patch surface current amplitudes for various feed positions. The current amplitude is seen to decrease as the feed position moves closer to the center of the circular patch. In general, there is no mode degeneracy problem for a circular microstrip antenna excited at the TM_{11} mode.

For the radiation characteristics, it is first noted that from the cross-polarization definition of (5.27)–(5.28), there is no cross-polarized field in the E plane ($\phi = 0°$ plane). Therefore, we study primarily cross-polarization results in the H plane ($\phi = 90°$ plane) of the spherical circular microstrip antenna with various sphere radii. Figure 5.5 presents typical results for a spherical circular patch antenna excited at the TM_{11} mode. It is seen that as the sphere radius increases, the cross-polarized radiation also increases. This suggests that the spherical circular patch antenna has a better linear polarization characteristic than the planar circular patch antenna.

The dependence of copolarized and cross-polarized radiation patterns on the feed position is also calculated. A typical case is shown in Figure 5.6. It is found that as the feed position moves toward the edge of the patch, the cross-polarized fields can be reduced. The cross-polarization results for different substrate

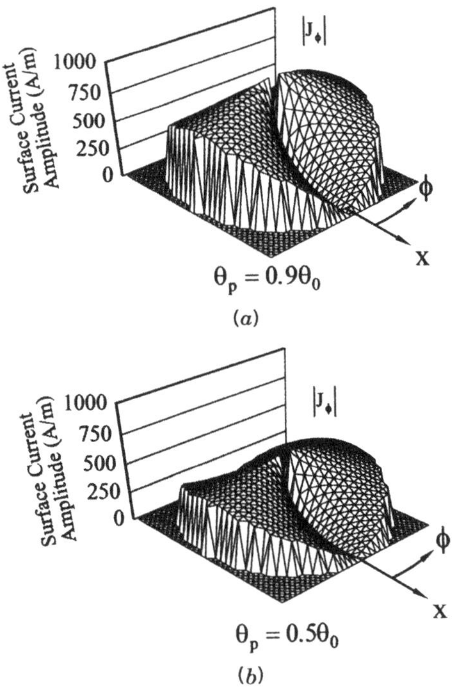

FIGURE 5.4 $\hat{\phi}$-Directed current distribution of the spherical circular microstrip antenna at 2.94 GHz (resonant frequency of the TM_{11} mode); antenna parameters are the same as in Figure 5.3. (*a*) $\theta_p = 0.9\theta_0$; (*b*) $\theta_p = 0.5\theta_0$.

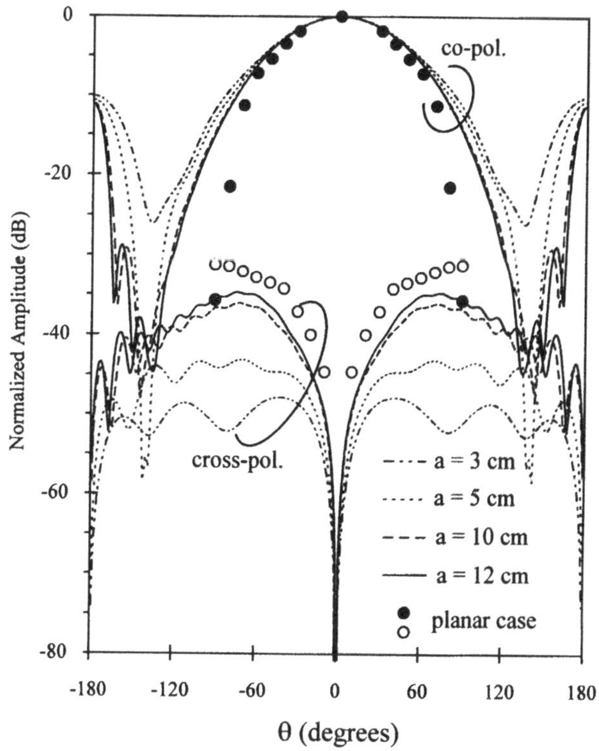

FIGURE 5.5 Radiation patterns of copolarized and cross-polarized fields in the H plane; $h = 1.59$ mm, $\epsilon_1 = 2.32$, $r_d = 10.68$ mm, $\theta_p = 0.9\theta_0$. (From Ref. [13], © 1993 John Wiley & Sons, Inc.)

thicknesses and permittivities are studied and cross-polarization levels are calculated. Figures 5.7 and 5.8 show the results as a function of resonant frequency for $a = 3$ and 5 cm, respectively. For fixed substrate thickness and permittivity, different resonant frequencies in the figure correspond to different circular patch radii. From the results it is observed that smaller substrate thickness and higher substrate permittivity can improve the cross-polarization level. Also, the cross-polarization level decreases with increasing resonant frequency. The results for several other planes are also calculated and show dependence similar to that presented here.

B. Annular-Ring Patch The input impedance for the annular-ring microstrip antenna excited at the TM_{12} mode is calculated and shown in Figure 5.9. For numerical computation to obtain good convergent solutions within the desired operating frequency band of the TM_{12} mode, the cavity-model basis functions [(3.122)–(3.123)] are selected to include TM_{mn} modes, $m = 0$ to 5, $n = 1$ to 8 (i.e., total 48 basis functions). The input impedances obtained for different sphere radii are presented in Figure 5.9. It is observed that variation of the input impedance

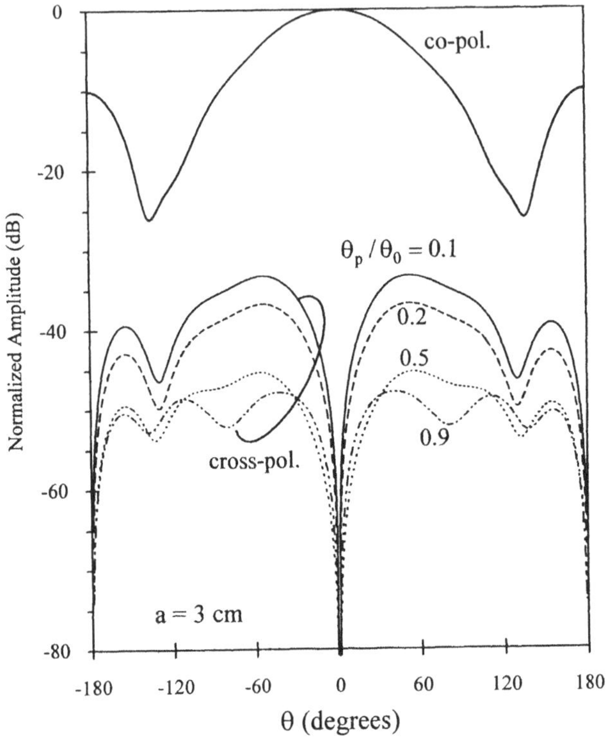

FIGURE 5.6 Radiation patterns of copolarized and cross-polarized fields in the H plane for $\theta_p/\theta_0 = 0.1$, 0.2, 0.5, 0.9; $a = 3$ cm, $h = 1.59$ mm, $\epsilon_1 = 2.32$, $r_d = 10.68$ mm, $f =$ 5.1 GHz. (From Ref. [13], © 1993 John Wiley & Sons, Inc.)

with sphere radius becomes small for $a > 30$ cm. The results of $a = 50$ cm obtained in the present method also agree in general with GTLM data [8]. Figure 5.10 also presents the resonant input resistance as a function of the feed position. From the results, the optimal feed position for impedance matching can easily be determined.

The surface current distributions on the annular-ring patch are also calculated and analyzed. Figures 5.11 and 5.12 show a typical case of patch surface current distributions due to, respectively, the contribution of all modes (TM_{0n} to TM_{5n}, $n = 1$ to 8) and the TM_{12} mode alone. The patch is excited at the resonant frequency of the TM_{12} mode ($f = 3.23$ GHz). From a comparison of the J_θ and J_ϕ components of all modes (Figure 5.11) and of the TM_{12} mode only (Figure 5.12), it is seen that there are several other modes excited with the excitation of the TM_{12} mode, and these degenerate modes also contribute significantly to the patch current. From examining the current distribution due to all individual modes, it is found that in addition to the TM_{12} mode, the TM_{02} and TM_{22} modes dominate the contribution to the $\hat{\theta}$ component of the patch current, and the TM_{51} and TM_{22} modes dominate the contribution to the $\hat{\phi}$ component of the patch current.

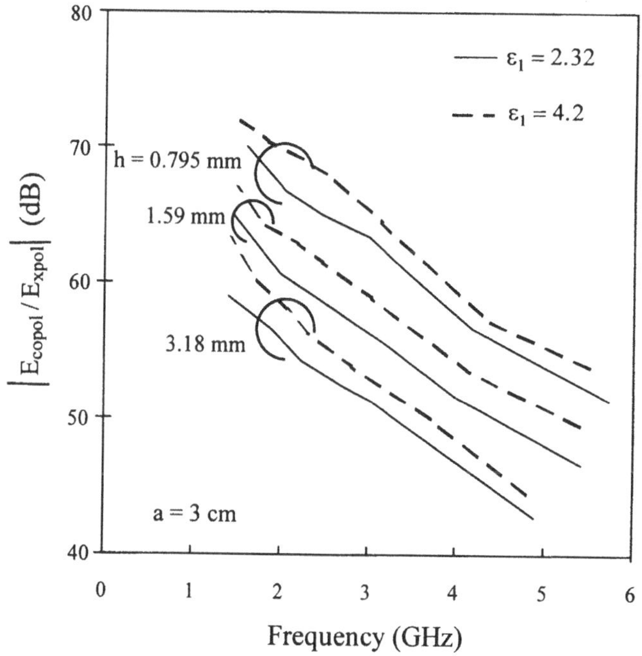

FIGURE 5.7 Cross-polarization level as a function of resonant frequency; $a = 3$ cm, $\epsilon_1 = 2.32$, 4.2, $h = 0.795$, 1.59, 3.18 mm, $\theta_p = 0.9\theta_0$. (From Ref. [13], © 1993 John Wiley & Sons, Inc.)

Contributions from other modes are relatively small. The current distributions of the TM_{02}, TM_{51}, and TM_{22} modes are shown in Figures 5.13 to 5.15, respectively. Note that the J_ϕ component of the TM_{02} mode is zero and is therefore not plotted in Figure 5.12. To show this more clearly, in Figure 5.16 we present the current distribution due to the contributions of the TM_{12}, TM_{02}, TM_{51}, and TM_{22} modes, which can be seen to be about the same as in Figure 5.11, due to the contribution of all modes. The presence of these degenerate modes may cause a deterioration in the radiated cross-polarization level and can cause significant effects in the input impedance level.

The copolarized and cross-polarized radiation patterns in the H plane are first calculated by assuming that the feed location is at $\phi_p = 0°$, $r_p = 1.1r_1$. A typical result is presented in Figure 5.17, where the copolarized and cross-polarized radiation patterns at the resonant frequency of the sphere radii of $a = 3$, 5, and 10 cm are shown. The field amplitudes in each case are normalized with respect to the copolarized field at $\theta = 0°$, and the frequencies $f = 15.8$, 15.4, and 15.2 GHz are the resonant frequencies for $a = 3$, 5, and 10 cm, respectively. From the results it is seen that the cross-polarization radiation increases with increasing sphere radius. This behavior is similar to that observed for a spherical circular microstrip antenna excited at the TM_{11} mode (see Section 5.2.1A).

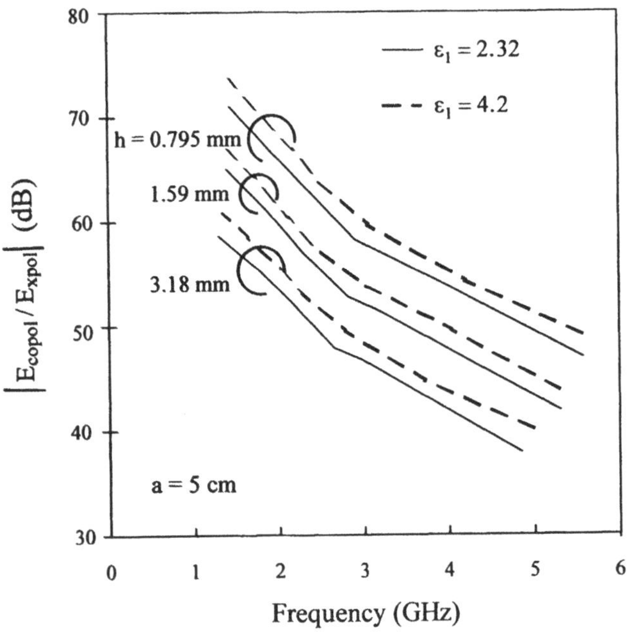

FIGURE 5.8 Cross-polarization level as a function of resonant frequency; $a = 5$ cm, $\epsilon_1 = 2.32$, 4.2, $h = 0.795$, 1.59, 3.18 mm, $\theta_p = 0.9\theta_0$. (From Ref. [13], © 1993 John Wiley & Sons, Inc.)

The copolarized and cross-polarized radiation patterns for various feed positions are also analyzed. The results are shown in Figure 5.18. It is observed that the cross-polarized field increases as the feed position moves toward the center of the annular-ring patch. The cross-polarization level as a function of the feed position is also calculated and presented in Figure 5.19. It can be seen that the cross-polarization level varies slightly for the feed position near the inner and outer edges of the patch, and decreases drastically when the feed position moves close to the patch center. This indicates that to have a better cross-polarization level, the feed position should be chosen to be close to the inner or outer edges of the patch. It should also be noted that for the feed position in the region of about 1.51 to $1.58r_1$ (see Figure 5.19), the cross-polarization level decreases more drastically and is below 0 dB; that is, the peak cross-polarized field is greater than the peak copolarized field. This seems to suggest that in this case other modes are more strongly excited than the TM_{12} mode. The cross-polarization characteristics in the planes of various values of ϕ are also discussed. The cross-polarization level obtained with $r_p = 1.1r_1$ is shown in Figure 5.20. It is seen that the cross-polarization level is minimum in the plane of $\phi = 45°$ and has a maximum value in the plane of $\phi = 0°$, where there is no cross-polarized radiation field.

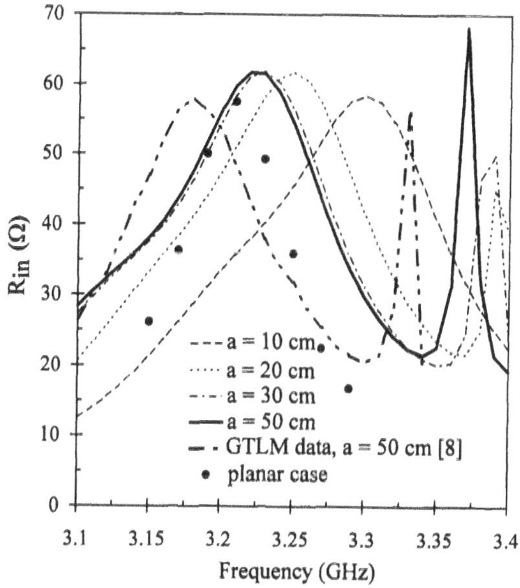

(*a*)

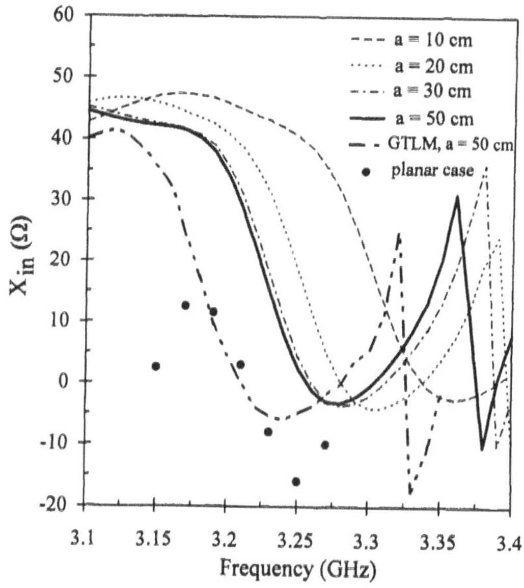

(*b*)

FIGURE 5.9 Input impedance versus frequency for a spherical annular-ring microstrip antenna; $h = 1.59$ mm, $\epsilon_1 = 2.2$, $r_1 = 3$ cm, $r_2 = 6$ cm, r_p $(=b\theta_p) = 3.4$ mm, $\phi_p = 0°$. (*a*) Input resistance; (*b*) input reactance. (From Ref. [3], © 1994 John Wiley & Sons, Inc.)

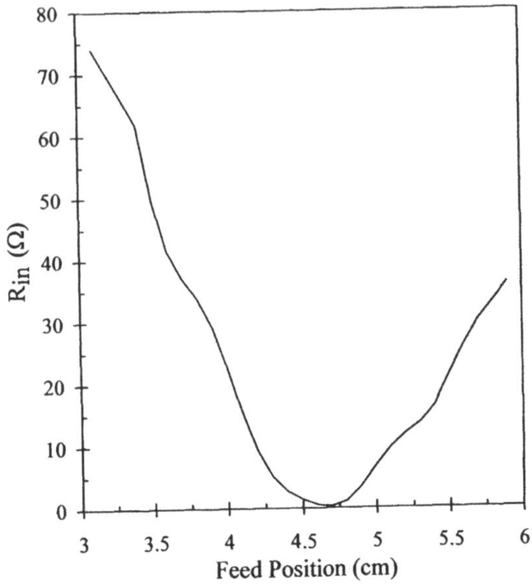

FIGURE 5.10 Resonant input resistance as a function of the feed position; $a = 30$ cm, $f = 3.23$ GHz. Other antenna parameters are given in Figure 5.5.

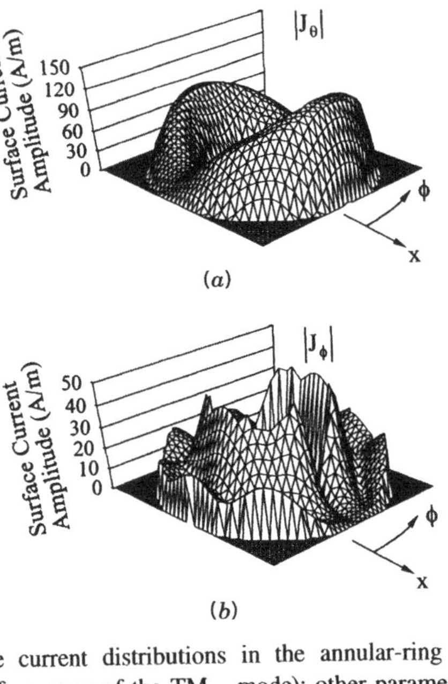

FIGURE 5.11 Surface current distributions in the annular-ring patch for $a = 30$ cm, $f = 3.23$ GHz (resonant frequency of the TM_{12} mode); other parameters are the same as in Figure 5.5. (*a*) $\hat{\theta}$-Directed current distribution due to contribution of all modes; (*b*) $\hat{\phi}$-directed current distribution due to contribution of all modes. (From Ref. [3], © 1994 John Wiley & Sons, Inc.)

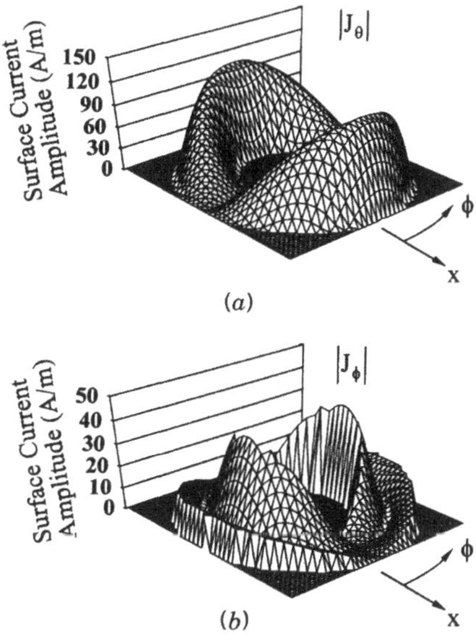

FIGURE 5.12 (*a*) $\hat{\theta}$-Directed and (*b*) $\hat{\phi}$-directed current distributions due to contribution of the TM_{12} mode for the case shown in Figure 5.7. (From Ref. [3], © 1994 John Wiley & Sons, Inc.)

5.2.2 Cavity-Model Solution

Cavity-model analysis of a probe-fed spherical circular microstrip antenna is described first. Expressions for the far-zone radiated fields and the input impedance are then derived. Numerical results for the radiation patterns and input impedance for the antenna excited at the TM_{11} mode are presented and discussed. We begin with the theoretical formulation described in Section 4.3 and the case of a source-free condition. By referring to the geometry in Figure 5.1, the cavity field inside the region (a spherical cavity) between the microstrip patch and the ground sphere is derived as [5,18]

$$E_r = E_0 J_m(k_{mp} b\theta) \cos m\phi \, , \tag{5.29}$$

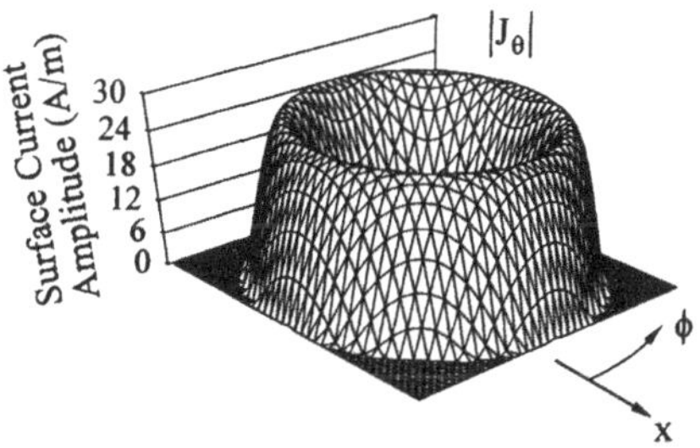

FIGURE 5.13 $\hat{\theta}$-Directed current distribution due to contribution of the TM_{02} mode for the case shown in Figure 5.11. (From Ref. [3], © 1994 John Wiley & Sons, Inc.)

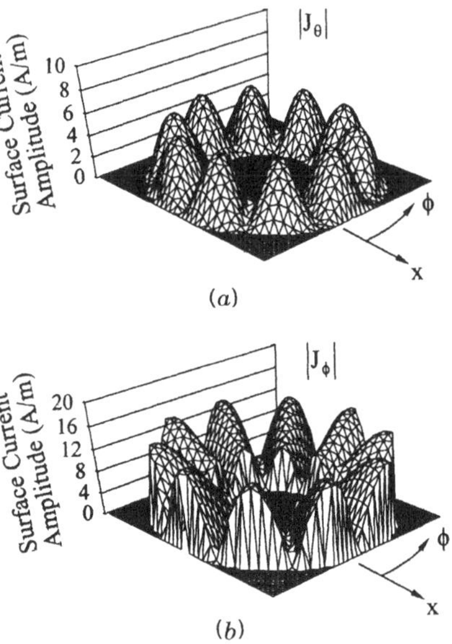

FIGURE 5.14 (*a*) $\hat{\theta}$-Directed and (*b*) $\hat{\phi}$-directed current distributions due to contribution of the TM_{51} mode for the case shown in Figure 5.11. (From Ref. [3], © 1994 John Wiley & Sons, Inc.)

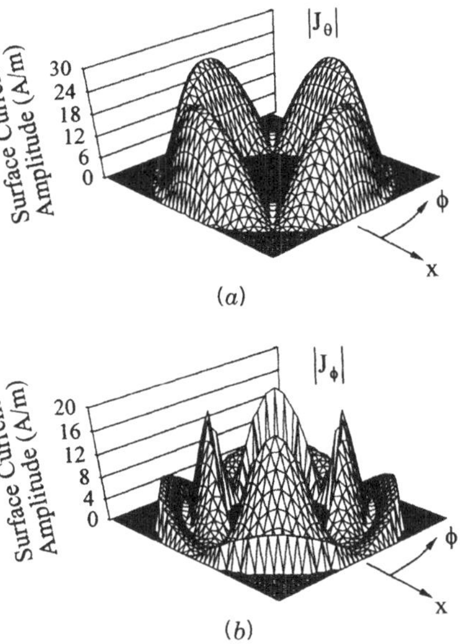

FIGURE 5.15 (*a*) $\hat{\theta}$-Directed and (*b*) $\hat{\phi}$-directed current distributions due to contribution of the TM_{22} mode for the case shown in Figure 5.11. (From Ref. [3], © 1994 John Wiley & Sons, Inc.)

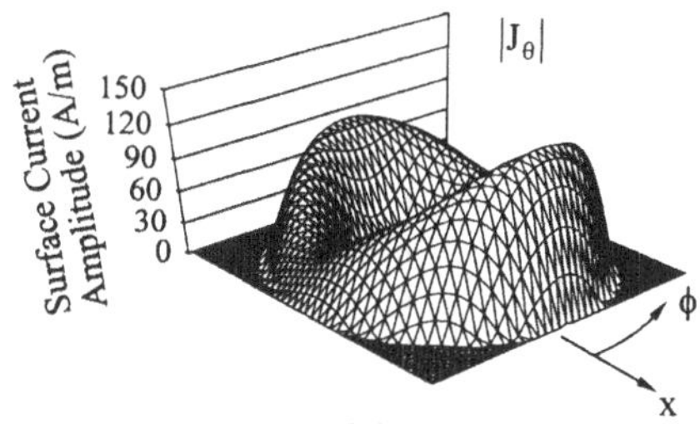

(a)

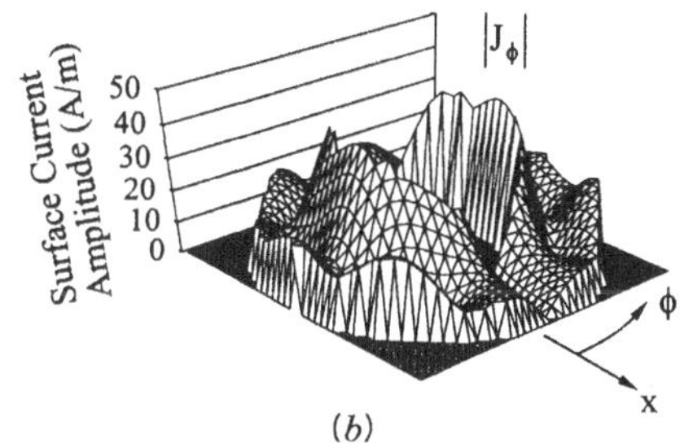

(b)

FIGURE 5.16 (a) $\hat{\theta}$-Directed and (b) $\hat{\phi}$-directed current distributions due to contribution of the TM_{12}, TM_{02}, TM_{51}, and TM_{22} mode for the case shown in Figure 5.11. (From Ref. [3], © 1994 John Wiley & Sons, Inc.)

with

$$J'_m(k_{mp}b\theta_0) = 0\,, \tag{5.30}$$

$$f_{mp} = \frac{c}{2\pi\sqrt{\epsilon_1}}\,k_{mp}\,, \tag{5.31}$$

where k_{mp} and f_{mp} are, respectively, the wavenumber and resonant frequency of the TM_{mp} mode; m denotes the azimuthal mode number and p is for the radial mode number. The far-zone radiated fields have also been derived by considering the equivalent magnetic current, obtained from (5.29), along the circumference of the cavity radiating in the presence of a spherical body and using an approximation for the spherical Hankel function [i.e., $\hat{H}_n^{(2)}(k_0 r) \simeq j^{n+1}e^{-jk_0 r}$]. The expressions are written as

$$E_\theta = \frac{-jE_0 h e^{-jk_0 r}\cos\phi}{2r}\sum_{n=1}^{\infty}\frac{j^n(2n+1)}{n^2(n+1)^2}\left[\frac{P_n^1(\cos\theta_0)}{\hat{H}_n^{(2)}(k_0 b)}\,\frac{P_n^1(\cos\theta)}{\sin\theta}\right.$$
$$\left. -\,\frac{j\sin\theta\,\sin^2\theta_0 P_n^{1'}(\cos\theta)P_n^{1'}(\cos\theta_0)}{\hat{H}_n^{(2)\prime}(k_0 b)}\right], \tag{5.32}$$

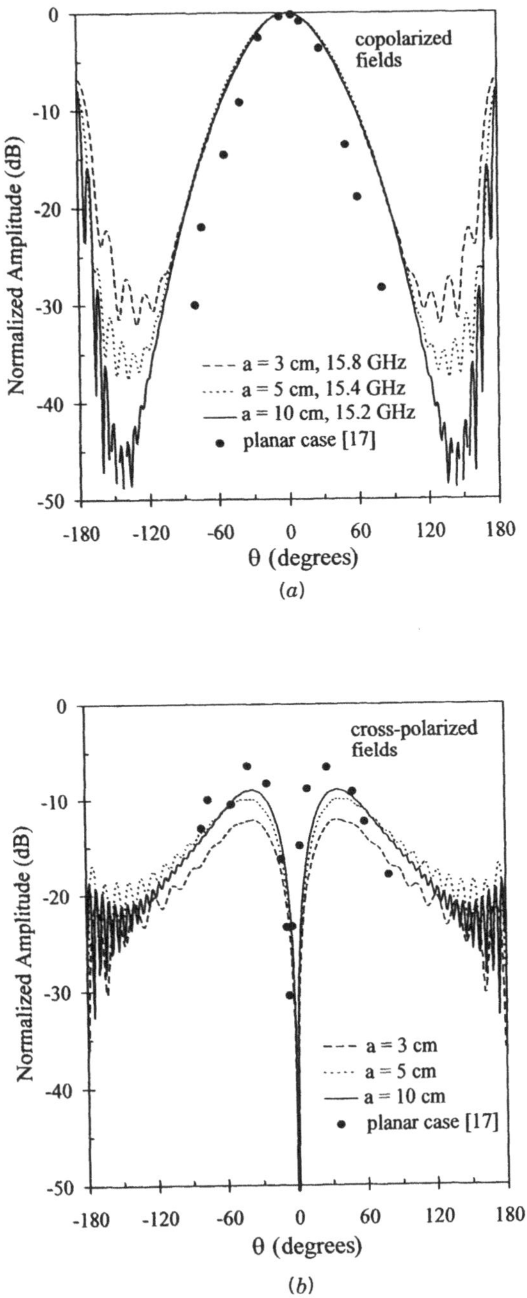

FIGURE 5.17 Radiation patterns of (*a*) copolarized and (*b*) cross-polarized fields in the *H* plane; $\epsilon_1 = 3.35$, $h = 0.76$ mm, $r_1 = 5$ mm, $r_2 = 9.5$ mm, $r_p = 1.1r_1$, $\phi_p = 0°$. (From Ref. [14], © 1994 John Wiley & Sons, Inc.)

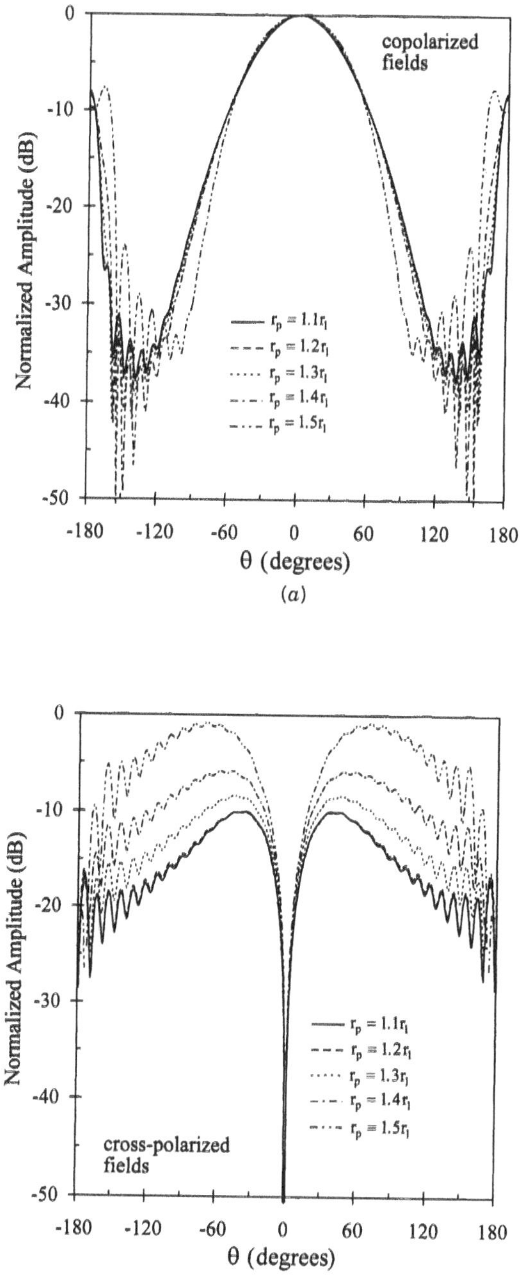

FIGURE 5.18 Radiation patterns of (*a*) copolarized and (*b*) cross-polarized fields in the H plane for various values of r_p/r_1; $a = 5$ cm, $\epsilon_1 = 3.35$, $h = 0.76$ mm, $r_1 = 5$ mm, $r_2 = 9.5$ mm, $\phi_p = 0°$, $f = 15.4$ GHz. (From Ref. [14], © 1994 John Wiley & Sons, Inc.)

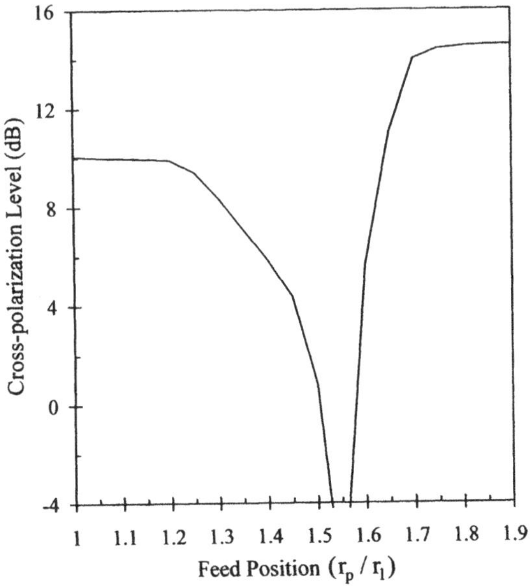

FIGURE 5.19 *H* plane cross-polarization level as a function of feed position; antenna parameters are as given in Figure 5.18. (From Ref. [14], © 1994 John Wiley & Sons, Inc.)

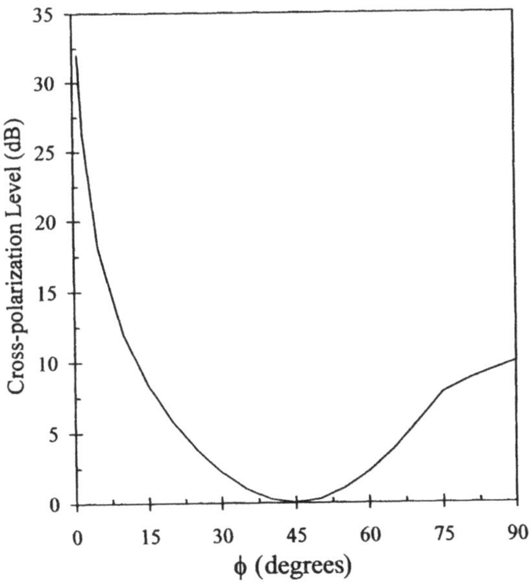

FIGURE 5.20 Cross-polarization level in the plane of various values of ϕ; $r_p = 1.1r_1$. Other antenna parameters are as given in Figure 5.18. (From Ref. [14], © 1994 John Wiley & Sons, Inc.)

$$E_\phi = \frac{jE_0 h e^{-jk_0 r} \sin\phi}{2r} \sum_{n=1}^{\infty} \frac{j^n(2n+1)}{n^2(n+1)^2} \left[\frac{P_n^1(\cos\theta_0) P_n^{1\prime}(\cos\theta) \sin\theta}{\hat{H}_n^{(2)}(k_0 b)} - \frac{jP_n^1(\cos\theta) P_n^{1\prime}(\cos\theta_0) \sin^2\theta_0}{\hat{H}_n^{(2)\prime}(k_0 b) \sin\theta} \right]. \tag{5.33}$$

Once the cavity field and the radiated field have been determined, the effective loss tangent δ_{eff} [Eq. (4.41)] of the substrate can be calculated by evaluating the radiation losses and the stored energy of the microstrip structure (see the formulation in Section 4.3). A determination of δ_{eff} is necessary for calculating the input impedance described below.

To calculate the input impedance, we consider the circular patch excited by a $\hat{r}$-directed current ribbon of effective arc length w_p at the feed position (θ_p, ϕ_p); that is,

$$\mathbf{J}_p(\theta, \phi) = \begin{cases} \hat{r}\delta(\theta - \theta_p), & \phi_p - \dfrac{w_p}{2b\theta_p} < \phi < \phi_p + \dfrac{w_p}{2b\theta_p}, \\ 0, & \text{elsewhere}. \end{cases} \tag{5.34}$$

In this case, the cavity field satisfies the wave equation

$$\frac{1}{b^2 \sin\theta} \frac{\partial}{\partial\theta}\left(\sin\theta \frac{\partial E_r}{\partial\theta}\right) + \frac{1}{b^2 \sin^2\theta} \frac{\partial^2 E_r}{\partial\phi^2} + k^2 E_r = -j\omega\mu_0 J_p \tag{5.35}$$

and can be expanded into an infinite summation of the source-free solution expressed in (5.29); that is,

$$E_r = \sum_{m=0}^{\infty} \sum_{p=1}^{\infty} C_{mp} J_m(k_{mp} b\theta) \cos m\phi, \tag{5.36}$$

with

$$C_{mp} = \frac{j\omega\mu_0 w \cos m\phi_p}{b(k_{\text{eff}}^2 - k_{mp}^2) U(m, p)} J_m(k_{mp} b\theta_p) \operatorname{sinc}\left(\frac{mw}{2b\theta_p}\right), \tag{5.37}$$

where sinc(x) is a sinc function ($= \sin x / x$), k_{eff} is the effective wavenumber $[= k_0\sqrt{\epsilon_1(1 - j\delta_{\text{eff}})}]$, and

$$U(m, p) = \frac{\sigma_m \pi \theta_0^2}{2}\left(1 - \frac{m^2}{k_{mp}^2 b^2 \theta_0^2}\right)[J_m(k_{mp} b\theta_0)]^2, \tag{5.38}$$

with

$$\sigma_m = \begin{cases} 2, & m = 0, \\ 1, & m > 0. \end{cases} \tag{5.39}$$

From (5.34) and (5.36), the input impedance can then be determined by

$$Z_{\text{in}} = -\frac{1}{I^2} \iiint_{v'} E_r J_p \, dv'$$

$$= \sum_{m=0}^{\infty} \sum_{p=1}^{\infty} \frac{-j\omega\mu_0 h \cos^2 m\phi_p}{b^2(k_{\text{eff}}^2 - k_{mp}^2)U(m, p)} J_m^2(k_{mp} b\theta_p) \operatorname{sinc}^2\left(\frac{mw_p}{2b\theta_p}\right), \quad (5.40)$$

where I is the input current amplitude and v' is the source region. By considering the equivalent magnetic current radiating in the presence of a spherical surface, we have the far-zone radiated field for the probe-fed case expressed as

$$E_\theta = \frac{-e^{-jk_0 r}}{k_0 r} \sum_{m=0}^{\infty} \cos m\phi \sum_{p=1}^{\infty} \left[b_{mn} j^{n+1} \frac{mP_n^m(\cos\theta)}{\sin\theta} - a_{mn} j^n \frac{dP_n^m(\cos\theta)}{d\theta} \right], \quad (5.41)$$

$$E_\phi = \frac{-e^{-jk_0 r}}{k_0 r} \sum_{m=0}^{\infty} \sin m\phi \sum_{p=1}^{\infty} \left[a_{mn} j^n \frac{mP_n^m(\cos\theta)}{\sin\theta} - b_{mn} j^{n+1} \frac{dP_n^m(\cos\theta)}{d\theta} \right], \quad (5.42)$$

where

$$a_{mn} = -\frac{(2n+1)(n-m)!}{2n(n+1)(n+m)!} \frac{M_m k_0 \sin\theta_0}{\hat{H}_n^{(2)\prime}(k_0 b)} \frac{dP_n^m(\cos\theta_0)}{d\theta}, \quad (5.43)$$

$$b_{mn} = \frac{(2n+1)(n-m)!}{2n(n+1)(n+m)!} \frac{M_m k_0 m}{\hat{H}_n^{(2)}(k_0 b)} P_n^m(\cos\theta_0), \quad (5.44)$$

$$M_m = \sum_{p=1}^{\infty} C_{mp} h J_m(k_{mp} b\theta_0). \quad (5.45)$$

Numerical results of (5.40)–(5.42) have also been calculated. Figure 5.21 shows typical results of input impedance versus frequency for the spherical circular microstrip antenna with different sphere radii excited at the TM_{11} mode. It is seen that the resonant frequency obtained from the zero crossing of the reactance curve is not affected by the sphere radius. This behavior is due to the approximation adopted in cavity-model analysis. However, the input resistance at the resonant frequency is seen to be a function of the sphere radius of the feed position. Figure 5.22 shows the resonant input resistance versus feed position for various sphere radii. The results are in good agreement with those obtained using GTLM theory [6], and the behavior also agrees with the full-wave solutions in Section 5.2.1. Typical radiation patterns are plotted in Figure 5.23. Results obtained using the full-wave analysis in Section 5.2.1 are shown in the figure for

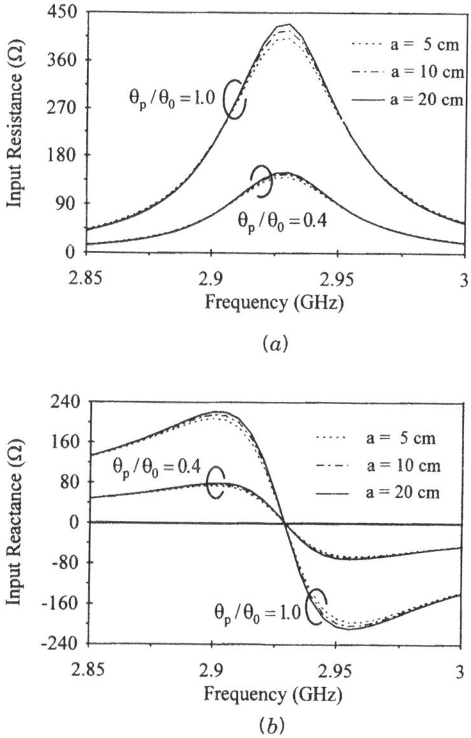

FIGURE 5.21 Input impedance at the TM_{11} mode versus frequency; $h = 1.6$ mm, $\epsilon_1 = 2.47$, $r_d = b\theta_0 = 19.1$ mm, $\phi_p = 0°$, $\theta_p/\theta_0 = 0.4$, 1.0. (*a*) Input resistance; (*b*) input reactance. (From Ref. [5], © 1994 John Wiley & Sons, Inc.)

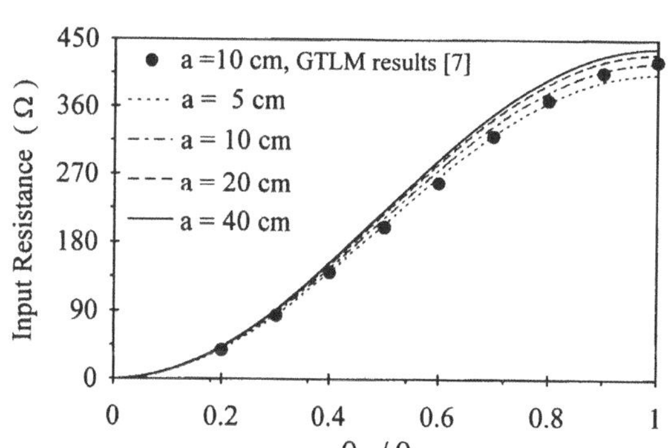

FIGURE 5.22 Resonant input resistance at the TM_{11} mode versus feed position; $h = 1.6$ mm, $\epsilon_1 = 2.47$, $r_d = b\theta_0 = 19.1$ mm, $\phi_p = 0°$. (From Ref. [5], © 1994 John Wiley & Sons, Inc.)

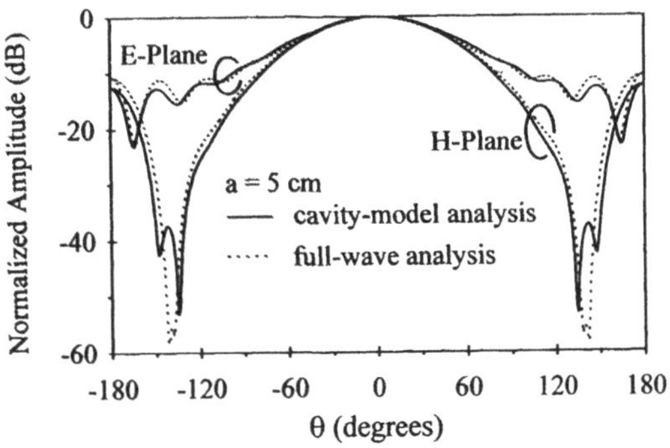

(*a*)

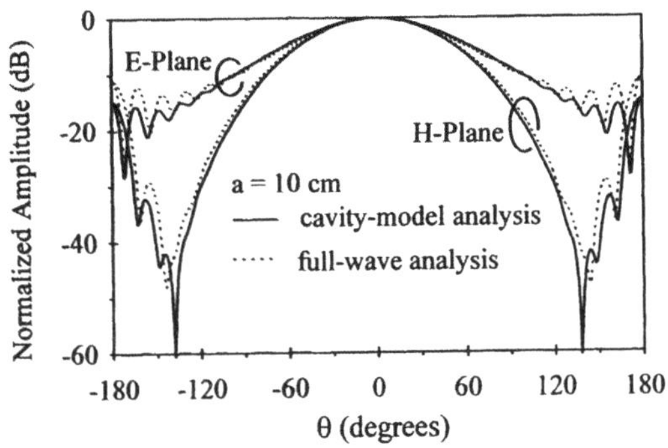

(*b*)

FIGURE 5.23 Radiation patterns for a spherical circular microstrip antenna with $h =$ 1.6 mm, $\epsilon_1 = 2.32$, $r_d = 10.68$ mm, $\phi_p = 0°$, $\theta_p/\theta_0 = 0.9$. (*a*) $a = 5$ cm; (*b*) $a = 10$ cm. (From Ref. [5], © 1994 John Wiley & Sons, Inc.)

comparison. Good agreement between cavity-model results and full-wave solutions is also observed. From the cavity-model solutions shown above, it can be concluded that the cavity-model formulation is useful for antenna designers to evaluate the radiation characteristics and resonant input resistance of spherical circular microstrip antennas.

For the spherical annular-ring microstrip antenna, the cavity-model formulation for far-zone radiated fields has also been derived [6,18]. Referring to the geometry in Figure 5.1 and considering the source-free case first, the cavity field in the substrate region under the annular-ring patch is written for the TM_{mp} mode as

$$E_r = E_0[J_m(k_{mp}b\theta)Y'_m(k_{mp}b\theta_1) - J'_m(k_{mp}b\theta_1)Y_m(k_{mp}b\theta)]\cos m\phi\,, \quad (5.46)$$

with

$$J'_m(k_{mp}b\theta_2)Y'_m(k_{mp}b\theta_1) - J'_m(k_{mp}b\theta_1)Y'_m(k_{mp}b\theta_2) = 0\,, \quad (5.47)$$

$$f_{mp} = \frac{ck_{mp}}{2\pi\sqrt{\epsilon_1}} . \tag{5.48}$$

Also, the far-zone radiated fields are expressed as

$$E_\theta = \frac{-jE_0 h \cos\phi e^{-jk_0 r}}{2r} \sum_{n=1}^{\infty} \frac{j^n(2n+1)}{n^2(n+1)^2} \left\{ \frac{P_n^1(\cos\theta_2)}{\hat{H}_n^{(2)}(k_0 b)\sin\theta} \right.$$
$$\times [P_n^1(\cos\theta_1) - S_{nm}(\theta_1, \theta_2)P_n^1(\cos\theta_2)]$$
$$\left. - \frac{jP_n^{1'}(\cos\theta)\sin\theta}{\hat{H}_n^{(2)}(k_0 b)} [P_n^{1'}(\cos\theta_1)\sin^2\theta_1 - S_{nm}(\theta_1, \theta_2)P_n^{1'}(\cos\theta_2)\sin^2\theta_2] \right\} , \tag{5.49}$$

$$E_\phi = \frac{jE_0 h \sin\phi e^{-jk_0 r}}{2r} \sum_{n=1}^{\infty} \frac{j^n(2n+1)}{n^2(n+1)^2} \left\{ \frac{-jP_n^1(\cos\theta)}{\hat{H}_n^{(2)}(k_0 b)\sin\theta} \right.$$
$$\times [P_n^{1'}(\cos\theta_1) - S_{nm}(\theta_1, \theta_2)P_n^{1'}(\cos\theta_2)]$$
$$\left. + \frac{P_n^{1'}(\cos\theta)\sin\theta}{\hat{H}_n^{(2)}(k_0 b)} [P_n^1(\cos\theta_1) - S_{nm}(\theta_1, \theta_2)P_n^1(\cos\theta_2)] \right\} , \tag{5.50}$$

with

$$S_{nm}(\theta_1, \theta_2) = \frac{J_m(k_{mp}b\theta_2)Y'_m(k_{mp}b\theta_1) - J'_m(k_{mp}b\theta_1)Y_m(k_{mp}b\theta_2)}{J_m(k_{mp}b\theta_1)Y'_m(k_{mp}b\theta_1) - J'_m(k_{mp}b\theta_1)Y_m(k_{mp}b\theta_1)} . \tag{5.51}$$

When the microstrip antenna with a probe feed given by (5.34) is considered, the cavity field of (5.46) is modified as [6]

$$E_r = \sum_{m=0}^{\infty} \sum_{p=1}^{\infty} B_{mp}[J_m(k_{mp}b\theta)Y'_m(k_{mp}b\theta_1) - J'_m(k_{mp}b\theta_1)Y_m(k_{mp}b\theta)] \cos m\phi , \tag{5.52}$$

where

$$B_{mp} = \frac{j\omega\mu_0 w \cos m\phi_p}{b(k_{\text{eff}}^2 - k_{mp}^2)V(m, p)} \Theta_{mp}(\theta_p) \operatorname{sinc}\left(\frac{mw}{2b\theta_p}\right) , \tag{5.53}$$

$$\Theta_{mp}(\theta) = J_m(k_{mp}b\theta) - Y_m(k_{mp}b\theta) \frac{J'_m(k_{mp}b\theta_1)}{Y'_m(k_{mp}b\theta_1)} , \tag{5.54}$$

$$V(m, p) = \frac{\sigma_m \pi}{2k_{mp}^2 b^2} [(k_{mp}^2 b^2 \theta_2^2 - m^2)\Theta_{mp}^2(\theta_2) - (k_{mp}^2 b^2 \theta_1^2 - m^2)\Theta_{mp}^2(\theta_1)] , \tag{5.55}$$

and the far-zone radiated fields of (5.49)–(5.50) are rewritten as

$$E_\theta = \frac{-e^{-jk_0 r}}{2k_0 r} \sum_{m=0}^{\infty} \cos m\phi \sum_{p=1}^{\infty} \left[b_{mn}^{\text{AR}} j^{n+1} \frac{mP_n^m(\cos\theta)}{\sin\theta} - a_{mn}^{\text{AR}} j^n \frac{dP_n^m(\cos\theta)}{d\theta} \right], \tag{5.56}$$

$$E_\phi = \frac{-e^{-jk_0 r}}{k_0 r} \sum_{m=0}^{\infty} \sin m\phi \sum_{p=1}^{\infty} \left[a_{mn}^{\text{AR}} j^n \frac{mP_n^m(\cos\theta)}{\sin\theta} - b_{mn}^{\text{AR}} j^{n+1} \frac{dP_n^m(\cos\theta)}{d\theta} \right], \tag{5.57}$$

where

$$a_{mn}^{\text{AR}} = -\frac{(2n+1)(n-m)!}{2n(n+1)(n+m)!} \frac{k_0}{\hat{H}_n^{(2)\prime}(k_0 b)} [N_m(\theta_2) \sin^2\theta_2 P_n^{m\prime}(\cos\theta_2) - N_m(\theta_1) \sin^2\theta_1 P_n^{m\prime}(\cos\theta_1)] , \tag{5.58}$$

$$b_{mn}^{\text{AR}} = -\frac{(2n+1)(n-m)!}{2n(n+1)(n+m)!} \frac{mk_0}{\hat{H}_n^{(2)\prime}(k_0 b)} [N_m(\theta_2) P_n^m(\cos\theta_2) - N_m(\theta_1) P_n^m(\cos\theta_1)] , \tag{5.59}$$

$$N_m = \sum_{p=1}^{\infty} B_{mp} h\Theta_{mp}(\theta) . \tag{5.60}$$

Numerical results of the expressions above for a probe-fed spherical annular-ring microstrip antenna excited at the TM_{12} mode have also been calculated and reported in [6]. The results obtained for the copolarized and cross-polarized radiation characteristics are found to be in good agreement with those obtained using full-wave analysis in Section 5.2.1.

5.2.3 GTLM Solution

GTLM theory has also been applied in an analysis of spherical microstrip antennas [7,8]. By following the GTLM formulation described in Section 4.4 and considering the geometry in Figure 5.1, the microstrip antenna is modeled as a transmission line in the $\hat{\theta}$ direction, and the modal voltage and the modal current are, respectively, defined as E_{rm} and $\pm a \sin\theta H_{\phi m}$ for wave propagation in the $\mp\hat{\theta}$ direction, where E_{rm} and $H_{\phi m}$ are the electric and magnetic fields inside the substrate layer under the patch. By considering the circular microstrip antenna first, E_{rm} and $H_{\phi m}$ for the TM_{mn} mode are expressed as

$$E_{rm}(\theta) = \begin{cases} E_0 P_\upsilon^m(\cos\theta)\cos m\phi , & 0 \le \theta \le \theta_p , \\ E_0[P_\upsilon^m(\cos\theta) + C_m Q_\upsilon^m(\cos\theta)]\cos m\phi , & \theta_p \le \theta \le \theta_0 , \end{cases} \tag{5.61}$$

and

$$H_{\phi m}(\theta)=\begin{cases}\dfrac{-\sin\theta}{j\omega\mu_0 a}E_0P_v^{m'}(\cos\theta)\cos m\phi\,, & 0\le\theta\le\theta_p\,,\\[2ex] \dfrac{-\sin\theta}{j\omega\mu_0 a}E_0[P_v^{m'}(\cos\theta)+C_mQ_v^{m'}(\cos\theta)]\cos m\phi\,, & \theta_p\le\theta\le\theta_0\,,\end{cases} \tag{5.62}$$

where $P_v^m(x)$ and $Q_v^m(x)$ are the associated Legendre functions of the first and second kinds, respectively; C_m are unknown coefficients. Similar to the case of a cylindrical circular microstrip antenna (Section 4.4.2), the equivalent circuit of the spherical circular microstrip antenna at the TM_m mode $(=\Sigma_n TM_{mn})$ can be constructed as shown in Figure 5.24, where the circular patch is also modeled as a π network (g_1, g_2, and g_3); y_p is the wall admittance at $\theta=\theta_p^-$ (just off the feed position toward the patch center) and y_w is the wall admittance at the patch edge; I_m is the feed current corresponding to the TM_m mode excitation, which is related to the total feed current I_t as shown by (4.160). These equivalent-circuit elements have been derived and can be written as follows: For the π network,

$$g_1=\frac{\sin^2\theta_p}{j\omega\mu_0}\left[\frac{\Pi_1(\theta_p,\theta_0)}{\Pi_2(\theta_p,\theta_0)}+\frac{\Pi_1(\theta_p,\theta_p)}{\Pi_2(\theta_0,\theta_p)}\right], \tag{5.63}$$

$$g_2=-\frac{\sin^2\theta_p}{j\omega\mu_0}\frac{\Pi_1(\theta_p,\theta_p)}{\Pi_2(\theta_0,\theta_p)}, \tag{5.64}$$

$$g_3=\frac{1}{j\omega\mu_0}\left[\sin^2\theta_p\frac{\Pi_1(\theta_p,\theta_p)}{\Pi_2(\theta_0,\theta_p)}-\sin^2\theta_0\frac{\Pi_1(\theta_0,\theta_p)}{\Pi_2(\theta_0,\theta_p)}\right], \tag{5.65}$$

where

$$\Pi_1(\alpha,\beta)=P_v^{m'}(\cos\alpha)Q_v^m(\cos\beta)-Q_v^{m'}(\cos\alpha)P_v^m(\cos\beta)\,, \tag{5.66}$$

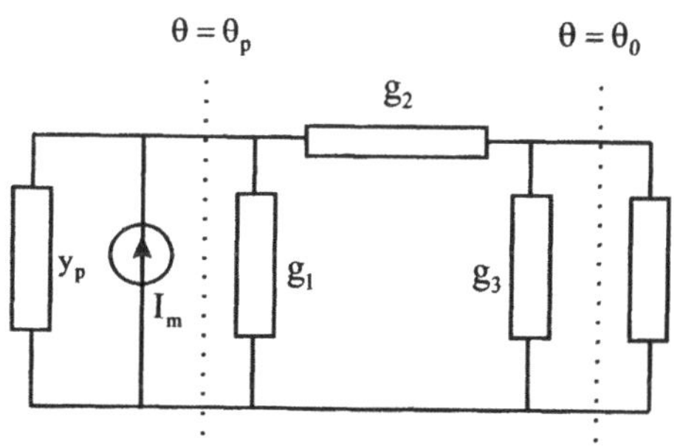

FIGURE 5.24 Equivalent circuit of a probe-fed spherical circular microstrip antenna, as shown in Figure 5.1, for GTLM analysis.

$$\Pi_2(\alpha, \beta) = P_v^m(\cos\alpha)Q_v^m(\cos\beta) - Q_v^m(\cos\alpha)P_v^m(\cos\beta), \tag{5.67}$$

and for the admittances at $\theta = \theta_p$ and θ_0,

$$y_p = -\frac{\sin^2\theta_p}{j\omega\mu_0}\frac{P_v^{m'}(\cos\theta_p)}{P_v^m(\cos\theta_p)}, \tag{5.68}$$

$$y_w = \frac{-1}{\pi h E_a^2}\iint_{S_a} H_\phi^a E_a \cos m\phi \, ds$$

$$= -j\frac{\sigma_m b^2}{h}\sqrt{\frac{\epsilon_1\epsilon_0}{\mu_0}}\sum_{n=m}^{\infty}\left[A_{mn}\frac{\hat{H}_n^{(2)'}(kb)}{\hat{H}_n^{(2)}(kb)} - B_{mn}\frac{\hat{H}_n^{(2)}(kb)}{\hat{H}_n^{(2)'}(kb)}\right], \tag{5.69}$$

where

$$A_{mn} = -\frac{(2n+1)}{2n(n+1)}\frac{(n-m)!}{(n+m)!}\int_{\theta_0}^{\theta_p} P_n^m(\cos\theta)\,d\theta, \tag{5.70}$$

$$B_{mn} = -\frac{(2n+1)}{2n(n+1)}\frac{(n-m)!}{(n+m)!}\int_{\theta_0}^{\theta_p}\frac{dP_n^m(\cos\theta)}{d\theta}\sin\theta\,d\theta. \tag{5.71}$$

In (5.69), $k = k_0\sqrt{\epsilon_1}$, $\sigma_m = 2$ for $m = 0$ and $\sigma_m = 1$ for $m > 0$ [see also (4.179)], S_a is the surface area (width h and length $2\pi r_d$) of the aperture around the edge of the circular patch, $E_a \cos m\phi$ is the electric field at the aperture ($\theta = \theta_p$), and $H_\phi^a(\theta)$ is the magnetic field due to the equivalent magnetic current at the aperture ($\theta = \theta_p$).

With all the elements in the equivalent circuit obtained, the input impedance of the antenna seen by the probe feed at the TM_{11} mode (i.e., $m = 1$) is given by

$$Z_{\text{in}} = \frac{h}{\pi}\left[y_p + g_1 + \frac{g_2(g_3 + y_w)}{g_2 + g_3 + y_w}\right]^{-1}. \tag{5.72}$$

It should also be noted that when the feed is placed at the patch edge (i.e., $\theta_p = \theta_0$), the input impedance can be determined by

$$Z_{\text{in}} = \frac{h}{\pi}(y_p + y_w)^{-1}. \tag{5.73}$$

Similarly, for the spherical annular-ring microstrip antenna shown in Figure 5.1, the equivalent circuit presented in Figure 5.25 for the antenna excited at the TM_m mode ($= \Sigma_n TM_{mn}$) is constructed. It is noted that due to the mode degeneracy problem associated with excitation of the TM_{12} mode for the annular-ring microstrip antenna [3], the input impedance calculation should include at least six modes (i.e., $Z_{\text{in}} = \Sigma_{m=0}^{5} Z_m$), where Z_m is the input impedance of the antenna seen by the total feed current at the TM_m mode. As shown in the equivalent circuit, the annular-ring microstrip antenna is modeled as a transmission line

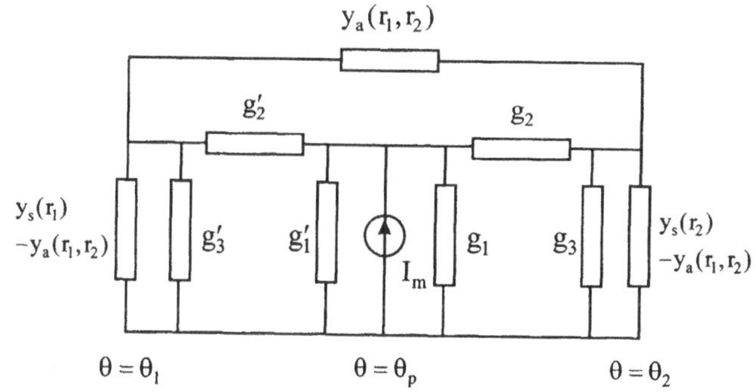

FIGURE 5.25 Equivalent circuit of a probe-fed spherical annular-ring microstrip antenna, as shown in Figure 5.1, for the GTLM analysis.

joining the radiating apertures at $\theta = \theta_1$ and θ_2. Two π networks [(g_1, g_2, g_3) and (g'_1, g'_2, g'_3)] are used to represent the sections of transmission line between the feed position ($\theta = \theta_p$) and the two apertures. The admittance y_a is the mutual admittance due to the interaction between two apertures, and y_s denotes the self-admittance of the apertures at the inner and outer edges of the annular-ring patch. I_m is again the feed current corresponding to TM_m-mode excitation, also related to the total feed current I_t, as shown by (4.160). These circuit elements in the equivalent circuit have also been derived and written as follows: For the π network between $\theta = \theta_p$ and θ_2,

$$g_1 = \frac{\sin^2\theta_p}{j\omega\mu_0}\left[\frac{\Pi_1(\theta_p, \theta_2)}{\Pi_2(\theta_p, \theta_2)} + \frac{\Pi_1(\theta_p, \theta_p)}{\Pi_2(\theta_2, \theta_p)}\right], \tag{5.74}$$

$$g_2 = -\frac{\sin^2\theta_p}{j\omega\mu_0}\frac{\Pi_1(\theta_p, \theta_p)}{\Pi_2(\theta_2, \theta_p)}, \tag{5.75}$$

$$g_3 = \frac{1}{j\omega\mu_0}\left[\sin^2\theta_p\frac{\Pi_1(\theta_p, \theta_p)}{\Pi_2(\theta_2, \theta_p)} - \sin^2\theta_2\frac{\Pi_1(\theta_2,\theta_p)}{\Pi_2(\theta_2,\theta_p)}\right], \tag{5.76}$$

where $\Pi_1(\alpha, \beta)$ and $\Pi_2(\alpha, \beta)$ are given in (5.66)–(5.67); for the π network between $\theta = \theta_p$ and θ_1,

$$g'_1 = \frac{\sin^2\theta_p}{j\omega\mu_0}\left[\frac{\Pi_1(\theta_1, \theta_p)}{\Pi_2(\theta_1, \theta_p)} + \frac{\Pi_1(\theta_1, \theta_1)}{\Pi_2(\theta_p, \theta_1)}\right], \tag{5.77}$$

$$g'_2 = -\frac{\sin^2\theta_p}{j\omega\mu_0}\frac{\Pi_1(\theta_1, \theta_1)}{\Pi_2(\theta_p, \theta_1)}, \tag{5.78}$$

$$g_3' = \frac{1}{j\omega\mu_0}\left[\sin^2\theta_p \frac{\Pi_1(\theta_1, \theta_1)}{\Pi_2(\theta_p, \theta_1)} - \sin^2\theta_1 \frac{\Pi_1(\theta_p, \theta_1)}{\Pi_2(\theta_p, \theta_1)}\right]; \tag{5.79}$$

and for the mutual admittance $y_a(\theta_1, \theta_2)$,

$$y_a = \frac{-1}{\pi h E_a E_b}\iint_{S_a} H_\phi^a E_b \cos m\phi \, ds$$

$$= j\sigma_m h\sqrt{\frac{\epsilon_1\epsilon_0}{\mu_0}}\sum_{n=m}^{\infty}\left[A'_{mn}(\theta_1)A'_{mn}(\theta_2)\frac{\hat{H}_n^{(2)\prime}(kb)}{\hat{H}_n^{(2)}(kb)} - B'_{mn}(\theta_1)B'_{mn}(\theta_2)\frac{\hat{H}_n^{(2)}(kb)}{\hat{H}_n^{(2)\prime}(kb)}\right], \tag{5.80}$$

where

$$A'_{mn}(\theta_i) = mP_n^m(\cos\theta_i), \qquad i = 1, 2, \tag{5.81}$$

$$B'_{mn}(\theta_i) = P_n^{m\prime}(\cos\theta_i)\sin^2\theta_i, \qquad i = 1, 2. \tag{5.82}$$

Also, the self-admittance $y_s(\theta_i)$, $i = 1, 2$, can be calculated from $y_a(\theta_i, \theta_i)$ in (5.80). With the forgiving equivalent-circuit elements obtained, the input impedance of the annular-ring microstrip antenna at the TM_m mode can be evaluated using (4.178). Numerical results calculated from (5.72) for the circular patch antenna and (4.178) for the annular-ring patch antenna have also been analyzed [7,8], and the results obtained show good agreement with those obtained using full-wave analysis (Section 5.2.1).

5.3 CONICAL MICROSTRIP ANTENNAS

In addition to the cylindrical and spherical microstrip antennas, which have received much attention, microstrip antennas mounted on a conical surface have also been studied [9–12]. This kind of conical microstrip antenna is of particular importance in aircraft and missiles, portions of whose bodies are conical in shape. Typical configurations of conical microstrip antennas are shown in Figures 5.26 to 5.29. In Figures 5.26 and 5.27, a circular microstrip antenna is mounted inside or outside a conical ground plane, to modify the radiation pattern of the microstrip antenna in the elevation plane. Results have shown that for the case shown in Figure 5.26, a relatively small conical ground plane can improve the front-to-back ratio of the elevation-plane radiation pattern by at least 8 dB, compared to the ratio using a same-size planar ground plane [9]. However, it should be noted that for the case shown by Figure 5.27, radiation in the lower hemisphere can be enhanced significantly compared to that of a planar microstrip antenna.

Also, due to the nonplanar ground plane, the resonant frequency of a conical microstrip antenna is also affected. For the conical circular microstrip antenna discussed here, an approximate equation for the resonant frequency is given as [9]

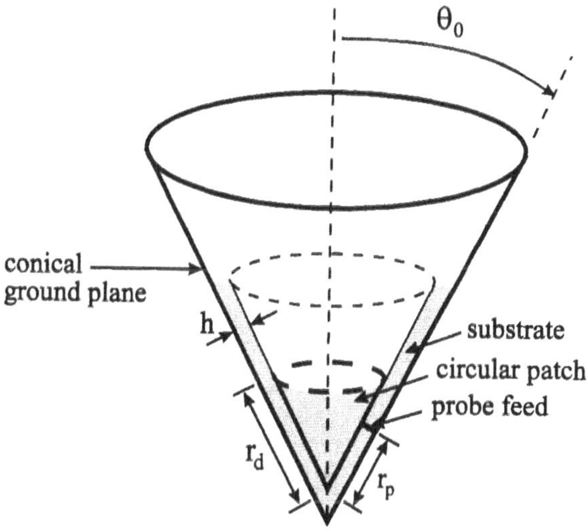

FIGURE 5.26 Geometry of a conical circular microstrip antenna; the circular microstrip patch is mounted inside a conical ground surface.

$$\frac{f_{\text{cone}} - f_{\text{plane}}}{(f_{\text{cone}} + f_{\text{plane}})/2} \approx 1 - \sin\frac{\theta_c}{2}, \tag{5.83}$$

where f_{cone} and f_{plane} are, respectively, the resonant frequencies of the circular microstrip antenna mounted on conical ground and planar ground planes; θ_c

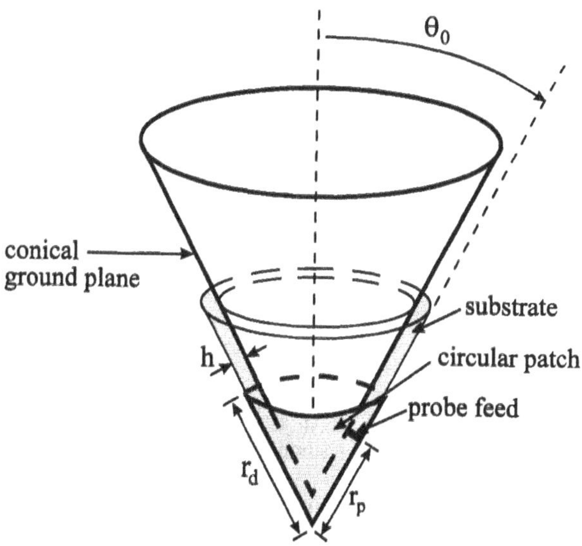

FIGURE 5.27 Geometry of a circular microstrip antenna mounted outside a conical ground surface.

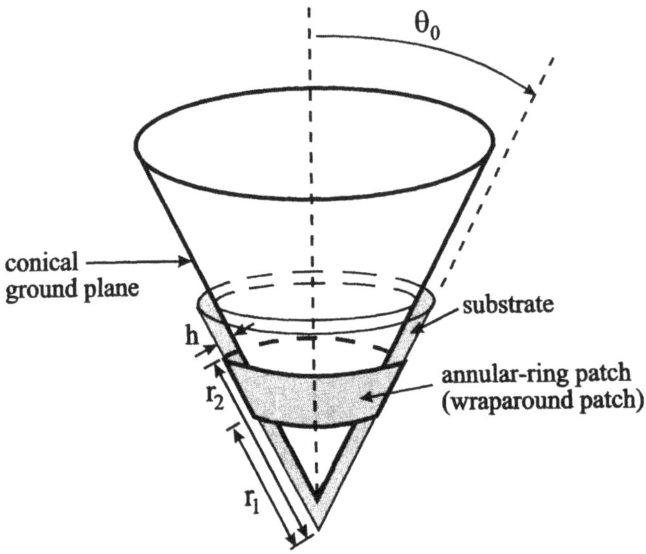

FIGURE 5.28 Geometry of a conical annular-ring (wraparound) microstrip antenna.

($= 2\theta_0$; see Figure 5.26) is the flare angle of the conical ground plane. From (5.83) it is clear that the resonant frequency increases with decreasing θ_c.

Figure 5.28 shows a conical annular-ring (wraparound) microstrip antenna [10]. In such a case, an omnidirectional radiation in the azimuthal direction can be obtained. The conical annular-ring-segment microstrip antenna shown in Figure 5.29 has also been studied [11,12]. The far-zone radiated fields and the input impedance have been derived using a cavity-model analysis. The formulation is described briefly below.

As shown by the geometry in Figure 5.29, an annular-ring-segment microstrip patch, having an angular width of $2\phi_0$ and a length of $r_2 - r_1$, is mounted on a ground cone of flare angle $2\theta_0$. The substrate is of thickness h and relative permittivity ϵ_1. The patch is centered with respect to the x–z plane ($\phi = 0°$) and excited by a probe feed at (r_p, ϕ_p), which is modeled here as a unit-amplitude $\hat{\theta}$-directed current ribbon of effective width w_p given as

$$J_\theta = J(\phi)\delta(r - r_p)\,, \tag{5.84}$$

with

$$J(\phi) = \begin{cases} 1\,, & -\dfrac{w_p}{2} \leq (r_p \sin\theta_0)\phi \leq \dfrac{w_p}{2}\,, \\ 0\,, & \text{elsewhere}\,. \end{cases} \tag{5.85}$$

To utilize cavity-model analysis, the substrate thickness is assumed to be very small and the inner and outer radii (r_1, r_2) of the annular-ring-segment patch and the curvature radius ($\sim r_1 \sin\theta_0$) of the ground conical surface are also assumed to

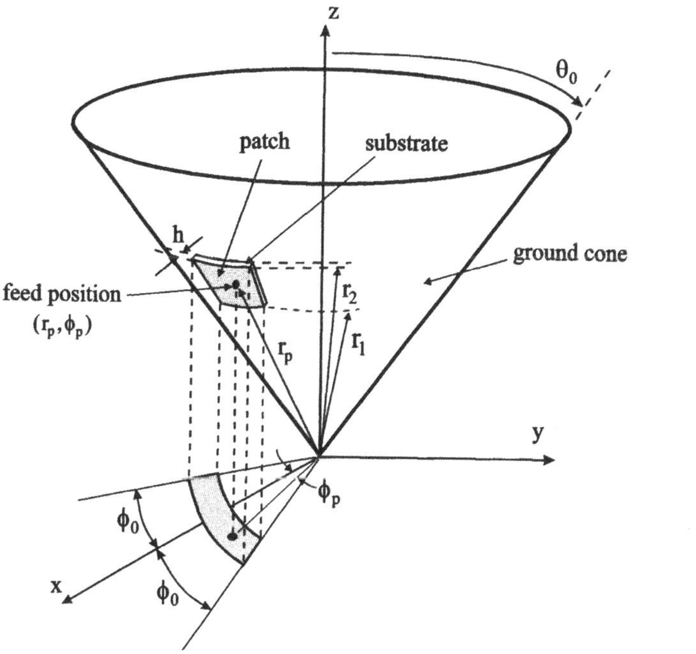

FIGURE 5.29 Geometry of a conical annular-ring-segment microstrip antenna.

be large compared to the operating wavelength. In such a case, the region between the microstrip patch and the ground conical surface can be treated as a cavity bounded by electric walls on the top and bottom and magnetic walls around the circumference of the cavity. By expanding the cavity fields in terms of spherical wave functions and imposing the boundary conditions of the cavity, the following eigenvalue equations are obtained:

$$P_n^{m'}(\cos\theta_0)Q_n^{m'}[\cos(\theta_0+\Delta\theta)] - P_n^{m'}[\cos(\theta_0+\Delta\theta)]Q_n^{m'}(\cos\theta_0) = 0\,, \quad (5.86)$$

$$\hat{J}'_n(k_l r_1)\hat{N}'_n(k_l r_2) - \hat{J}'_n(k_l r_2)\hat{N}'_n(k_l r_1) = 0\,, \quad (5.87)$$

$$\sin(2m\phi_0) = 0\,, \qquad m = \frac{q\pi}{2\phi_0}\,, \quad q = 0, 1, 2, \ldots, \quad (5.88)$$

where k_l $[= l\pi/(r_2 - r_1)$ with $l = 1, 2, 3, \ldots]$ are the eigenvalues that determine the resonant frequencies of the cavity; $\hat{J}_n(x)$ and $\hat{N}_n(x)$ are, respectively, spherical Bessel functions of the first and second kinds [see (3.47) for modified spherical Bessel functions with $\nu = n + \frac{1}{2}$]; and $P_n^m(x)$ and $Q_n^m(x)$ are the associated Legendre functions of the first and second kinds, respectively. By knowing that only modes with index $n = 0$ can exist in the cavity, due to the thin-substrate ($h \ll \lambda_0$) condition, and assuming that an annular-ring-segment patch on the conical surface is close to a rectangular shape, the resonant frequencies of the conical annular-ring-segment microstrip antenna can be written as follows: For the (l, q) mode,

$$f_{lq} = \frac{c}{2\sqrt{\epsilon_1}}\left[\left(\frac{l}{2L}\right)^2 + \left(\frac{q}{2W}\right)^2\right]^{1/2}, \tag{5.89}$$

with

$$2W = [(r_2 - r_1)\sin\theta_0]\phi_0\,, \tag{5.90}$$

$$2L = r_2 - r_1\,, \tag{5.91}$$

where $2L$ and $2W$ are, respectively, the effective length and width of the patch, similar to the expression of (1.4). It is also noted that from the expressions above, the mode with $n = q = 0$, $l = 1$ is the fundamental resonant mode of the cavity. The electric field inside the cavity can also be approximated to be independent of θ and have only a θ component; that is, we have the cavity field for the (l, q) mode given by [12]

$$E_\theta = \frac{-1}{r\sin\theta}\sum_{l,q} C_{lq}\cos[k_l(r - r_1)]\cos\left[\frac{q\pi}{2\phi_0}(\phi - \phi_0)\right], \tag{5.92}$$

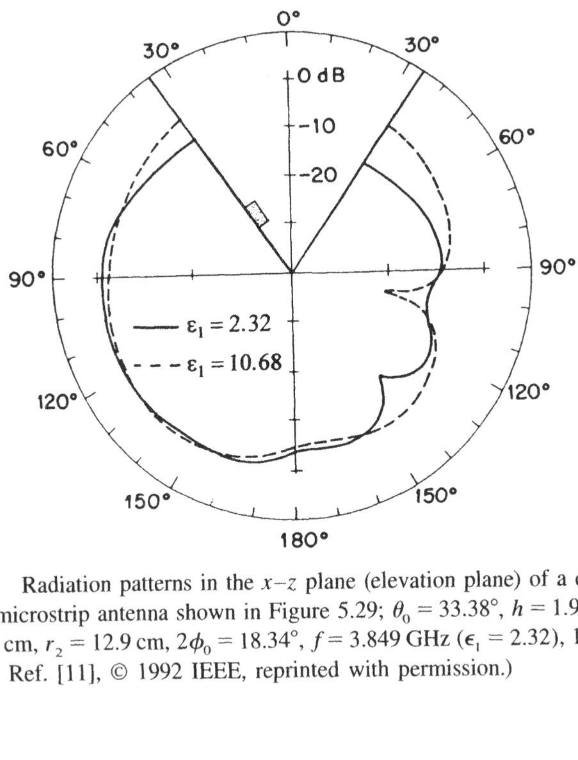

FIGURE 5.30 Radiation patterns in the x–z plane (elevation plane) of a conical annular-ring-segment microstrip antenna shown in Figure 5.29; $\theta_0 = 33.38°$, $h = 1.9$ mm, $\epsilon_1 = 2.32$, 10.68, $r_1 = 10$ cm, $r_2 = 12.9$ cm, $2\phi_0 = 18.34°$, $f = 3.849$ GHz ($\epsilon_1 = 2.32$), 1.794 GHz ($\epsilon_1 = 10.68$). (From Ref. [11], © 1992 IEEE, reprinted with permission.)

where C_{lq} are coefficients. Based on (5.84) and (5.92), the input impedance can also be evaluated from

$$Z_{\text{in}} = \frac{-h}{w_p^2} \int_{\phi_p - \Delta\phi}^{\phi_p - \Delta\phi} E_\theta \big|_{r=r_p} J_\theta r_p \sin\theta_0 \, d\phi , \tag{5.93}$$

where $\Delta\phi = w_p / 2r_p \sin\theta_0$. Numerical results of the input impedance have also been calculated. Variations of the input impedance with feed position [excited at the $(l, q) = (1, 0)$ mode] are found to be similar to those of a rectangular microstrip antenna at the TM_{01} mode [11]; that is, a null input impedance occurs for the feed position at about the center of the patch, and the maximum value of input impedance is obtained when the feed position is at about the patch edge ($r = r_1, r_2$). The far-zone radiated fields have also been derived by evaluating the radiation from four equivalent magnetic currents (two $\hat{\phi}$-directed magnetic currents and two $\hat{r}$-directed magnetic currents) along the edges of the cavity. Results have been given in [11,12], and a typical radiation pattern is shown in Figure 5.30, where the radiation pattern is seen to be broadened and a nearly omnidirectional radiation in the elevation plane is observed.

REFERENCES

1. H. T. Chen and Y. T. Cheng, "Full-wave analysis of a disk-loaded spherical annular-ring microstrip antenna," *Microwave Opt. Technol. Lett.*, vol. 12, pp. 353–358, Aug. 20, 1996.
2. H. T. Chen, H. D. Chen, and Y. T. Cheng, "Full-wave analysis of the annular-ring-loaded spherical circular microstrip antenna," *IEEE Trans. Antennas Propagat.*, vol. 45, pp. 1581–1583, Nov. 1997.
3. H. D. Chen and K. L. Wong, "Full-wave analysis of input impedance and patch current distribution of spherical annular-ring microstrip antennas excited by a probe feed," *Microwave Opt. Technol. Lett.*, vol. 11, pp. 524–528, Aug. 5, 1994.
4. W. Y. Tam, A. K. Y. Lai, and K. M. Luk, "Input impedance of spherical microstrip antenna," *IEE Proc.—Micro. Antennas Propag.*, vol. 142, pp. 285–288, June 1995.
5. H. T. Chen and K. L. Wong, "Analysis of probe-fed spherical-circular microstrip antennas using cavity-model theory," *Microwave Opt. Technol. Lett.*, vol. 7, pp. 309–312, May 1994.
6. H. T. Chen, W. S. Chen, and Y. T. Cheng, "Cavity-model analysis for radiation characteristics of a probe-fed spherical annular-ring microstrip antenna," *Proc. 1995 International Symposium on Communications*, pp. 930–934.
7. B. Ke and A. A. Kishk, "Analysis of spherical circular microstrip antennas," *IEE Proc.* pt. H, vol. 138, pp. 542–548, Dec. 1991.
8. A. A. Kishk, "Analysis of spherical annular-ring microstrip antennas," *IEEE Trans. Antennas Propagat.*, vol. 41, pp. 338–343, Mar. 1993.

9. N. Fayyaz, A. Abbaspour-Tamijani, S. Safavi-Naeini, and N. Hodjat, "Design and analysis of a circular patch antenna on a finite conical ground plane," *1996 IEEE AP-S International Symposium Digest*, pp. 680–683.
10. D. N. Meeks and P. F. Wahid, "Input impedance of a wraparound microstrip antenna on a conical surface," *1996 IEEE AP-S International Symposium Digest*, pp. 676–679.
11. J. R. Descardeci and A. J. Giarola, "Microstrip antenna on a conical surface," *IEEE Trans. Antennas Propagat.*, vol. 40, pp. 460–463, Apr. 1992.
12. R. Shavit, "Circular polarization microstrip antenna on a conical surface," *IEEE Trans. Antennas Propagat.*, vol. 45, pp. 1086–1092, July 1997.
13. H. T. Chen and K. L. Wong, "Cross-polarization characteristics of a probe-fed spherical-circular microstrip patch antenna," *Microwave Opt. Technol. Lett.*, vol. 6, pp. 705–710, Sept. 20, 1993.
14. H. D. Chen and K. L. Wong, "Cross-polarization characteristics of spherical annular-ring microstrip antennas," *Microwave Opt. Technol. Lett.*, vol. 7, pp. 616–619, Sept. 1994.
15. A. C. Ludwig, "The definition of cross polarization," *IEEE Trans. Antennas Propagat.*, vol. 21, pp. 116–119, Jan. 1973.
16. K. F. Lee, K. M. Luk, and P. Y. Tam, "Crosspolarization characteristics of circular patch antennas," *Electron. Lett.*, vol. 28, pp. 587–589, Mar. 12, 1992.
17. N. S. Nurie and R. J. Langley, "Crosspolarization performance of annular-ring microstrip antenna with novel mode suppression," *Electron. Lett.*, vol. 25, pp. 656–657, May 1, 1989.
18. K. M. Luk and W. Y. Tam, "Patch antenna on a spherical body," *IEE Proc.*, pt. H, vol. 138, pp. 103–108, Feb. 1991.

CHAPTER SIX

Coupling between Conformal Microstrip Antennas

6.1 INTRODUCTION

Coupling characteristics of cylindrical and spherical microstrip antennas have also been investigated [1–13]. The mutual coupling coefficients between two cylindrical microstrip antennas of rectangular [1–3,5,7], triangular [4], and circular [6,8] patches have been analyzed theoretically and experimentally. Various theoretical approaches using full-wave analysis, cavity-model method, and GTLM theory are described in this chapter, and the data measured are compared with the theoretical results. The broadband characteristics of cylindrical microstrip antennas with gap-coupled parasitic elements have also been reported [9,10]. On the other hand, for microstrip antennas mounted on a spherical ground surface, the coupling behavior between concentric circular and annular-ring microstrip antennas has also been shown [11–13]. These related coupling studies are presented in subsequent sections.

6.2 MUTUAL COUPLING OF CYLINDRICAL MICROSTRIP ANTENNAS

6.2.1 Full-Wave Solution of Rectangular Patches

The mutual coupling characteristics between two cylindrical rectangular microstrip antennas have been analyzed extensively using various theoretical models. The theoretical formulation using a full-wave approach and a moment-method calculation is described first. Numerical results for E- and H-plane mutual coupling coefficients for cylindrical microstrip antennas with different cylinder radii are presented, and the curvature effects on the mutual coupling are studied.

Figure 6.1 shows the geometry of two cylindrical rectangular microstrip

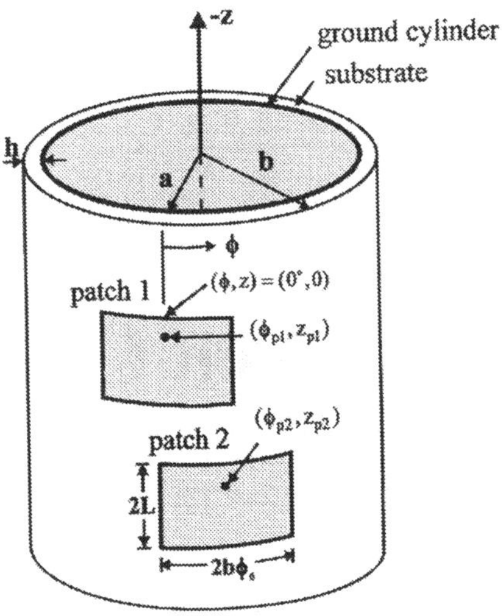

FIGURE 6.1 Geometry of two cylindrical rectangular microstrip antennas.

antennas. The two patches are probe fed at (ϕ_{p1}, z_{p1}) and (ϕ_{p2}, z_{p2}), respectively, and are of the same size $(2L \times 2b\phi_0)$; $2\phi_0$ is the angle subtended by the rectangular patch. The probe feed is modeled as a line source with a unit-amplitude current density written as

$$\mathbf{J}_p^{(i)} = \frac{\hat{\rho}\delta(\phi' - \phi_{pi})\delta(z' - z_{si})}{\rho'}, \qquad a \leq \rho' \leq b, \quad i = 1, 2, \tag{6.1}$$

where the superscript i denotes the source current at antenna i. Other parameters used here are of the same meaning as in other chapters. To begin with the mutual coupling study, the two probe-fed microstrip antennas are treated as a two-port network with a 2×2 port matrix $[Z^P]$, as shown in Figure 6.2. The relation between the port voltages and currents are defined as

$$\begin{bmatrix} V_1^P \\ V_2^P \end{bmatrix} = \begin{bmatrix} Z_{11}^P & Z_{12}^P \\ Z_{21}^P & Z_{22}^P \end{bmatrix} \begin{bmatrix} I_1^P \\ I_2^P \end{bmatrix}, \tag{6.2}$$

where the superscript p denotes the port quantities, and it is necessary to differentiate port impedance Z_{ij}^P from the impedance element Z_{ij} in (4.8). To determine the port impedance matrix, the input impedance Z_{11}^P of the excited antenna with the presence of the other antenna open-circuited and the mutual impedance Z_{12}^P between the two antennas must be calculated. It should also be noted that since measurement of voltages and currents at microwave frequencies is difficult, a scattering matrix related to the direct measurement of incident, reflected, and transmitted waves is used to describe the multiport network problem

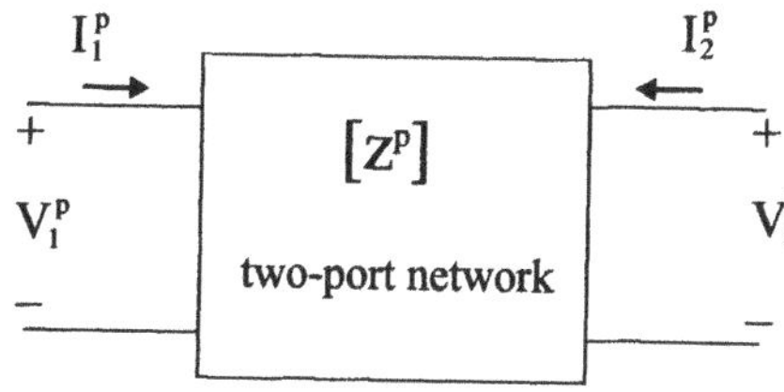

FIGURE 6.2 Two-port network representation of two coupled probe-fed rectangular microstrip antennas.

completely. For a two-port network shown in Figure 6.2, the scattering matrix $[S]$ can be evaluated from

$$[S] = ([Z^p] - Z_c[U])([Z^p] + Z_c[U])^{-1}, \tag{6.3}$$

where $[U]$ is a unit matrix of order 2×2 and Z_c is the characteristic impedance of the feeding coax (assumed to be 50 Ω here). To solve Z_{11}^p and Z_{21}^p, the following formulas can be used:

$$Z_{11}^p = -\int_{v_{p1}} \mathbf{E}^{(1)} \cdot \mathbf{J}_p^{(1)}\, dv, \tag{6.4}$$

$$Z_{21}^p = -\int_{v_{p2}} \mathbf{E}^{(1)} \cdot \mathbf{J}_p^{(2)}\, dv, \tag{6.5}$$

where $\mathbf{E}^{(1)}$ is the total electric field inside the substrate layer due to the probe current at port 1, and v_{pi} is the volume over the probe at port i. By using the full-wave analysis and Galerkin's moment-method technique discussed in Section 4.2, (6.4) and (6.5) are rewritten as

$$Z_{11}^p = -\int_a^b E_\rho^{(1)}(\rho, \phi_{p1}, z_{p1})\, d\rho, \tag{6.6}$$

$$Z_{21}^p = -\int_a^b E_\rho^{(1)}(\rho, \phi_{p2}, z_{p2})\, d\rho, \tag{6.7}$$

with

$$E_\rho^{(1)}(\rho, \phi_{pi}, z_{pi}) = \frac{1}{2\pi} \sum_{n=-\infty}^{\infty} e^{jn\phi_{pi}} \int_{-\infty}^{\infty} dk_z\, e^{jk_z z_{pi}} \bar{\bar{\tilde{G}}}_\rho^{\mathrm{T}}(n, k_z) \begin{bmatrix} \tilde{J}_\phi^{(i)}(n, k_z) \\ \tilde{J}_z^{(i)}(n, k_z) \end{bmatrix},$$

$$i = 1, 2, \tag{6.8}$$

where $\bar{\bar{\tilde{G}}}_\rho(n, k_z)$ relates the $\hat{\rho}$ component of the spectral-domain electric field inside the substrate layer to the patch surface current densities $\tilde{J}_\phi^{(i)}$ and $\tilde{J}_z^{(i)}$ of patch i; the superscript T denotes a transpose of the matrix, and the tilde again represents a Fourier transform. The matrix $\bar{\bar{\tilde{G}}}_\rho(n, k_z)$ in (6.8) is also derived as

$$\bar{\bar{\tilde{G}}}_\rho = \begin{bmatrix} \tilde{G}_{\rho\phi} \\ \tilde{G}_{\rho z} \end{bmatrix} = \frac{1}{X_{11}X_{22} - X_{12}X_{21}} \begin{bmatrix} a_1 X_{22} - a_2 X_{21} \\ -a_1 X_{12} + a_2 X_{11} \end{bmatrix}, \qquad a < \rho < b, \quad (6.9)$$

with

$$a_1 = \frac{jk_z}{k_{1\rho}} \left[H_n^{(1)\prime}(k_{1\rho}\rho) - \frac{H_n^{(1)}(k_{1\rho}a)}{J_n(k_{1\rho}a)} J_n'(k_{1\rho}\rho) \right], \quad (6.10)$$

$$a_2 = -\frac{n\omega\mu_0}{\rho k_{1\rho}^2} \left[H_n^{(1)}(k_{1\rho}\rho) - \frac{H_n^{(1)\prime}(k_{1\rho}a)}{J_n'(k_{1\rho}a)} J_n(k_{1\rho}\rho) \right], \quad (6.11)$$

$$k_{1\rho}^2 = k_1^2 - k_z^2, \qquad k_1 = \omega\sqrt{\mu_0 \epsilon_0 \epsilon_1}, \quad (6.12)$$

where the expressions of X_{ij} in (6.9) can be found in (2.35)–(2.38). Once Z_{11}^P and Z_{21}^P are determined, the mutual coupling coefficient can be calculated from (6.3) and given as

$$S_{12} = 2Z_{21}^P \frac{Z_c}{(Z_{11}^P + Z_c)^2 - (Z_{21}^P)^2}. \quad (6.13)$$

From the formulation above, the mutual coupling characteristics can be analyzed. Typical results versus edge spacing *S*, the distance between edges of two patches, for the *E*- and *H*-plane coupling cases are presented in Figures 6.3 to 6.7. In the study the input impedance of the isolated rectangular microstrip antenna with various cylinder radii of interest is adjusted to be 50 Ω at the TM_{01}-mode excitation (patch excited in the $\hat{z}$ direction). For this purpose, the feed positions for the cases studied in Figures 6.3 to 6.6 with $a = 5$, 10, and 20 cm are selected to be at $(\phi_p, z_p) = (0°, 1.79\text{ cm})$, (0°, 1.89 cm), and (0°, 1.97 cm), respectively, and the corresponding operating frequencies are at $f = 1419$, 1417, and 1415 MHz. The Z_{11}^P value obtained is found to vary slightly with 50 Ω and the Z_{21}^P value is strongly affected by the cylinder radius and edge spacing, as shown in Figure 6.3 for the *E*-plane case. The results of *E*-plane coupling coefficients versus edge spacing are shown in Figure 6.4. The results for larger cylinder radii are seen to approach those for the planar case [14–16]. It is also observed that for a large edge spacing, the coupling decreases more rapidly with increasing cylinder radius. Another case of thicker substrate thickness ($h = 0.32$ cm) is also studied. Similar coupling results are observed, and the coupling level for $h = 0.32$ cm is about 0.5–1 dB higher than that for $h = 0.1575$ cm, shown in Figure 6.4.

For the *H*-plane coupling case, mutual impedance Z_{21}^P and mutual coupling coefficient S_{12} are shown in Figures 6.5 and 6.6, respectively. As compared to the *E*-plane case, different curvature effects on the mutual coupling are observed. The coupling is decreased with decreasing cylinder radius. This is probably because for fixed edge spacing, the two patches on a cylindrical body with smaller radius are subtended by a large angle, which results in attenuation of the space wave and thus

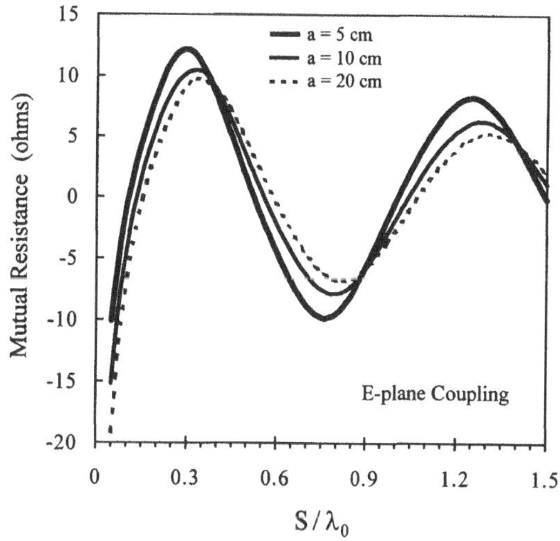

(a)

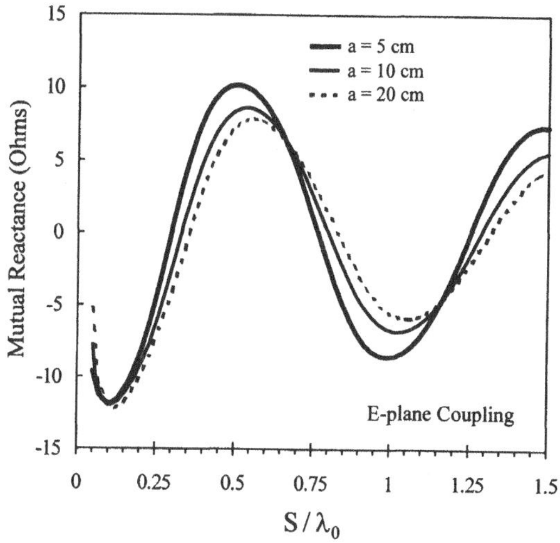

(b)

FIGURE 6.3 (*a*) Mutual resistance and (*b*) mutual reactance for the *E*-plane coupling case; $h = 1.575$ mm, $\epsilon_1 = 2.5$, $2L = 6.55$ cm, $2b\phi_0 = 10.57$ cm. For $a = 5$ cm, $f = 1.419$ GHz; $a = 10$ cm, $f = 1.417$ GHz; $a = 20$ cm, $f = 1.415$ GHz.

weakens the coupling. It should also be noted that the coupling curve for $a = 5$ cm is symmetrical with respect to $S/\lambda_0 \approx 0.26$. In this case, the two patches are on opposite sides of the cylindrical host, which results in a minimal coupling level.

Mutual coupling characteristics have also been studied experimentaly [2]. Rectangular microstrip antennas are fabricated on a flexible copper-clad substrate of thickness 0.762 mm and relative permittivity 2.98, which can easily be made conformal to cylindrical ground surfaces of different radii. The mutual coupling

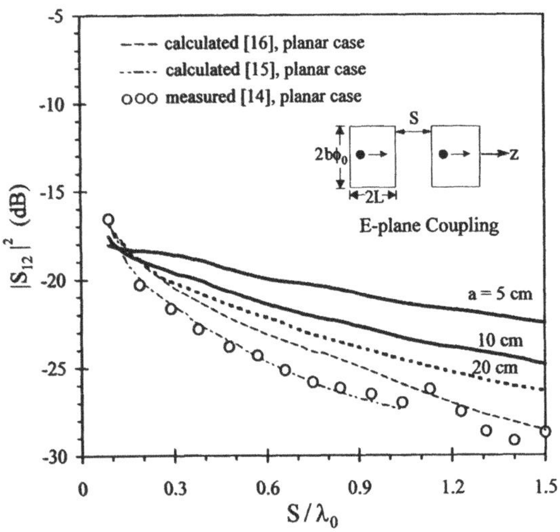

FIGURE 6.4 Mutual coupling coefficient for the *E*-plane coupling case; antenna parameters are as in Figure 6.3. (From Ref. [1], © 1994 John Wiley & Sons, Inc.)

coefficient S_{12} was measured as a function of edge spacing between two patch antennas, and the feed position for each patch antenna is adjusted to achieve a 50-Ω input impedance. A grounded dielectric substrate, with the same parameters as the patch substrate, was inserted between the two patch antennas to provide an approximation as good as that of a continuous substrate. The two-element microstrip array was then mounted on a large conducting ground plane or conducting cylinders of different radii. Different lengths of the grounded dielectric substrate represent different edge spacings between two microstrip antennas. The S_{12} coefficient was then measured as the transmission coefficient between the two antennas, with a 50-Ω source and a 50-Ω detector connected, respectively, at the feed positions of antennas 1 and 2. The data measured are presented in Figure 6.7. The results show the same behavior as theoretical prediction from the full-wave solution, and good agreement between experiment and theory is also seen.

6.2.2 Full-Wave Solution of Triangular Patches

In this section, a study of the mutual coupling between two cylindrical triangular microstrip antennas, as shown in Figure 6.8, is presented. Typical results for the mutual coupling coefficient are calculated using the full-wave approach as described in Section 6.2.1. Experiments are also performed to verify theoretical results. For the geometry shown in Figure 6.8, the two triangular patches are also of the same size and are assumed to be equilateral and probe fed at (ϕ_{p1}, z_{p1}) and (ϕ_{p2}, z_{p2}), respectively. The probe feed is also modeled as a line source, shown by (6.1). Knowing the fact that the resonant frequency of the fundamental mode (TM_{10} mode) of the triangular patch is far below the other higher-order modes, we

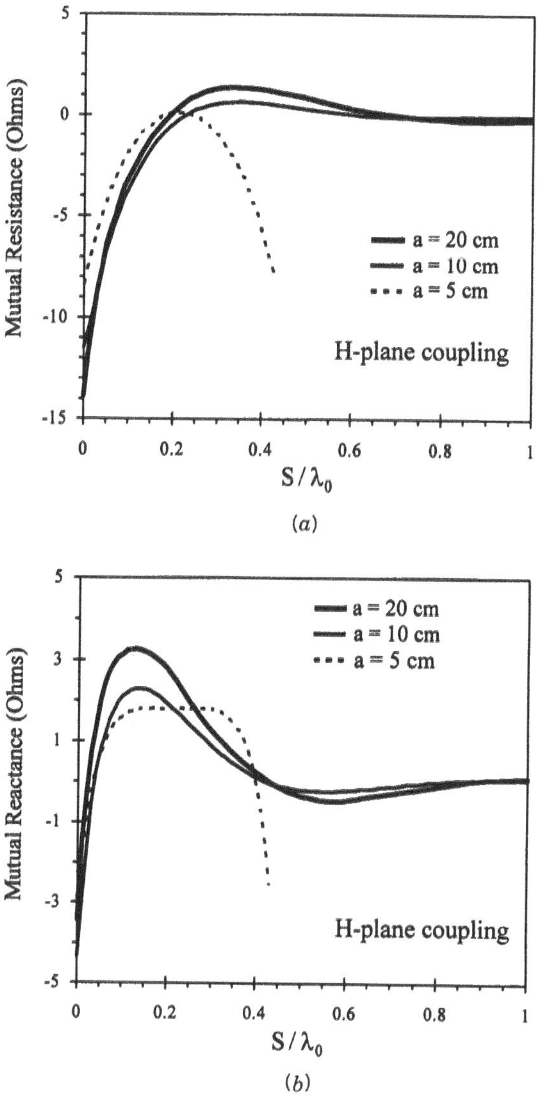

FIGURE 6.5 (*a*) Mutual resistance and (*b*) mutual reactance for the *H*-plane coupling case; antenna parameters are as in Figure 6.3.

can select only one expansion basis function [(2.99)–(2.100)] for the unknown surface current density on the two triangular patches to simplify the formulation; that is,

$$\mathbf{J}^{(i)}(\phi, z) = \hat{\phi} I_{\phi}^{(i)} J_{\phi}^{(i)}(\phi, z) + \hat{z} I_{z}^{(i)} J_{z}^{(i)}(\phi, z)\,, \qquad i = 1, 2\,, \tag{6.14}$$

with

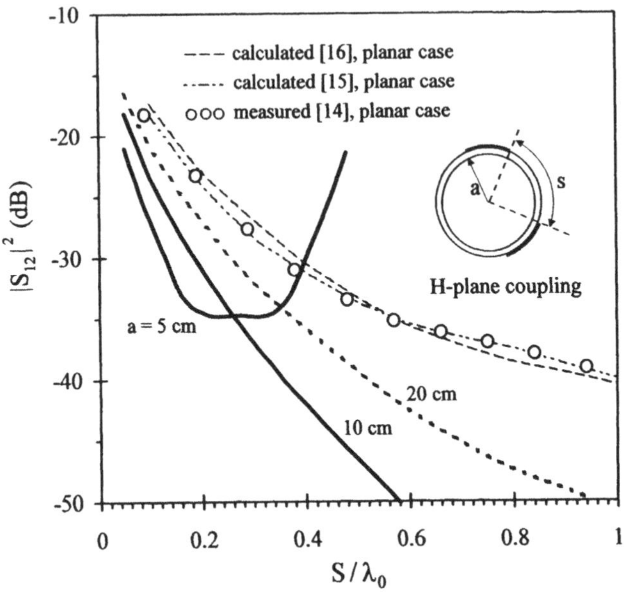

FIGURE 6.6 Mutual coupling coefficient for the *H*-plane coupling case; antenna parameters are as in Figure 6.3. (From Ref. [1], © 1994 John Wiley & Sons, Inc.)

$$J_\phi^{(1)}(\phi, z) = \frac{b\phi}{z - d_h/2} \sin \frac{(z + d_h/2)\pi}{d_h}, \tag{6.15}$$

$$J_z^{(1)}(\phi, z) = \sin \frac{(z + d_h/2)\pi}{d_h}, \tag{6.16}$$

$$J_\phi^{(2)}(\phi, z) = \frac{b(\phi - \phi_{sp})}{z - z_{sp} - d_h/2} \sin \frac{(z - z_{sp} + d_h/2)\pi}{d_h}, \tag{6.17}$$

$$J_z^{(2)}(\phi, z) = \sin \frac{(z - z_{sp} + d_h/2)\pi}{d_h}, \tag{6.18}$$

where the subscript i (=1, 2) denotes patches 1 and 2, respectively; $I_\phi^{(i)}$ and $I_z^{(i)}$ are unknown coefficients to be solved; d_h is the distance from the tip to the bottom side of the triangle; and ϕ_{sp} and z_{sp} are the spacings between two triangular patches in the $\hat{\phi}$ and $\hat{z}$ directions, respectively (see Figure 6.8). From (4.8), $I_\phi^{(i)}$ and $I_z^{(i)}$ can be solved. In turn, the input impedance, Z_{11}^p, of the excited antenna with the presence of the other antenna open-circuited and the mutual impedance, Z_{21}^p, between the two antennas can be evaluated. With Z_{11}^p and Z_{21}^p obtained, the mutual coupling coefficient can also be calculated from (6.13).

The calculated and measured *E*-plane ($\phi_{sp} = 0$) and *H*-plane ($z_{sp} = 0$) mutual

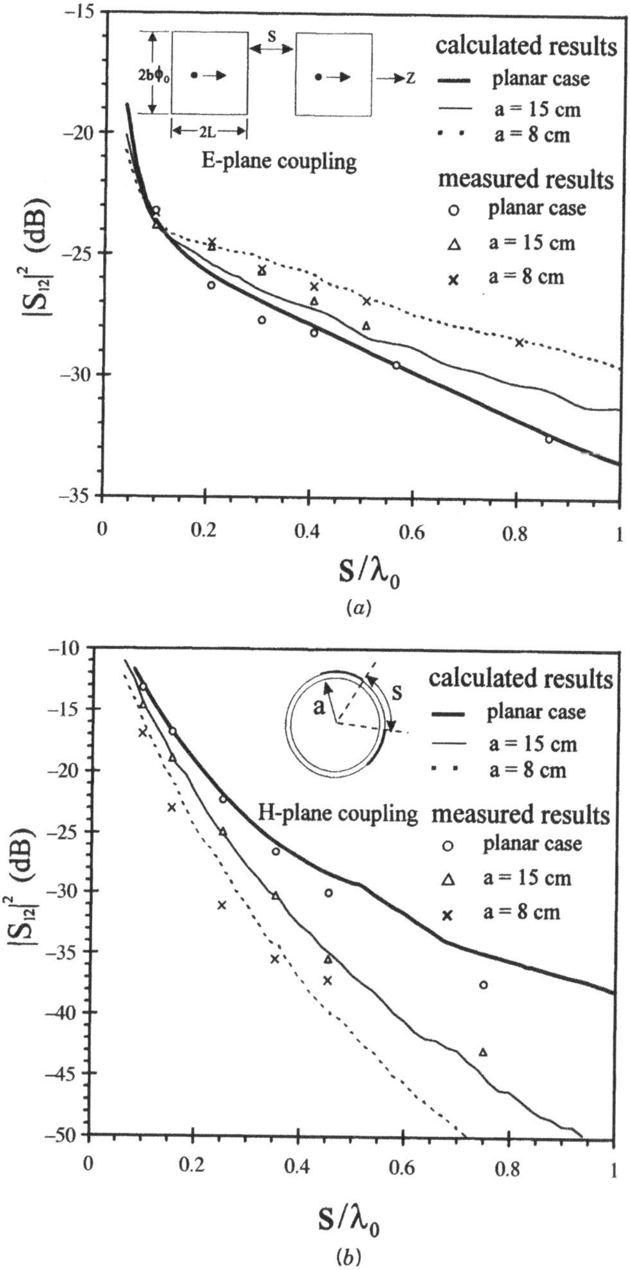

FIGURE 6.7 Measured mutual coupling coefficients versus edge spacing between two cylindrical rectangular microstrip antennas; $h = 0.762$ mm, $\epsilon_1 = 2.98$, $2L = 6$ cm, $2b\phi_0 = 4$ mm. (*a*) *E*-plane coupling case; (*b*) *H*-plane coupling case. (From Ref. [2], © 1996 John Wiley & Sons, Inc.)

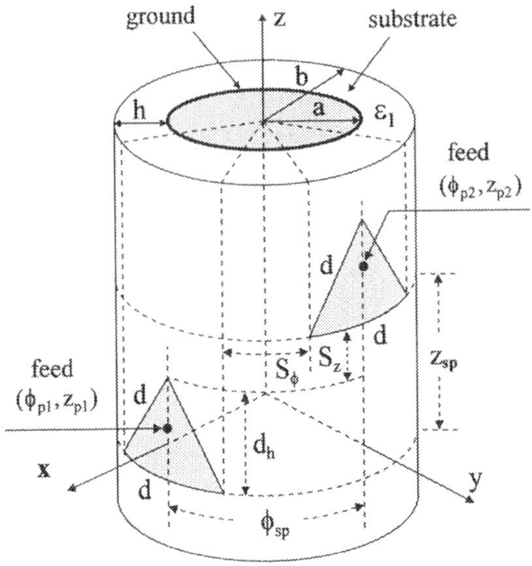

FIGURE 6.8 Geometry of two cylindrical triangular microstrip antennas. (From Ref. [4], © 1997 IEE, reprinted with permission.)

coupling coefficients versus the edge spacing between two patch antennas are presented in Figures 6.9 and 6.10. The spacings S_z and S_ϕ are all normalized with respect to the free-space wavelength, λ_0. In the study the TM_{10} mode is excited at 1.9 GHz. It is also noted that since the input impedance is affected by the curvature variation, the feed positions of the antennas with different cylinder radii are adjusted such that a 50-Ω input impedance is obtained and impedance matching to the 50-Ω coax is maintained. From the results shown, good agreement between theory and experiment is obtained. For the E-plane coupling case, mutual

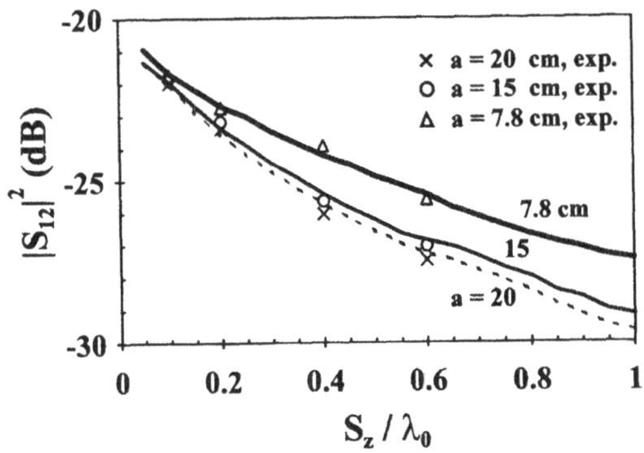

FIGURE 6.9 Mutual coupling coefficients calculated and measured for the E-plane coupling case; $h = 0.762$ mm, $\epsilon_r = 3.0$, $d = 6$ cm. (From Ref. [4], © 1997 IEE, reprinted with permission.)

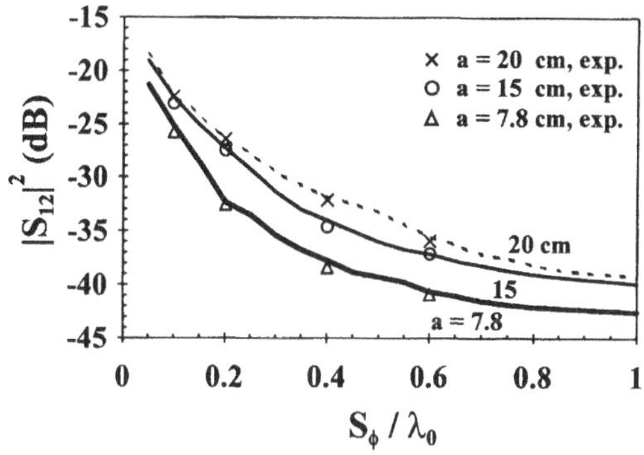

FIGURE 6.10 Mutual coupling coefficients calculated and measured for the H-plane coupling case; $h = 0.762$ mm, $\epsilon_1 = 3.0$, $d = 6$ cm. (From Ref. [4], © 1997 IEE, reprinted with permission.)

coupling is seen to decrease with increasing cylinder radius. As for the H-plane coupling case, the mutual coupling increases with increasing cylinder radius, which is in contrast to the E-plane coupling behavior. This observation is similar to those obtained for cylindrical rectangular (see Sections 6.2.1, 6.2.3, and 6.2.5) and cylindrical circular (see Sections 6.2.4 and 6.2.6) microstrip antennas. From the study of a case of different substrate thickness ($h = 1.524$ mm), it is observed that similar coupling behavior is obtained, with the coupling level for $h = 1.524$ mm case about 0.5–2 dB larger than that for $h = 0.762$ mm, shown in Figures 6.9 and 6.10. This increase in mutual coupling is probably due to the increase in surface-wave excitation in thicker substrates, which enhances the mutual coupling between microstrip antennas.

6.2.3 Cavity-Model Solution of Rectangular Patches

Theoretical formulation using a cavity-model method has also been used for the study of mutual coupling between cylindrical microstrip antennas [5,6]. Expressions for mutual impedance and the mutual coupling coefficient have been derived, and typical numerical results for the E- and H-plane coupling cases have been reported. Cavity-model solutions for rectangular microstrip antennas mounted on a cylindrical ground surface are presented and discussed below.

By considering the geometry shown in Figure 6.1 for two probe-fed cylindrical rectangular microstrip antennas, the probe currents of antennas 1 and 2 are modeled as a $\hat{\rho}$-directed current ribbon of effective arc length w_p with unit-amplitude current density written as

$$\mathbf{J}^{(i)} = \hat{\rho} J(a\phi)\delta(z' - z_{pi}), \qquad i = 1, 2, \tag{6.19}$$

with

$$J(a\phi) = \begin{cases} 1\,, & a\phi_{pi} - \frac{w_p}{2} \le a\phi \le a\phi_{pi} + \frac{w_p}{2}\,, \\ 0\,, & \text{elsewhere}\,. \end{cases} \tag{6.20}$$

To adopt the cavity-model theory, a thin-substrate condition is assumed and the substrate thickness is assumed to be much smaller than the cylinder radius and operating wavelength. In this case, the input impedance of an isolated antenna excited at the TM_{01} mode has been derived in (4.61). To study mutual coupling between two antennas, the input impedance of each antenna is adjusted to 50 Ω by choosing a suitable feed position from (4.61). Due to the thin-substrate assumption, the surface-wave effect on mutual coupling can be ignored. It is also assumed that the mutual interaction does not disturb the interior field distribution inside the cavity below the microstrip patches. Based on these assumptions, mutual impedance between the two antennas can be calculated from

$$Z_{21} = \frac{1}{I_1 I_2} \oint \mathbf{H}_1 \cdot \mathbf{M}_2 \, dl\,, \tag{6.21}$$

where I_1 and I_2 are the feed currents of the two antennas, $\mathbf{H}_1$ is the magnetic field set up by antenna 1 on antenna 2, and $\mathbf{M}_2$ is the equivalent magnetic current of antenna 2. Following the theoretical formulation in Section 4.3, mutual impedance for the E- and H-plane coupling cases can be derived. The formulation is shown in the following.

A. For the E-plane Coupling Case In this case, the feed positions of the two antennas can be represented by (ϕ_{p1}, z_{p1}) and $(\phi_{p1}, z_{p1} - 2L - S)$. The equivalent magnetic current along the edges of the cavity can be obtained from

$$\mathbf{M} = hE_\rho \hat{\rho} \times \hat{n}\,, \tag{6.22}$$

where $\hat{n}$ is a unit vector pointing outward at the edge of the cavity. From (6.22) we have the magnetic current around patch 1 for the TM_{01} mode written as follows: For $z = \pm L$ and $-\phi_0 \le \phi \le \phi_0$,

$$M^E_{\phi 1} = -hE_{01}\,, \tag{6.23}$$

and for $\phi = \pm\phi_0$ and $-L \le z \le L$,

$$M^E_{z1} = \pm hE_{01} \cos\left[\frac{\pi}{2L}(z - L)\right], \tag{6.24}$$

with

$$E_{01} = \frac{j\omega\mu_0}{k^2_{\text{eff}} - (\pi/2L)^2} \frac{w_p}{2aL\phi_0} \cos\frac{\pi z_{p1}}{2L}\,, \tag{6.25}$$

where E_{01} $(=j\omega\mu_0 C_{01})$ is the amplitude of the excited field at the TM_{01} mode and can be obtained from (4.54). By further taking the Fourier transform of (6.23) and (6.24), we have

$$\tilde{M}^{E}_{\phi 1}(p,u) = -\frac{hE_{01}\sin p\phi_0}{\pi p}(1+e^{j2Lu}), \tag{6.26}$$

$$\tilde{M}^{E}_{z1}(p,u) = -\frac{hE_{01}\sin p\phi_0}{\pi a}\frac{u(1+e^{j2Lu})}{(\pi/2L)^2+u^2}. \tag{6.27}$$

In the expressions above, the superscript E denotes the case of two cylindrical rectangular microstrip antennas with an E-plane coupling arrangement; that is, the two antennas are both excited and aligned in the axial direction.

Then, to calculate the magnetic field radiated by antenna 1, we define the following magnetic and electric vector potentials:

$$A_z = \frac{1}{2\pi}\sum_{p=-\infty}^{\infty} e^{jp\phi}\int_{-\infty}^{\infty} f_p(u)H_p^{(2)}(\rho\sqrt{k^2-u^2})e^{juz}\,du, \tag{6.28}$$

$$F_z = \frac{1}{2\pi}\sum_{p=-\infty}^{\infty} e^{jp\phi}\int_{-\infty}^{\infty} g_p(u)H_p^{(2)}(\rho\sqrt{k^2-u^2})e^{juz}\,du, \tag{6.29}$$

where $f_p(u)$ and $g_p(u)$, defined in (4.33) and (4.34), respectively, are functions of the equivalent magnetic currents of (6.26) and (6.27). From (6.28) and (6.29), we can have the magnetic field evaluated from [17]

$$\mathbf{H} = \nabla\times\mathbf{A} - j\omega\epsilon_0\epsilon_1\mathbf{F} + \frac{1}{j\omega\mu_0}\nabla(\nabla\cdot\mathbf{F}), \tag{6.30}$$

with

$$\mathbf{A} = A_z\hat{z}, \qquad \mathbf{F} = F_z\hat{z}. \tag{6.31}$$

From (6.30), the magnetic field set up by antenna 1 on antenna 2 is derived as follows:

1. At $\phi = -\phi_0$, $-L-S \geq z \geq -3L-S$,

$$H_z^E = -j\omega\epsilon_0 F_z + \frac{1}{j\omega\mu_0}\frac{\partial^2 F_z}{\partial z^2}$$
$$= \frac{1}{j2\pi\omega\mu_0}\sum_{p=-\infty}^{\infty} e^{-jp\phi_0}\int_{-\infty}^{\infty}(k_0^2-u^2)g_p(u)H_p^{(2)}(a\sqrt{k_0^2-u^2})e^{juz}\,du. \tag{6.32}$$

2. At $\phi = \phi_0$, $-L-S \geq z \geq -3L-S$,

$$H_z^E = \frac{1}{j2\pi\omega\mu_0} \sum_{p=-\infty}^{\infty} e^{jp\phi_0} \int_{-\infty}^{\infty} (k_0^2 - u^2) g_p(u) H_p^{(2)}(a\sqrt{k_0^2 - u^2}) e^{juz}\, du\,. \tag{6.33}$$

3. At $z = -L - S$, $-\phi_0 \leq \phi \leq \phi_0$,

$$\begin{aligned} H_\phi^E &= -\frac{\partial A_z}{\partial \rho} + \frac{1}{j\omega\mu_0\rho} \frac{\partial^2 F_z}{\partial \phi\, \partial z} \\ &= \frac{-1}{2\pi} \sum_{p=-\infty}^{\infty} e^{jp\phi} \int_{-\infty}^{\infty} (k_0^2 - u^2)^{1/2} f_p(u) H_p^{(2)\prime}(a\sqrt{k_0^2 - u^2}) e^{-ju(L+S)}\, du \\ &\quad - \frac{1}{j2\pi\omega\mu_0 a} \sum_{p=-\infty}^{\infty} p e^{jp\phi} \int_{-\infty}^{\infty} u g_p(u) H_p^{(2)}(a\sqrt{k_0^2 - u^2}) e^{-ju(L+S)}\, du\,. \end{aligned} \tag{6.34}$$

4. At $z = -3L - S$, $-\phi_0 \leq \phi \leq \phi_0$,

$$\begin{aligned} H_\phi^E &= \frac{-1}{2\pi} \sum_{p=-\infty}^{\infty} e^{jp\phi} \int_{-\infty}^{\infty} (k_0^2 - u^2)^{1/2} f_p(u) H_p^{(2)\prime}(a\sqrt{k_0^2 - u^2}) e^{-ju(3L+S)}\, du \\ &\quad - \frac{1}{j2\pi\omega\mu_0 a} \sum_{p=-\infty}^{\infty} p e^{jp\phi} \int_{-\infty}^{\infty} u g_p(u) H_p^{(2)}(a\sqrt{k_0^2 - u^2}) e^{-ju(3L+S)}\, du\,. \end{aligned} \tag{6.35}$$

Also, using (6.22), we have the equivalent magnetic current of antenna 2 expressed as follows: For $z = -L - S$, $-3L - S$, and $-\phi_0 \leq \phi \leq \phi_0$,

$$M_{\phi 2}^E = -hE_{01}\,, \tag{6.36}$$

and for $\phi = \pm\phi_0$ and $-L - S \geq z \geq -3L - S$,

$$M_{z2}^E = \pm hE_{01} \cos\left[\frac{\pi}{2L}(z + L + S)\right]. \tag{6.37}$$

By substituting (6.32)–(6.37) into (6.21) and after some lengthy manipulation, the mutual impedance between two antennas 1 and 2 can be derived as

$$Z_{21}^E = \left(\frac{hE_{01}}{2\pi w_p}\right)^2 \left\{R_1 + \sum_{p=1}^{\infty} [R_2(p) + R_3(p)]\right\}, \tag{6.38}$$

where

$$R_1 = -j8\omega\epsilon_0 a^2 \phi_0^2 \int_0^{\infty} \frac{A_1 H_0^{(2)\prime}(\xi)}{\xi H_0^{(2)}(\xi)}\, du\,, \tag{6.39}$$

$$R_2(p) = \frac{j8\omega\epsilon_0 a^2(\cos 2p\phi_0 - 1)}{p^2} \int_0^\infty \frac{A_1 H_p^{(2)\prime}(\xi)}{\xi H_p^{(2)}(\xi)}\, du\,, \tag{6.40}$$

$$R_3(p) = \frac{j8(\cos 2p\phi_0 - 1)}{\omega\mu_0 a} \int_0^\infty u^2 A_1 A_2 \frac{H_p^{(2)}(\xi)}{H_p^{(2)\prime}(\xi)} \left\{ \frac{a}{\xi} - \frac{\xi}{a[(\pi/2L)^2 - u^2]} \right\} du\,, \tag{6.41}$$

with

$$\xi = a\sqrt{k_0^2 - u^2}\,, \tag{6.42}$$

$$A_1 = \cos uS + 2\cos[u(L+S)] + \cos[u(3L+S)]\,, \tag{6.43}$$

$$A_2 = \frac{1}{(\pi/2L)^2 - u^2} - \frac{1}{k_0^2 - u^2}\,. \tag{6.44}$$

B. For the H-plane Coupling Case As for the H-plane coupling case, the feed positions of the two antennas are arranged at (ϕ_{p1}, z_{p1}) and $(\phi_{p1} + 2\phi_0 + S/b, z_{p1})$; that is, the two antennas are aligned in the $\hat{\phi}$ direction, with the two patches still excited in the axial direction. The equivalent magnetic current of antenna 1 is shown by (6.23)–(6.24). By following the same procedure as that described for the E-plane coupling case, the magnetic field set up by antenna 1 on antenna 2 can be written as follows:

1. At $\phi = \phi_0 + S/b$, $-L \le z \le L$,

$$H_z^H = \frac{1}{j2\pi\omega\mu_0} \sum_{p=-\infty}^{\infty} e^{jp(\phi_0 + S/b)} \int_{-\infty}^{\infty} (k_0^2 - u^2) g_p(u) H_p^{(2)}(a\sqrt{k_0^2 - u^2}) e^{juz}\, du\,. \tag{6.45}$$

2. At $\phi = 3\phi_0 + S/b$, $-L \le z \le L$,

$$H_z^H = \frac{1}{j2\pi\omega\mu_0} \sum_{p=-\infty}^{\infty} e^{jp(3\phi_0 + S/b)} \int_{-\infty}^{\infty} (k_0^2 - u^2) g_p(u) H_p^{(2)}(a\sqrt{k_0^2 - u^2}) e^{juz}\, du\,. \tag{6.46}$$

3. At $z = L$, $\phi_0 + S/b \le \phi \le 3\phi_0 + S/b$,

$$H_\phi^H = \frac{-1}{2\pi} \sum_{p=-\infty}^{\infty} e^{jp\phi} \int_{-\infty}^{\infty} (k_0^2 - u^2)^{1/2} f_p(u) H_p^{(2)\prime}(a\sqrt{k_0^2 - u^2}) e^{juL}\, du$$
$$- \frac{1}{j2\pi\omega\mu_0 a} \sum_{p=-\infty}^{\infty} p e^{jp\phi} \int_{-\infty}^{\infty} u g_p(u) H_p^{(2)}(a\sqrt{k_0^2 - u^2}) e^{juL}\, du\,. \tag{6.47}$$

4. At $z = -L$, $\phi_0 + S/b \leq \phi \leq 3\phi_0 + S/b$,

$$H_\phi^H = \frac{-1}{2\pi} \sum_{p=-\infty}^{\infty} e^{jp\phi} \int_{-\infty}^{\infty} (k_0^2 - u^2)^{1/2} f_p(u) H_p^{(2)\prime}(a\sqrt{k_0^2 - u^2}) e^{-juL} \, du$$
$$- \frac{1}{j2\pi\omega\mu_0 a} \sum_{p=-\infty}^{\infty} p e^{jp\phi} \int_{-\infty}^{\infty} u g_p(u) H_p^{(2)}(a\sqrt{k_0^2 - u^2}) e^{-juL} \, du . \tag{6.48}$$

The equivalent magnetic current of antenna 2 can be expressed as follows: For $z = \pm L$ and $\phi_0 + S/b \leq \phi \leq 3\phi_0 + S/b$,

$$M_{\phi 2}^H = -hE_{01} , \tag{6.49}$$

and for $-L \leq z \leq L$,

$$M_{z2}^H = \begin{cases} -hE_{01} \cos\left[\dfrac{\pi}{2L}(z - L)\right], & \phi = \phi_0 + \dfrac{S}{b}, \\ hE_{01} \cos\left[\dfrac{\pi}{2L}(z - L)\right], & \phi = 3\phi_0 + \dfrac{S}{b}, \end{cases} \tag{6.50}$$

where E_{01} is as given in (6.25).

Substitution of (6.45)–(6.50) in (6.21) gives the H-plane mutual impedance expressed as

$$Z_{21}^H = \left(\frac{hE_{01}}{2\pi w_p}\right)^2 \left\{X_1 + \sum_{p=1}^{\infty} [X_2(p) + X_3(p) + X_4(p)]\right\}, \tag{6.51}$$

with

$$X_1 = -j16\omega\epsilon_0 a^2 \phi_0^2 \int_0^\infty \frac{B_1 H_0^{(2)\prime}(\xi)}{\xi H_0^{(2)}(\xi)} \, du , \tag{6.52}$$

$$X_2(p) = \frac{j16\omega\epsilon_0 a^2(1 - \cos 2p\phi_0)\cos[p(2\phi_0 + S/b)]}{p^2} \int_0^\infty \frac{B_1 H_p^{(2)\prime}(\xi)}{\xi H_p^{(2)}(\xi)} \, du , \tag{6.53}$$

$$X_3(p) = \frac{j16(\cos 2p\phi_0 - 1)\cos[p(2\phi_0 + S/b)]}{\omega\mu_0} \int_0^\infty \frac{u^2 B_1 B_2 H_p^{(2)}(\xi)}{\xi H_p^{(2)\prime}(\xi)} \, du , \tag{6.54}$$

$$X_4(p) = \frac{-j8}{\omega\mu_0} \int_0^\infty \frac{\xi H_p^{(2)}(\xi)}{H_p^{(2)\prime}(\xi)} \frac{u^2 B_1 B_2}{a^2[(\pi/2L)^2 - u^2]} \, du , \tag{6.55}$$

where ξ is given in (6.42) and

$$B_1 = 1 + \cos 2uL , \tag{6.56}$$

$$B_2 = \cos\left[p\left(\frac{S}{b}\right)\right] - 2\cos\left[p\left(2\phi_0 + \frac{S}{b}\right)\right] + \cos\left[p\left(4\phi_0 + \frac{S}{b}\right)\right]. \quad (6.57)$$

With the mutual impedance obtained from (6.38) and (6.51), the mutual coupling coefficient S_{12} for the E- and H-plane coupling cases can then be calculated from (6.13), with Z_{21}^E or Z_{21}^H in place of Z_{21}^p. Due to the assumption that the mutual interaction does not disturb the interior field distribution of the antenna, the self-impedance Z_{11} is here equal to the input impedance [Eq. (4.61)] of the isolated antenna. Thus, in this case, (6.13) can be reduced to

$$S_{12} = \frac{100Z_{21}}{10^4 - Z_{21}^2}. \quad (6.58)$$

Numerical results calculated from (6.58) have also been analyzed. The mutual impedance calculated versus edge spacing for the E-plane coupling case is presented in Figure 6.11. The resonant frequency f_{01} is at 1413 MHz, and for the antenna parameters of $\epsilon_r = 2.5$, $h = 0.1575$ cm, $2L = 6.55$ cm, $2b\phi_0 = 10.57$ cm studied here, the feed positions for antenna 1 with $a = 5$, 10, and 20 cm are selected to be, respectively, at $(\phi_{p1}, z_{p1}) = (0°, -1.8 \text{ cm})$, $(0°, -1.91 \text{ cm})$, and $(0°, -1.97 \text{ cm})$ to obtain a 50-Ω input impedance. Results show that the mutual resistance and reactance have an oscillatory behavior. The mutual coupling coefficient is presented in Figure 6.12. The full-wave solutions, obtained from Section 6.2.1, are plotted in the figure for comparison. Good agreement between the cavity-model and full-wave solutions is observed. The results for the H-plane coupling case are presented in Figures 6.13 and 6.14. It is seen that the cavity-model solutions of mutual coupling coefficient obtained again agree well with the full-wave solutions. Since the computation of one cavity-model solution of S_{12} requires less time than that of the full-wave solution, this makes cavity-model analysis useful for antenna engineers in the related designs of microstrip antenna arrays.

6.2.4 Cavity-Model Solution of Circular Patches

The geometry shown in Figure 6.15 for two probe-fed cylindrical circular microstrip antennas is considered. The two circular patches are of the same radius r_d and are separated by an edge spacing S. The centers of the two patches are at (b, ϕ_1, z_1) and (b, ϕ_2, z_2). Also, to facilitate the analysis, the $(\hat{\rho}, \hat{l}, \hat{\beta})$ coordinates defined in (4.73) are adopted. In this case the feed positions of the two patches can be denoted to be at (l_{p1}, β_{p1}) and (l_{p2}, β_{p2}), respectively. Similar to the formulation in Section 4.3.3, the feed current for each patch is also modeled as a unit-amplitude current ribbon of effective length w_p. For an isolated cylindrical circular microstrip antenna, the resonant frequency and input impedance at the TM_{11} mode can be written, respectively, as (see Section 4.3.3)

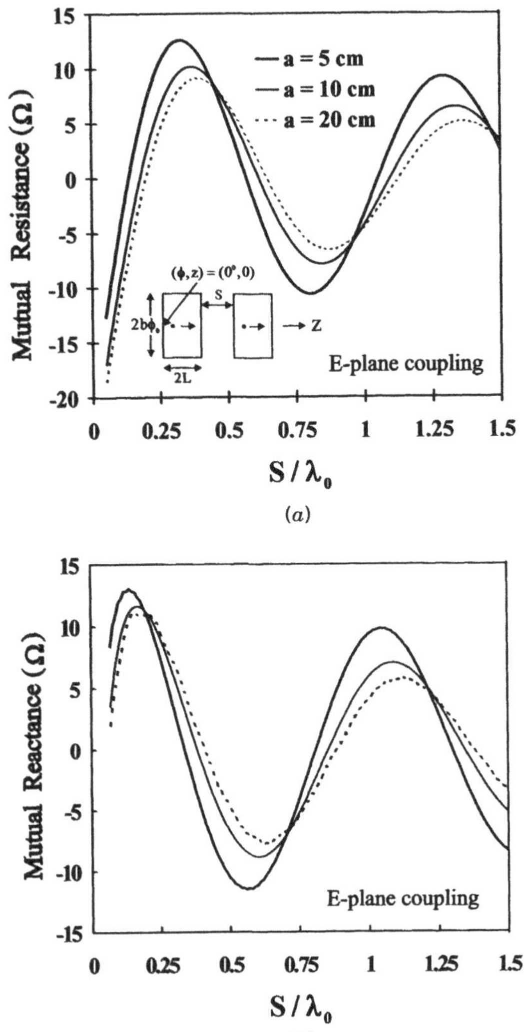

FIGURE 6.11 (*a*) Mutual resistance and (*b*) mutual reactance for the *E*-plane coupling of two cylindrical rectangular microstrip antennas; $h = 1.575$ mm, $\epsilon_1 = 2.5$, $2L = 6.55$ mm, $2b\phi_0 = 10.57$ cm. (From Ref. [5], © 1995 John Wiley & Sons, Inc.)

$$f_{11} = \frac{k_{11}c}{2\pi\sqrt{\epsilon_1}}\left[1 + \frac{2h}{\pi r_d \epsilon_1}\left(\ln\frac{\pi r_d}{2h} + 1.7726\right)\right]^{-1/2}, \tag{6.59}$$

$$Z_{\text{in}} = \frac{hk_{11}^2}{\pi^2\epsilon_0\epsilon_1}\,\frac{J_1^2(k_{11}l_p)\,\text{sinc}^2(w_p/2)}{J_1^2(k_{11}r_d)(k_{11}^2 r_d^2 - 1)}\,\frac{\delta_{\text{eff}}f^3 - jf(f^2 - f_{11}^2)}{\delta_{\text{eff}}^2 f^4 + (f^2 - f_{11}^2)^2}, \tag{6.60}$$

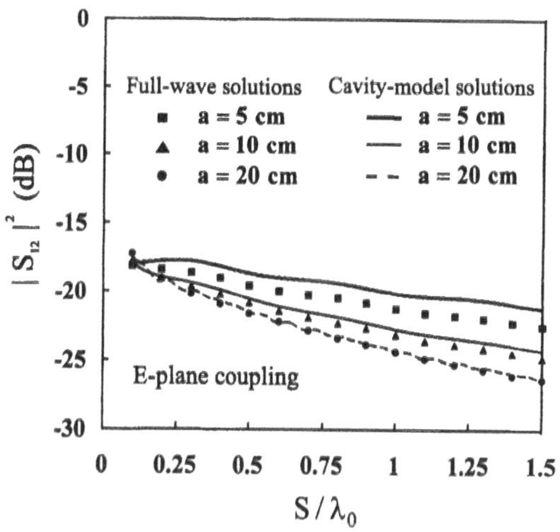

FIGURE 6.12 Mutual coupling coefficients for the E-plane coupling case; the antenna parameters are as in Figure 6.11. (From Ref. [5], © 1995 John Wiley & Sons, Inc.)

where k_{11} satisfies $J_1'(k_{11}r_d) = 0$. To study the mutual coupling, we first select suitable feed positions from (6.60) to obtain a 50-Ω input impedance for each antenna. For mutual impedance, the formulation for the feed positions in both the x–z plane ($\beta_{p1} = \beta_{p2} = 90°$) and x–y plane ($\beta_{p1} = \beta_{p2} = 0°$) is discussed.

A. Feed Positions in the x–z Plane ($\beta_{p1} = \beta_{p2} = 90°$) For the E-plane coupling case, the patch center and feed position of patch 1 are chosen to be at $(\phi_1, z_1) = (0°, 0)$ and $(l_{p1}, \beta_{p1}) = (l_p, 90°)$, respectively. For patch 2, the patch center and feed position are chosen to be, respectively, at $(\phi_2, z_2) = (0°, -2r_d - S)$ and $(l_{p2}, \beta_{p2}) = (l_p, 90°)$. It should be noted that the coordinates of (l_{p2}, β_{p2}) shown above are with respect to the center of patch 2. The equivalent magnetic current along the edge of patch 1 is written as

$$\begin{aligned}\mathbf{M}_1^E &= M_{z1}^E\hat{z} + M_{\phi 1}^E\hat{\phi} \\ &= hE_{11}J_1(k_{11}r_d)\sin\beta(\cos\beta\hat{z} - \sin\beta\hat{\phi}),\end{aligned} \tag{6.61}$$

with

$$E_{11} = \frac{-j2\omega\mu_0 k_{11}^2}{(k_{\text{eff}}^2 - k_{11}^2)} \frac{w_p J_1(k_{11}l_p)\,\text{sinc}(w_p/2)}{\pi J_1^2(k_{11}r_d)(k_{11}^2 r_d^2 - 1)}, \tag{6.62}$$

where E_{11} ($=j\omega\mu_0 C_{11}$) is the amplitude of the excited field at the TM_{11} mode and can be obtained from (4.83). Then, by taking the Fourier transform of (6.61) and

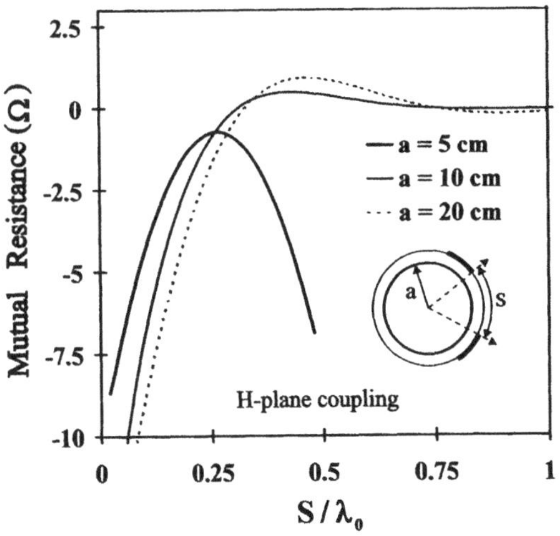

(a)

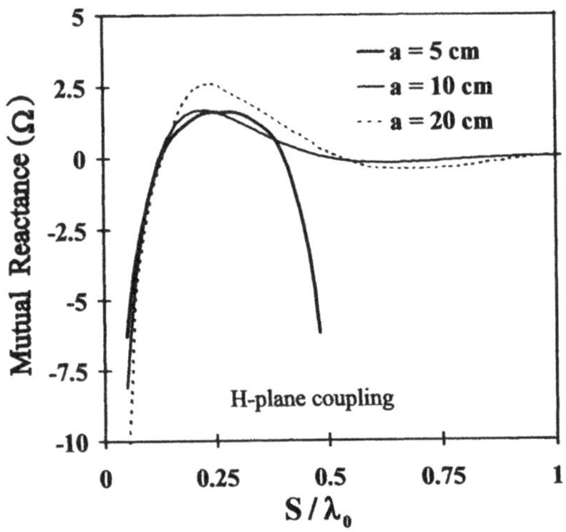

(b)

FIGURE 6.13 (*a*) Mutual resistance and (*b*) mutual reactance for the *H*-plane coupling of two cylindrical rectangular microstrip antennas; the parameters are as in Figure 6.11. (From Ref. [5], © 1995 John Wiley & Sons, Inc.)

applying to (6.30) the magnetic and electric vector potentials defined in (6.28)–(6.29), the magnetic field radiated by antenna 1 on antenna 2 can be expressed as

$$H_z^E = \frac{1}{j2\pi\omega\mu_0} \sum_{p=-\infty}^{\infty} e^{jpr_d \cos\beta/a} \int_{-\infty}^{\infty} (k_0^2 - u^2) g_p(u) H_p^{(2)}(a\sqrt{k_0^2 - u^2})$$
$$\times e^{-ju(S+2r_d)} e^{jur_d \sin\beta}\, du\,, \tag{6.63}$$

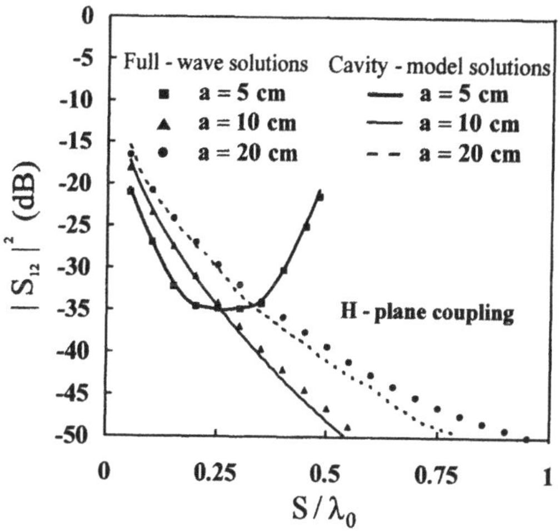

FIGURE 6.14 Mutual coupling coefficients for the H-plane coupling case; the antenna parameters are as in Figure 6.11. (From Ref. [5], © 1995 John Wiley & Sons, Inc.)

$$H_{\phi}^{E} = \frac{-1}{2\pi} \sum_{p=-\infty}^{\infty} e^{jpr_d \cos \beta / a} \int_{-\infty}^{\infty} (k_0^2 - u^2)^{1/2} f_p(u) H_p^{(2)\prime}(a\sqrt{k_0^2 - u^2}) \times e^{-ju(S+2r_d)} e^{jur_d \sin \beta} \, du - \frac{1}{j2\pi\omega\mu_0 a} \sum_{p=-\infty}^{\infty} e^{jpr_d \cos \beta / a} \int_{-\infty}^{\infty} u g_p(u) H_p^{(2)}(a\sqrt{k_0^2 - u^2}) \times e^{-ju(S+2r_d)} e^{jur_d \sin \beta} \, du . \quad (6.64)$$

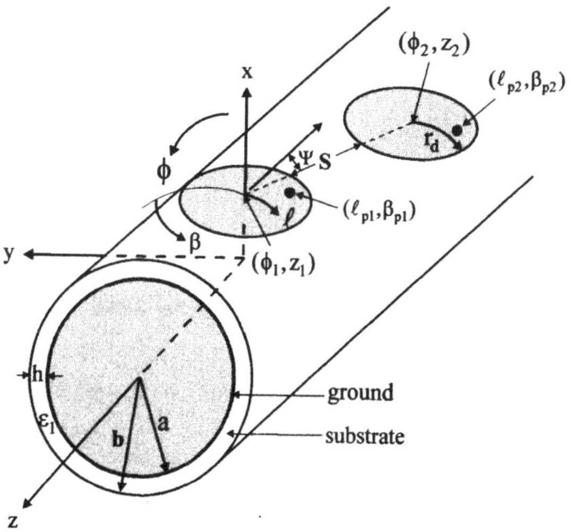

FIGURE 6.15 Geometry of two cylindrical circular microstrip antennas.

Then, similar to (6.61), the equivalent magnetic current of antenna 2 is written as

$$\mathbf{M}_2^E = M_{z2}^E \hat{z} + M_{\phi 2}^E \hat{\phi}$$
$$= hE_{11} J_1(k_{11} r_d) \sin \beta (\cos \beta \hat{z} - \sin \beta \hat{\phi}) . \quad (6.65)$$

From (6.63)–(6.65), the mutual impedance can be evaluated from (6.21) and is written as

$$Z_{21}^E = \left[\frac{r_d h E_{11} J_1(k_{11} r_d)}{2\pi w_p} \right]^2 \left\{ R_1 + \sum_{p=1}^{\infty} [R_2(p) + R_3(p)] \right\} , \quad (6.66)$$

with

$$R_1 = -j2\omega\epsilon_0 \int_0^\infty \frac{A H_0^{(2)\prime}(\xi)}{\xi H_0^{(2)}(\xi)} I_a^2(0, u)\, du , \quad (6.67)$$

$$R_2(p) = -j4\omega\epsilon_0 \int_0^\infty \frac{A H_p^{(2)\prime}(\xi)}{\xi H_p^{(2)}(\xi)} I_a^2(p, u)\, du , \quad (6.68)$$

$$R_3(p) = \frac{-4p}{j\omega\mu_0 a} \int_0^\infty \frac{A u H_p^{(2)}(\xi)}{\xi H_p^{(2)\prime}(\xi)} \left[2I_a(p,u) I_b(p,u) + \frac{apu}{\xi^2} I_a^2(p,u) + \frac{\xi^2 I_b^2(p,u)}{apu} \right] du , \quad (6.69)$$

where ξ is in (6.42) and

$$A = \cos[u(S + 2r_d)] , \quad (6.70)$$

$$I_a(p, u) = \pi J_2 \left[r_d \sqrt{\left(\frac{p}{a}\right)^2 + u^2} \right] \cos\left(2 \tan^{-1} \frac{au}{p}\right) + \pi J_0 \left[r_d \sqrt{\left(\frac{p}{a}\right)^2 + u^2} \right] , \quad (6.71)$$

$$I_b(p, u) = -\pi J_2 \left[r_d \sqrt{\left(\frac{p}{a}\right)^2 + u^2} \right] \sin\left(2 \tan^{-1} \frac{au}{p}\right) . \quad (6.72)$$

For the H-plane coupling case, with the patch center and feed position of patch 1 given at $(\phi_1, z_1) = (0°, 0)$ and $(l_{p1}, \beta_{p1}) = (l_p, 90°)$, the patch center and feed position of patch 2 are chosen, respectively, to be at $(\phi_2, z_2) = [(S + 2r_d)/b, 0]$ and $(l_{p2}, \beta_{p2}) = (l_p, 90°)$. The equivalent magnetic current of antenna 1 is the same as that of the E-plane coupling case [Eq. (6.61)]. In this case, the magnetic field set up by antenna 1 on antenna 2 is derived as

$$H_z^H = \frac{1}{j2\pi\omega\mu_0} \sum_{p=-\infty}^{\infty} e^{jp(S+2r_d+r_d\cos\beta)/a} \int_{-\infty}^{\infty} (k_0^2 - u^2) g_p(u)$$
$$\times H_p^{(2)}(a\sqrt{k_0^2 - u^2}) e^{jur_d \sin\beta}\, du\,, \tag{6.73}$$

$$H_\phi^H = \frac{-1}{2\pi} \sum_{p=-\infty}^{\infty} e^{jp(S+2r_d+r_d\cos\beta)/a} \int_{-\infty}^{\infty} (k_0^2 - u^2)^{1/2} f_p(u)$$
$$\times H_p^{(2)\prime}(a\sqrt{k_0^2 - u^2}) e^{jur_d \sin\beta}\, du$$
$$- \frac{1}{j2\pi\omega\mu_0 a} \sum_{p=-\infty}^{\infty} p e^{jp(S+2r_d+r_d\cos\beta)/a} \int_{-\infty}^{\infty} u g_p(u)$$
$$\times H_p^{(2)}(a\sqrt{k_0^2 - u^2}) e^{jur_d \sin\beta}\, du\,. \tag{6.74}$$

From (6.61), (6.71), and (6.74), the expression of the mutual impedance for the H-plane coupling case can be obtained by evaluating (6.21), and the result is the same as that [Eq. (6.66)] of the E-plane coupling case with $A = \cos[p(S + 2r_d)/a]$ in place of (6.70).

B. Feed Positions in the x–y Plane ($\beta_{p1} = \beta_{p2} = 0°$) For the E-plane coupling case, the patch center and feed position of patch 1 are chosen to be at $(\phi_1, z_1) = (0°, 0)$ and $(l_{p1}, \beta_{p1}) = (l_p, 0°)$. For patch 2, the patch center and feed position are at $(\phi_2, z_2) = [(2r_d + S)/b, 0]$ and $(l_{p2}, \beta_{p2}) = (l_p, 0°)$. Following the formulation in Section 6.2.4A, the mutual impedance is derived as

$$Z_{21}^E = \left[\frac{r_d h E_{11} J_1(k_{11} r_d)}{2\pi w_p}\right]^2 \left\{X_1 + \sum_{p=1}^{\infty} [X_2(p) + X_3(p)]\right\}, \tag{6.75}$$

with

$$X_1 = \frac{-2}{j\omega\mu_0 a} \int_0^\infty \frac{A\xi H_0^{(2)}(\xi)}{\xi H_0^{(2)\prime}(\xi)} I_b^2(0, u)\, du\,, \tag{6.76}$$

$$X_2(p) = -j4\omega\epsilon_0 \int_0^\infty \frac{AH_p^{(2)\prime}(\xi)}{\xi H_p^{(2)}(\xi)} I_b^2(p, u)\, du\,, \tag{6.77}$$

$$X_3(p) = \frac{-4p}{j\omega\mu_0 a} \int_0^\infty \frac{AuH_p^{(2)}(\xi)}{\xi H_p^{(2)\prime}(\xi)} \left[2I_a(p, u) I_b(p, u) + \frac{apu}{\xi^2} I_b^2(p, u)\right.$$
$$\left. + \frac{\xi^2 I_a^2(p, u)}{apu}\right] du\,, \tag{6.78}$$

where ξ is as in (6.42), $I_a(p, u)$ as in (6.71), $I_b(p, u)$ as in (6.72), and $A = \cos[p(S + 2r_d)/a]$.

For the H-plane coupling case, the patch center and feed position of patch 2 are at $(\phi_2, z_2) = (0°, -2r_d - S)$ and $(l_{p2}, \beta_{p2}) = (l_p, 0°)$, with the coordinates of patch 1 unchanged. The expression of the mutual impedance is derived to be given in (6.75) with $A = \cos[u(S + 2r_d)]$.

With the mutual impedance obtained, the mutual coupling coefficient can readily be calculated from (6.58). Numerical results of the feed positions in the x–z plane are first calculated and shown in Figures 6.16 to 6.19. The antenna parameters considered are $r_d = 3.85$ cm, $\epsilon_1 = 2.5$, and $h = 0.1575$ cm. The resonant frequency f_{11} is calculated at 1405 MHz. Due to the dependence of the input impedance on the cylinder radius, the feed positions for patch 1 with $a = 5$, 10, 20, and 30 cm are selected to be, respectively, at $(l_p, \beta_p) = (1.06\text{ cm}, 90°)$, $(0.95\text{ cm}, 90°)$, $(0.89\text{ cm}, 90°)$, and $(0.86\text{ cm}, 90°)$. The E- and H-plane mutual impedances are shown in Figures 6.16 and 6.18, and the mutual coupling coefficients are presented in Figures 6.17 and 6.19. It is observed that the dependence of mutual coupling on the edge spacing is the same as discussed for the cylindrical rectangular and triangular microstrip antennas. It is also seen that the coupling curves of $a = 5$ cm and 10 cm show a minimum coupling level around $S/\lambda_0 \simeq 0.37$ and 1.1, respectively. This is because in this case, the two circular patches are about on opposite sides of the cylinder host, which gives minimal mutual coupling between the two patches. This behavior is similar to that observed in Figures 6.6 and 6.14.

For the x–y plane, the feed positions of patch 1 with $a = 5$, 10, and 30 cm are chosen to be, respectively, at $(l_p, \beta_p) = (0.75\text{ cm}, 90°)$, $(0.8\text{ cm}, 90°)$, and $(0.85\text{ cm}, 90°)$ for 50-Ω impedance matching. The calculated mutual impedance and mutual coupling coefficients are shown in Figures 6.20 to 6.23. It is observed that the variation of E-plane mutual coupling coefficients with the cylinder radius (Figure 6.21) is in contrast to that for feed positions in the x–z plane. This is probably because in this case, the edge spacing of the two antennas is subtended by a large angle for the ground cylinder of a smaller cylinder radius, which reduces the radiated space wave set up by antenna 1 on antenna 2 and thus weakens the mutual interaction.

6.2.5 GTLM Solution of Rectangular Patches

Mutual coupling between two cylindrical rectangular microstrip antennas has also been studied using GTLM theory [7]. Referring to the geometry in Figure 6.1 and considering the patches to be excited at the TM_{01} mode, the two rectangular microstrip antennas can be represented by an equivalent circuit, shown in Figure 6.24, where the two rectangular patches are modeled as sections of transmission lines and are replaced by the equivalent π networks of g_1, g_2, g_3 [(4.126)–(4.127)] and g_1', g_2', g_3' [(4.128)–(4.129)]. The circuit element $y_s - y_m$ is the total wall admittance at the radiating edges, y_s denotes the self-admittance, and y_m denotes the mutual admittance [see (4.115)]. Then to solve the equivalent circuit,

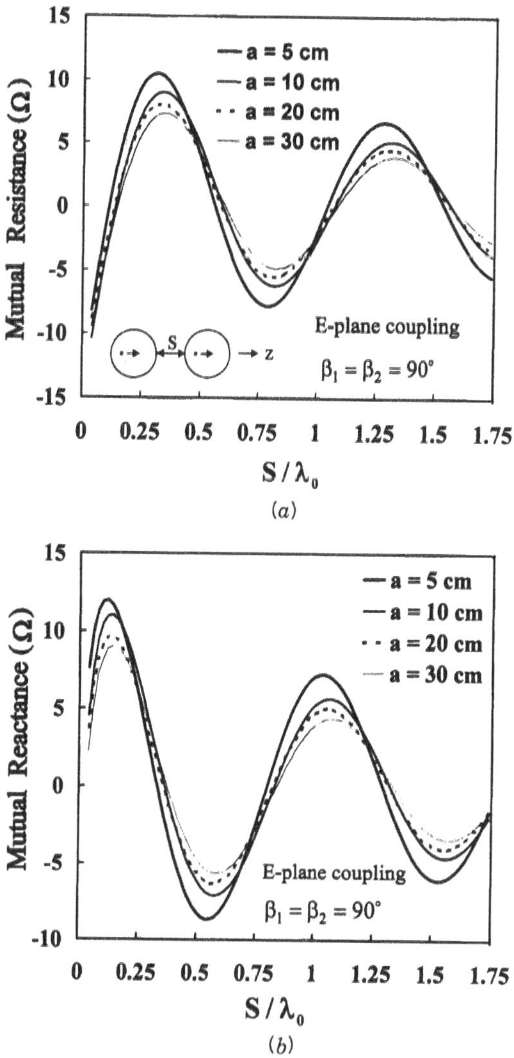

FIGURE 6.16 (*a*) Mutual resistance and (*b*) mutual reactance for the *E*-plane coupling of two cylindrical circular microstrip antennas; $h = 1.575$ mm, $\epsilon_1 = 2.5$, $r_d = 3.85$ cm, $\beta_{p1} = \beta_{p2} = 90°$.

we decompose the circuit into three cascade connections of two-port networks. The *ABCD* matrix for a two-port network [18] is used in the analysis of the cascade networks with the following relations:

$$\begin{bmatrix} V_1 \\ I_1 \end{bmatrix} = \begin{bmatrix} A^{(1)} & B^{(1)} \\ C^{(1)} & D^{(1)} \end{bmatrix} \begin{bmatrix} V_2 \\ -I_2 \end{bmatrix}, \qquad (6.79)$$

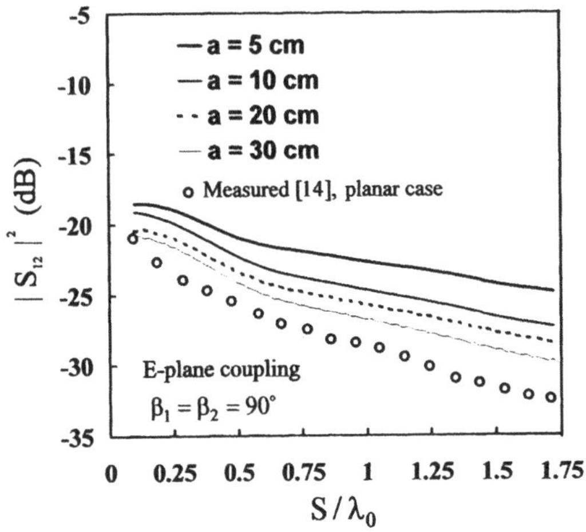

FIGURE 6.17 Mutual coupling coefficients for the *E*-plane coupling case; the antenna parameters are as in Figure 6.16. (From Ref. [6], © 1995 John Wiley & Sons, Inc.)

$$\begin{bmatrix} V_2 \\ -I_2 \end{bmatrix} = \begin{bmatrix} A^{(2)} & B^{(2)} \\ C^{(2)} & D^{(2)} \end{bmatrix} \begin{bmatrix} V_3 \\ -I_3 \end{bmatrix}, \tag{6.80}$$

$$\begin{bmatrix} V_3 \\ -I_3 \end{bmatrix} = \begin{bmatrix} A^{(3)} & B^{(3)} \\ C^{(3)} & D^{(3)} \end{bmatrix} \begin{bmatrix} V_4 \\ -I_4 \end{bmatrix}, \tag{6.81}$$

where I_1 and I_4 are the probe input currents and V_1 and V_4 are the probe excitation voltages; I_2 and I_3 are the currents flowing into ports 2 or 3, and V_2 and V_3 are voltages at ports 2 or 3. Therefore, the *ABCD* matrix of the cascade connection can be written in terms of the individual *ABCD* matrix as follows:

$$\begin{aligned} \begin{bmatrix} V_1 \\ I_1 \end{bmatrix} &= \begin{bmatrix} A & B \\ C & D \end{bmatrix} \begin{bmatrix} V_4 \\ -I_4 \end{bmatrix} \\ &= \begin{bmatrix} A^{(1)} & B^{(1)} \\ C^{(1)} & D^{(1)} \end{bmatrix} \begin{bmatrix} A^{(2)} & B^{(2)} \\ C^{(2)} & D^{(2)} \end{bmatrix} \begin{bmatrix} A^{(3)} & B^{(3)} \\ C^{(3)} & D^{(3)} \end{bmatrix} \begin{bmatrix} V_4 \\ -I_4 \end{bmatrix}. \end{aligned} \tag{6.82}$$

The relationship between the *ABCD* and *Z* parameters can then be obtained from

$$\begin{bmatrix} Z_{11} & Z_{12} \\ Z_{21} & Z_{22} \end{bmatrix} = \begin{bmatrix} A/C & (AD - BC)/C \\ 1/C & D/C \end{bmatrix}. \tag{6.83}$$

With the [Z] matrix determined, the mutual coupling coefficient can be calculated from (6.13) with Z_c set to 50 Ω.

Typical results of GTLM solutions calculated for *E*- and *H*-plane mutual coupling coefficients versus edge spacing *S* are presented in Figures 6.25 and 6.26.

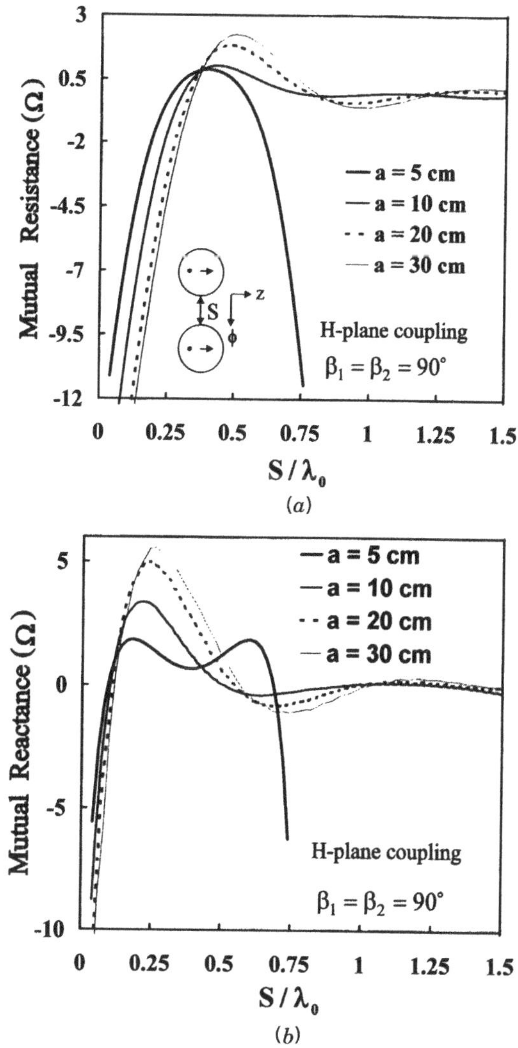

FIGURE 6.18 (*a*) Mutual resistance and (*b*) mutual reactance for the *H*-plane coupling of two cylindrical circular microstrip antennas; $h = 1.575$ mm, $\epsilon_1 = 2.5$, $r_d = 3.85$ cm, $\beta_{p1} = \beta_{p2} = 90°$.

The edge spacing is normalized to the free-space wavelength and the operating frequency is at 1441 MHz, where the TM_{01} mode is excited. The measured data are also shown in the figure, for comparison. From the results it is seen that good agreement between GTLM theory and experiment is observed, and the coupling behavior is the same as that observed using the full-wave approach (Section 6.2.1) and cavity-model method (Section 6.2.3).

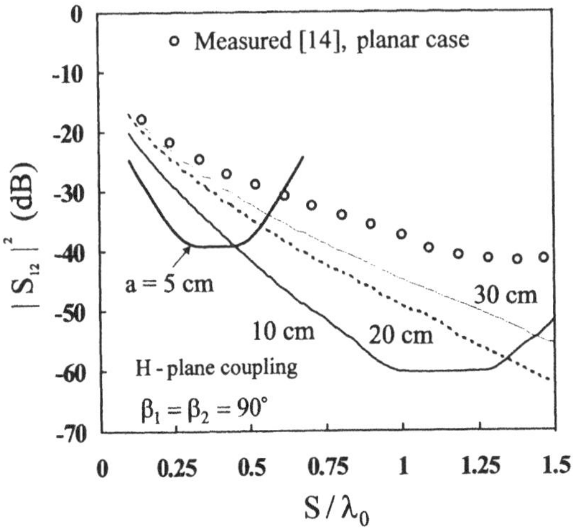

FIGURE 6.19 Mutual coupling coefficients for the *H*-plane coupling case; the antenna parameters are as in Figure 6.18. (From Ref. [6], © 1995 John Wiley & Sons, Inc.)

6.2.6 GTLM Solution of Circular Patches

GTLM analysis for mutual coupling between two cylindrical circular microstrip antennas, shown in Figure 6.15, is described here. Following the formulation of coupling between two planar circular microstrip antennas in [19], the equivalent circuit shown in Figure 6.27 is obtained. Compared to the equivalent circuit shown in Figure 4.26 for a single circular microstrip antenna, an additional mutual admittance, y_{AB}, between the radiating edges of the two circular patches is introduced in Figure 6.27. The mutual admittance has been derived as

$$y_{AB} = \frac{hr_d^2}{4\pi^3 j\omega\mu_0 a} \sum_{p=-\infty}^{\infty} \int_{-\infty}^{\infty} \left\{ A_p B_p(\xi) \left[\frac{\xi}{a} I_b^2(p,u) + \frac{2pu}{\xi} I_a(p,u) I_b(p,u) \right.\right.$$

$$\left.\left. + \frac{ap^2u^2}{\xi^3} I_a^2(p,u) \right] - A_p \frac{\omega^2 \mu_0 \epsilon_0 a}{\xi B_p(\xi)} I_b^2(p,u) \right\} du\,, \tag{6.84}$$

with

$$A_p = \exp\left[-j \frac{p(S+2r_d)}{a} \sin\Psi - ju(S+2r_d)\cos\Psi \right], \tag{6.85}$$

$$B_p(\xi) = \frac{H_p^{(2)}(\xi)}{H_p^{(2)\prime}(\xi)}, \tag{6.86}$$

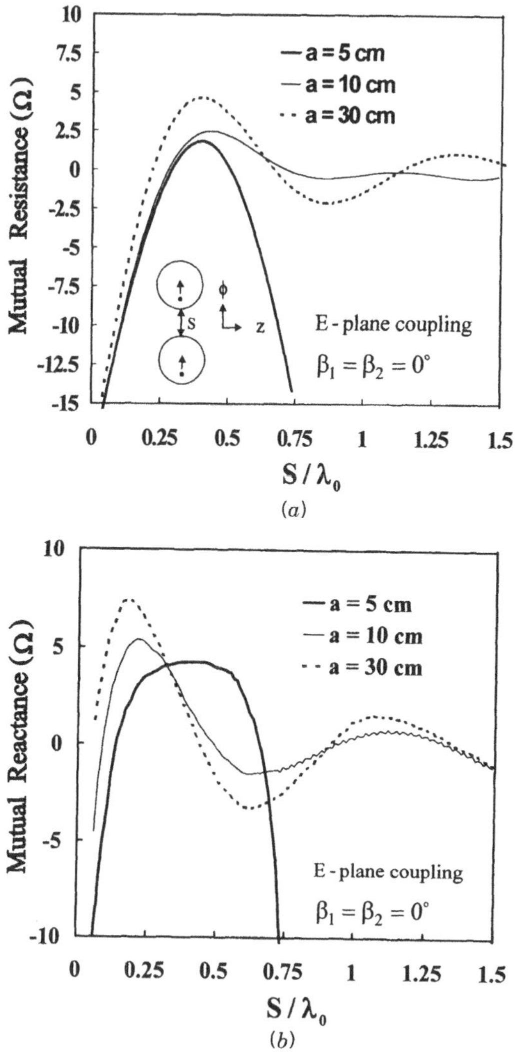

FIGURE 6.20 (*a*) Mutual resistance and (*b*) mutual reactance for the *E*-plane coupling of two cylindrical circular microstrip antennas; $h = 1.575$ mm, $\epsilon_1 = 2.5$, $r_d = 3.85$ cm, $\beta_{p1} = \beta_{p2} = 0°$.

where ξ is as in (6.42), $I_a(p, u)$ as in (6.71), and $I_b(p, u)$ as in (6.72); Ψ is an angle representing the orientation of patch 2 with respect to patch 1 (see Figure 6.15). Expressions of other parameters in Figure 6.27 are the same as in Section 4.4.2. Next, by simplifying the equivalent circuit, a two-port network with a 2×2 impedance matrix $[Z]$ can be obtained and is given as

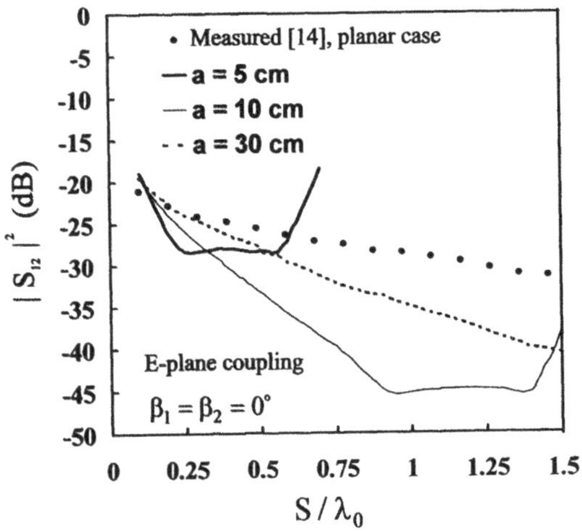

FIGURE 6.21 Mutual coupling coefficients for the *E*-plane coupling case; the antenna parameters are as in Figure 6.20.

$$\begin{bmatrix} V_1 \\ V_2 \end{bmatrix} = \begin{bmatrix} Z_{11} & Z_{12} \\ Z_{21} & Z_{22} \end{bmatrix} \begin{bmatrix} I_1 \\ I_2 \end{bmatrix}, \tag{6.87}$$

with

$$Z_{11} = Z_{22} = \frac{\Delta_1}{\Delta}, \tag{6.88}$$

$$Z_{12} = Z_{21} = \frac{g_2^2 y_w}{\Delta}, \tag{6.89}$$

$$\Delta_1 = \frac{y_0 + g_1 + g_2}{(g_2 + g_3 + y_w)^2 - y_{AB}^2} - g_2^2(g_2 + g_3 + y_w), \tag{6.90}$$

$$\Delta = \frac{\Delta_1^2 - g_2^4 y_{AB}^2}{(g_2 + g_3 + y_w)^2 - y_{AB}^2}, \tag{6.91}$$

where I_1 and I_2 are the probe input currents and V_1 and V_2 are the probe excitation voltages. Also, with the [Z] matrix obtained, the mutual coupling coefficient is readily evaluated from (6.13).

Figure 6.28 shows measured and calculated mutual coupling coefficients versus edge spacing for the *E*-plane ($\Psi = 0°$) and *H*-plane ($\Psi = 90°$) coupling cases. The calculated resonant frequency of TM_{11} excitation is at about 1563 MHz, while the measured resonant frequencies are at about $f = 1559$, 1558, and 1557 MHz for

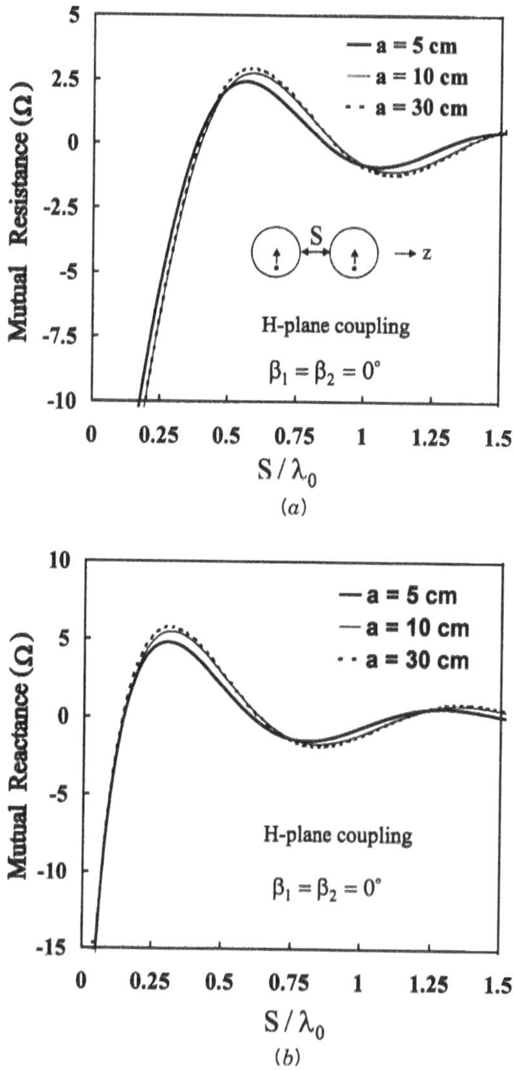

FIGURE 6.22 (*a*) Mutual resistance and (*b*) mutual reactance for the *H*-plane coupling of two cylindrical circular microstrip antennas; $h = 1.575$ mm, $\epsilon_1 = 2.5$, $r_d = 3.85$ cm, $\beta_{p1} = \beta_{p2} = 0°$.

$a = 8$, 15, and 50 cm, respectively. The difference between the calculated and measured resonant frequencies is thus within 6 MHz (about 0.4%) for the case studied here. From the results shown, good agreement is observed between the data measured and GTLM solutions calculated, and the dependence of mutual coupling on the cylinder radius is the same as that obtained in Section 6.2.4 using cavity-model solutions.

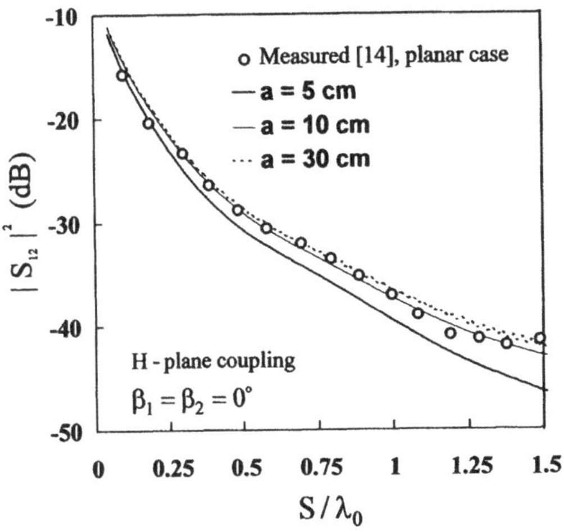

FIGURE 6.23 Mutual coupling coefficients for the *H*-plane coupling case; the antenna parameters are as in Figure 6.22.

6.3 CYLINDRICAL MICROSTRIP ANTENNAS WITH PARASITIC PATCHES

Using parasitic patches gap coupled to the radiating edges of a rectangular microstrip antenna has been shown to be capable of significantly broadening the antenna bandwidth [20], which improves the narrow-bandwidth characteristic of microstrip antennas. The applicability of such a method to cylindrical rectangular microstrip antennas has also been studied using a full-wave approach [9,10]. The theoretical formulation is described here, and numerical results for the antenna bandwidth, determined to be the frequency range over which the voltage standing-

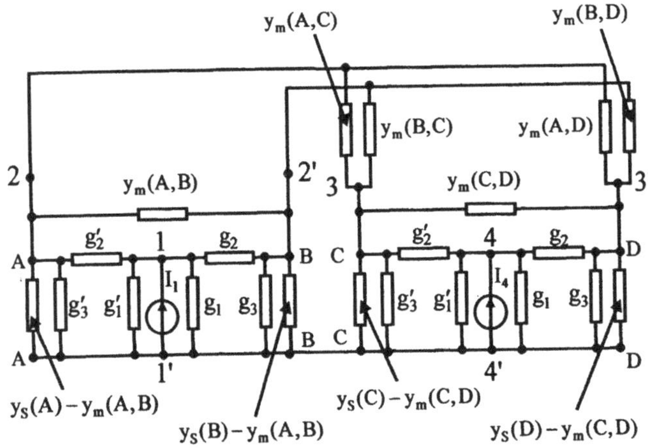

FIGURE 6.24 Equivalent circuit of the two rectangular microstrip antennas shown in Figure 6.1. (From Ref. [7], © 1997 IEE, reprinted with permission.)

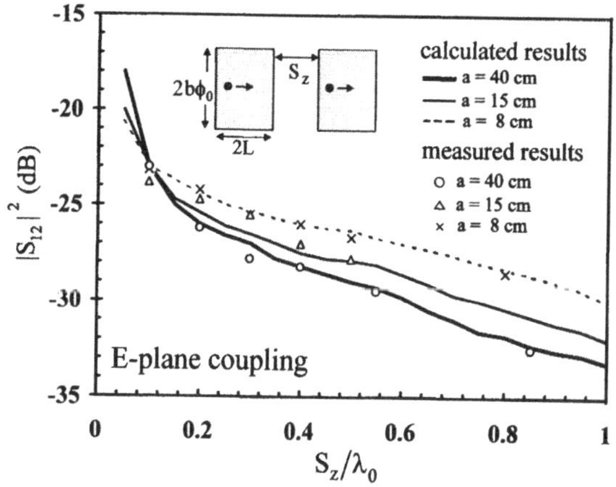

FIGURE 6.25 Mutual coupling coefficient for the *E*-plane coupling case; $\epsilon_1 = 2.98$, $h = 0.762$ mm, $2L = 6$ cm, $2b\phi_0 = 4$ cm. (From Ref. [7], © 1997 IEE, reprinted with permission.)

wave ratio (VSWR) is less than 2, are presented and analyzed. The curvature effects on the antenna bandwidth improvement using this method are also investigated.

Figure 6.29 shows the geometry of a cylindrical rectangular microstrip antenna with two parasitic patches gap coupled to its radiating edges. All the patches have a gap spacing of S. The driven patch (patch 2) has a length of $2L_2$ and a width of $2W$ ($=2b\phi_0$) and is excited by a probe feed at (ϕ_p, z_p). The dimensions of the two parasitic patches (patch 1 and patch 3) are chosen to be $2L_1 \times 2W$ and $2L_3 \times 2W$,

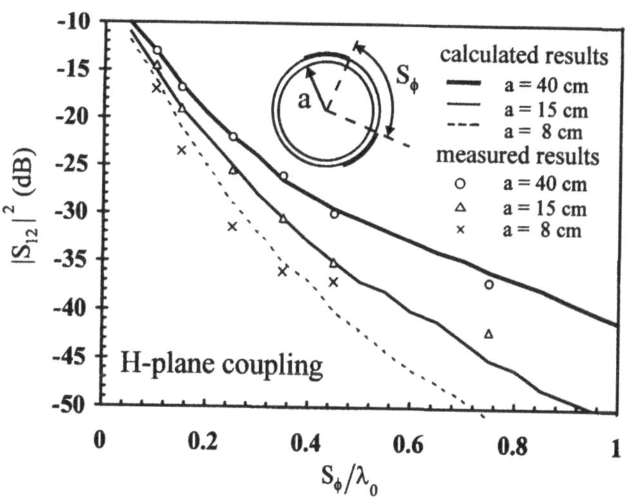

FIGURE 6.26 Mutual coupling coefficient for the *H*-plane coupling case; antenna parameters are as in Figure 6.25. (From Ref. [7], © 1997 IEE, reprinted with permission.)

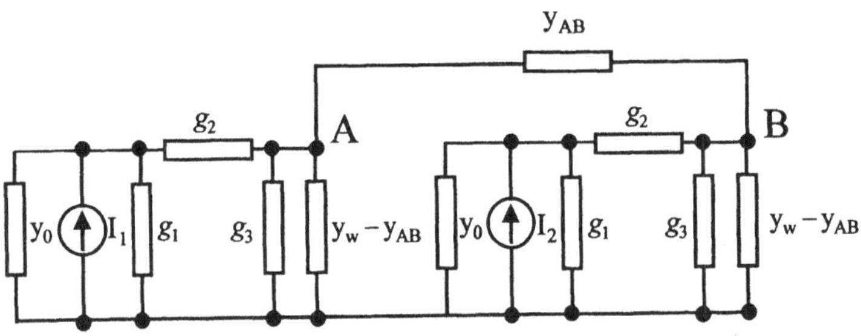

FIGURE 6.27 Equivalent circuit of the two circular microstrip antennas shown in Figure 6.15.

respectively, where the patch widths are taken to be equal to that of the driven patch and the patch lengths are slightly different from that of the driven patch. Thus, at the center operating frequency, the driven patch is at resonance, and at nearby frequencies, the two parasitic patches can also become resonant. This staggering of resonances can make the antenna bandwidth wider.

To solve the problem, the probe feed is first modeled as a unit-amplitude current source of (4.1), as discussed in Section 4.2. Next, by imposing the boundary condition that the total electric field tangential to the surface of the driven and parasitic patches must be zero and following the theoretical formulation in Section 4.2, we can have the following integral equation on the patches:

$$\sum_{q=-\infty}^{\infty} e^{jq\phi} \int_{-\infty}^{\infty} dk_z\, e^{jk_z z} \tilde{\bar{\bar{G}}}(q, k_z) \begin{bmatrix} \sum_{i=1}^{3} \tilde{J}_{\phi}^{(i)}(q, k_z) \\ \sum_{i=1}^{3} \tilde{J}_{z}^{(i)}(q, k_z) \end{bmatrix}$$

$$= -\sum_{q=-\infty}^{\infty} e^{q(\phi-\phi_p)} \int_{-\infty}^{\infty} dk_z\, e^{jk_z(z-z_p)} \begin{bmatrix} T_{11} & T_{12} \\ T_{21} & T_{22} \end{bmatrix} \begin{bmatrix} R_1 \\ R_2 \end{bmatrix}, \qquad (6.92)$$

where the superscript i $(=1, 2, 3)$ denotes patch i; $\tilde{\bar{\bar{G}}}(q, k_z)$ is given in (4.6), with the tilde denoting the spectral domain; the elements in the $[T]$ and $[R]$ matrices have been derived in [21]. The integral equation of (6.92) is then solved using Galerkin's moment method. The surface current density on patch i is expanded in terms of linear combinations of cavity-model basis functions [(2.68)–(2.69)]; that is,

$$\mathbf{J}^{(i)}(\phi, z) = \sum_{n=1}^{N_i} I_{\phi n}^{(i)} \mathbf{J}_{\phi n}^{(i)}(\phi, z) + \sum_{m=1}^{M_i} I_{zm}^{(i)} \mathbf{J}_{zm}^{(i)}(\phi, z)\,, \qquad (6.93)$$

where $I_{\phi n}^{(i)}$ and $I_{zm}^{(i)}$ are unknown expansion coefficients of the basis functions $J_{\phi n}^{(i)}$ and $J_{zm}^{(i)}$, respectively. For the configuration studied here, the cavity-model basis

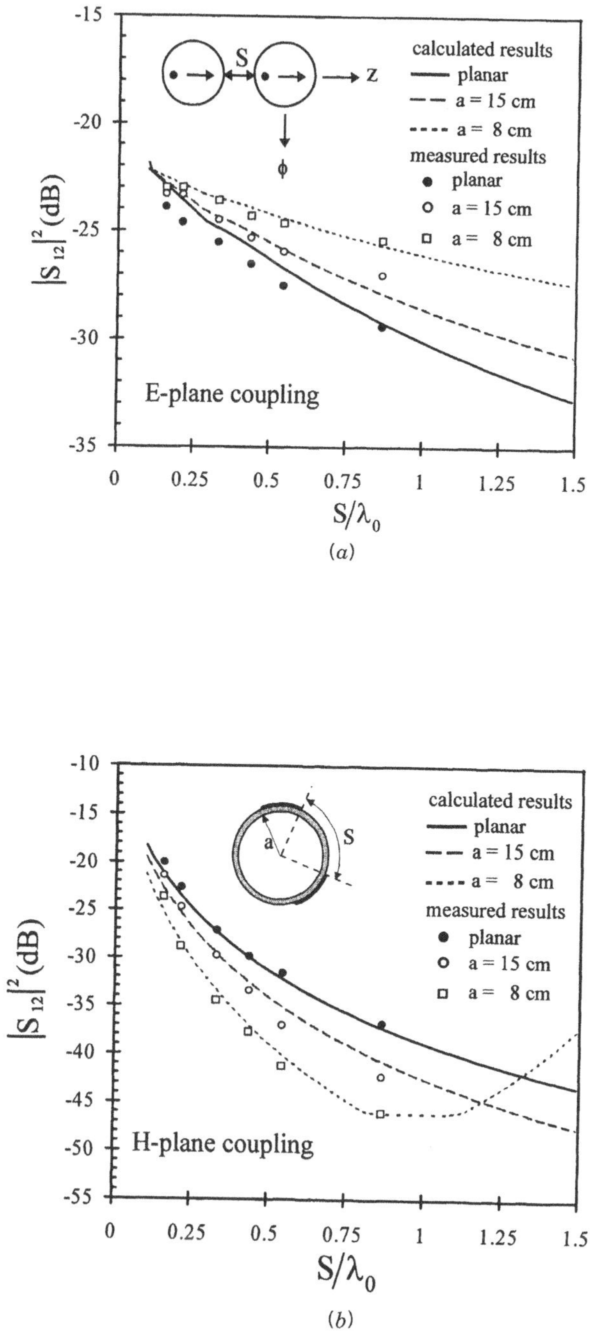

FIGURE 6.28 Mutual coupling coefficients measured and calculated versus edge spacing between two circular microstrip antennas; $h = 0.762$ mm, $\epsilon_1 = 3.0$, $r_d = 3.2$ cm. (*a*) *E*-plane coupling case; (*b*) *H*-plane coupling case. (From Ref. [8], © 1996 John Wiley & Sons, Inc.)

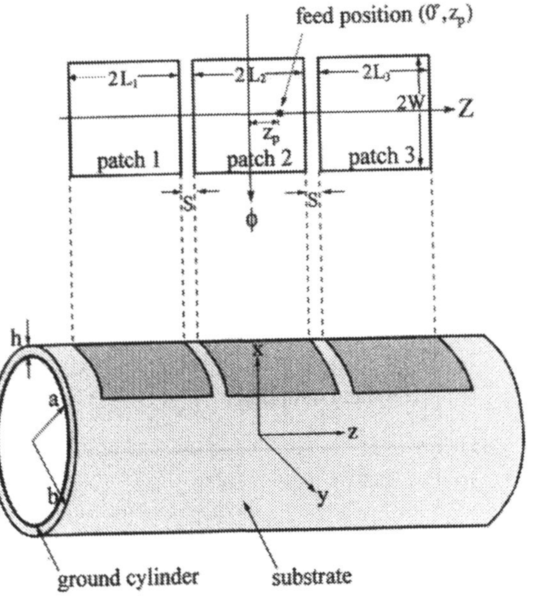

FIGURE 6.29 Configuration of a cylindrical rectangular microstrip antenna with two parasitic patches gap coupled to its radiating edges.

functions in the spectral domain for patch i $(=1, 2, 3)$ are written as

$$\tilde{J}^{(i)}_{\phi n}(q, k_z) = j^{[p_1^{(i)}+p_2^{(i)}+1]} e^{-jk_z z_i} \frac{p_1^{(i)} k_z}{\phi_0} \frac{\sin(p_1^{(i)}\pi/2 - q\phi_0)}{q^2 - (p_1^{(i)}\pi/2\phi_0)^2} \frac{\sin(p_2^{(i)}\pi/2 - k_z L_i)}{k_z^2 - (p_2^{(i)}\pi/2L_i)^2}, \tag{6.94}$$

$$\tilde{J}^{(i)}_{zm}(q, k_z) = j^{[r_1^{(i)}+r_2^{(i)}+1]} e^{-jk_z z_i} \frac{q r_2^{(i)}}{L_i} \frac{\sin(r_1^{(i)}\pi/2 - q\phi_0)}{q^2 - (r_1^{(i)}\pi/2\phi_0)^2} \frac{\sin(r_2^{(i)}\pi/2 - k_z L_i)}{k_z^2 - (r_2^{(i)}\pi/2L_i)^2}, \tag{6.95}$$

where $z_1 = -(S + L_1 + L_2)$, $z_2 = 0$, and $z_3 = S + L_2 + L_3$; p_1, p_2, r_1, and r_2 are integers and are dependent on the mode numbers n and m. By substituting (6.94)–(6.95) into (6.92), applying Galerkin's moment-method procedure, and integrating over each patch area, the following matrix equation is obtained:

$$\begin{bmatrix} Z_{11} & Z_{12} & Z_{13} \\ Z_{21} & Z_{22} & Z_{23} \\ Z_{31} & Z_{32} & Z_{33} \end{bmatrix} \begin{bmatrix} I_1 \\ I_2 \\ I_3 \end{bmatrix} = \begin{bmatrix} V_1 \\ V_2 \\ V_3 \end{bmatrix}, \tag{6.96}$$

with

$$Z_{ij} = \begin{pmatrix} (Z_{ln}^{\phi\phi})_{N_i \times N_j} & (Z_{lm}^{\phi z})_{N_i \times M_j} \\ (Z_{kn}^{z\phi})_{M_i \times N_j} & (Z_{km}^{zz})_{M_i \times M_j} \end{pmatrix}, \tag{6.97}$$

$$I_i = \begin{pmatrix} (I_{\phi n}^{(i)})_{N_i \times 1} \\ (I_{zm}^{(i)})_{M_i \times 1} \end{pmatrix}, \tag{6.98}$$

$$V_i = \begin{pmatrix} (V_{\phi n}^{(i)})_{N_i \times 1} \\ (V_{zm}^{(i)})_{M_i \times 1} \end{pmatrix}, \qquad i = 1, 2, 3\,, \quad j = 1, 2, 3\,, \tag{6.99}$$

where the expressions of $Z_{ln}^{\phi\phi}$, $Z_{lm}^{\phi z}$, $Z_{kn}^{z\phi}$, and Z_{km}^{zz} in (6.97) are given in (2.73)–(2.76), and $V_{\phi n}$ and V_{zm} are expressed in (4.9)–(4.10). The terms Z_{11}, Z_{22}, and Z_{33} are contributed solely from patches 1, 2, and 3, respectively, and the term Z_{ij}, $i \neq j$, accounts for the coupling effect between patch i and patch j. By solving (6.96), the unknown expansion coefficients $I_{\phi n}^{(i)}$ and $I_{zm}^{(i)}$ for patch i can be obtained. By neglecting the self-impedance of the probe feed, which is much smaller than the impedance contributed from the patch current for thin-substrate cases, the input impedance of a cylindrical rectangular microstrip antenna with gap-coupled parasitic patches can be evaluated from

$$Z_{\text{in}} = -\int_a^b E_\rho(\rho, \phi_p, z_p)\, d\rho\,, \tag{6.100}$$

with

$$E_\rho(\rho, \phi, z) = \frac{1}{2\pi} \sum_{q=-\infty}^{\infty} e^{jq\phi} \int_{-\infty}^{\infty} dk_z\, e^{jk_z z} \bar{\bar{G}}_\rho^{\mathrm{T}}(q, k_z) \left[\sum_{i=1}^{3} \tilde{J}^{(i)}(q, k_z) \right], \tag{6.101}$$

where E_ρ is the $\hat{\rho}$ component of the electric field in the substrate layer due to the surface currents on the driven and parasitic patches, and $\bar{\bar{G}}_\rho(q, k_z)$, given in (4.13), relates the $\hat{\rho}$ component of the electric field in the spectral domain inside the substrate layer to a patch surface current at $\rho = b$.

From (6.100), the antenna bandwidth, defined here to be the frequency range over which the VSWR is less than 2, can be determined. To obtain numerical convergence for the moment-method calculation in the analysis, the unknown surface current density on the driven and parasitic patches are all expanded with 12 cavity-model basis functions ($N_i = M_i = 6$, $i = 1, 2, 3$); that is, a total of 36 basis functions are used in the calculation. In this case good convergence results can be achieved for the input impedance. With the characteristic impedance of the feeding coax set to 50 Ω, a typical result for VSWR versus frequency for various gap spacings with $a = 5$ cm is shown in Figure 6.30. The feed position is at $(\phi_p, z_p) = (0°, 1.15\text{ cm})$, where a maximum antenna bandwidth is obtained. This optimal feed position is selected from various positions along the z axis, with

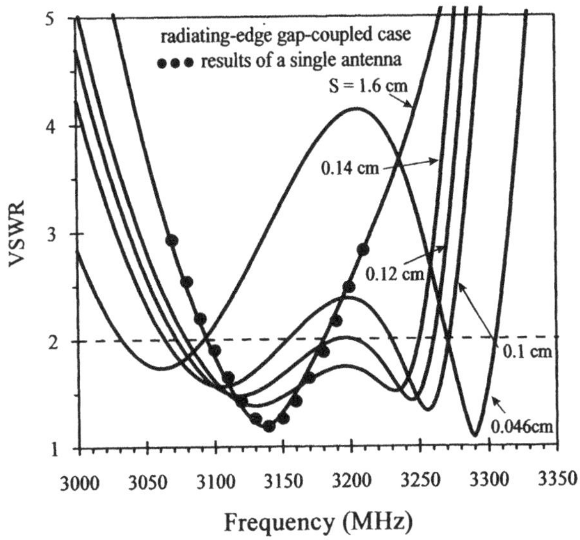

FIGURE 6.30 VSWR versus frequency for different gap spacings; $a = 5$ cm, $h = 1.59$ mm, $\epsilon_1 = 2.55$, $2L_2 = 29$ mm, $2L_1 = 2L_3 = 27.5$ mm, $2W = 40$ mm, $\phi_p = 0°$, $z_p = 11.5$ mm. (From Ref. [9], © 1994 John Wiley & Sons, Inc.)

$0 < z_p < L_2$. It is found that the optimal feed position depends strongly on the resonant lengths of the parasitic patches; however, it is insensitive to variations in cylinder radius. From the result in Figure 6.30 it can be seen that with an optimal gap spacing of $S = 0.12$ cm, the antenna bandwidth can reach 190 MHz (about 6%), which is about 2.1 times that (88 MHz) of a single rectangular microstrip antenna (shown by the solid circles in the figure and calculated by using the single patch formulation in Section 4.2). The results for a much larger gap spacing S are also plotted in the figure. For $S > 1.6$ cm, the results obtained show very small differences from those of $S = 1.6$ cm, which are very similar to those (shown by solid circles) for a single antenna.

The results for cylinder radii of $a = 5$, 10, and 15 cm with optimal gap spacings of $S = 0.12$, 0.15, and 0.16 cm, respectively, are also shown in Figure 6.31*a*. The results for a corresponding single antenna are calculated and presented in Figure 6.31*b*. It can be observed that the optimal gap spacing increases with increasing cylinder radius, and is on the order of one substrate thickness. The antenna bandwidths are 190, 166, and 160 MHz for $a = 5$, 10, and 15 cm, respectively, which are all about 2.1 times that of a corresponding single antenna [see Figure 6.31*b*]. It should be noted that more significant antenna bandwidth improvement can be expected if parasitic patches of unequal lengths are used [20] or the parasitic patches are short-circuited [22]. Also, the parasitic patches can be placed to be gap coupled to the nonradiating edges of the driven rectangular patch antenna [10]. However, since the coupling due to the nonradiating-edge gap-

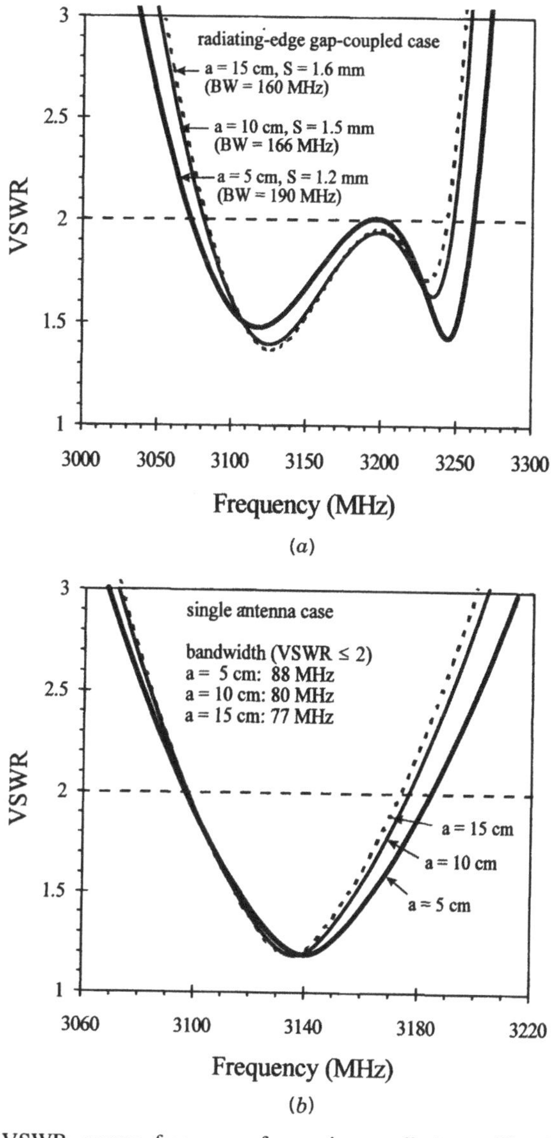

FIGURE 6.31 VSWR versus frequency for various cylinder radii with optimal gap spacings; other parameters are as in Figure 6.30: (*a*) antenna with radiating-edge gap-coupled patches; (*b*) single-antenna case. (From Ref. [9], © 1994 John Wiley & Sons, Inc.)

coupled parasitic patches is small because of the curvature of the ground cylinder (see Section 6.2.1), improvement in the antenna bandwidth is smaller for the nonradiating-edge coupling case [10] than for the radiating-edge coupling case studied here.

6.4 COUPLING BETWEEN CONCENTRIC SPHERICAL MICROSTRIP ANTENNAS

Figure 6.32 shows the geometry of concentric spherical circular and annular-ring microstrip antennas. The spacing between the two concentric patches is denoted as S $(=r_1 - r_d)$. Two cases are considered: One is the annular-ring patch as a parasitic patch gap coupled to the circular patch [12], and the other is the circular patch as a parasitic patch gap coupled to the annular-ring patch [11]. From the results obtained in Section 6.3, it is expected that by choosing the gap spacing to be about one substrate thickness or less, the antenna bandwidth of the spherical microstrip antenna can be enhanced. To study this problem, a rigorous Green's function formulation in the spectral domain and a Galerkin moment-method calculation have been utilized [11,12].

6.4.1 Annular-Ring Patch as a Parasitic Patch

By considering the annular-ring patch as a parasitic patch, the circular patch is assumed to be excited by a probe feed, expressed by (5.1), at (θ_p, ϕ_p). By following the theoretical formulation described in Section 5.2.1, we apply the boundary condition that the tangential electric field must vanish on all patches; that is,

$$\hat{r} \times (\mathbf{E}^D + \mathbf{E}^{\mathrm{AR}} + \mathbf{E}^P) = 0\,, \tag{6.102}$$

where $\mathbf{E}^D$ is the electric field due to the surface current density $\mathbf{J}^D$ on the circular patch, $\mathbf{E}^{\mathrm{AR}}$ the electric field due to the surface current density $\mathbf{J}^{\mathrm{AR}}$ on the annular-ring patch, and $\mathbf{E}^P$ the electric field due to the probe current with all the patches being absent. The unknown surface current density on the two concentric patches are then expanded into different sets of cavity-model basis functions:

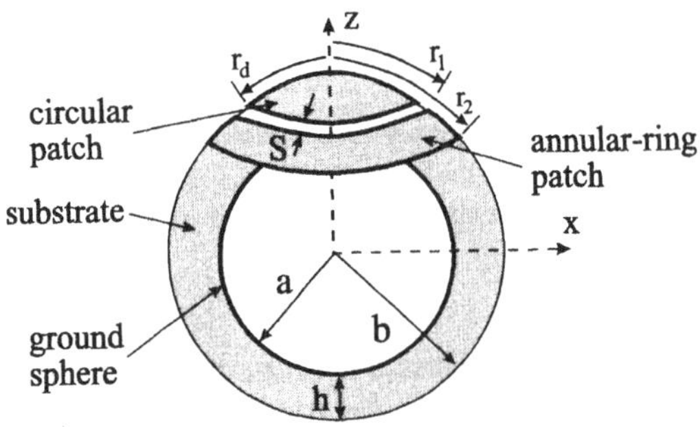

FIGURE 6.32 Geometry of concentric spherical circular and annular-ring microstrip antennas.

$$\mathbf{J}^{\mathrm{AR}} = \sum_{i=1}^{N_1} I_i^{\mathrm{AR}} \mathbf{J}_i^{\mathrm{AR}} , \tag{6.103}$$

$$\mathbf{J}^{D} = \sum_{i=1}^{N_2} I_i^{D} \mathbf{J}_i^{D} , \tag{6.104}$$

where I_i^{AR} is the unknown coefficient of the ith expansion function $\mathbf{J}_i^{\mathrm{AR}}$ for the annular-ring patch, whose expressions are listed in (3.121) and (3.123); and I_i^D is the unknown coefficient of the ith expansion function $\mathbf{J}_i^D$ [(3.92)–(3.93)] for the circular patch. To solve for the unknowns, Galerkin's procedure is used, applying weighting functions identical to the basis functions to (6.102), and we have the following matrix equation:

$$[Z]_{N\times N}[I]_{N\times 1} = [V]_{N\times 1} , \qquad N = N_1 + N_2 . \tag{6.105}$$

Once the surface current density on the circular (driven) and annular-ring (parasitic) patches are evaluated, the input impedance of the annular-ring-patch-loaded (ARL) circular microstrip antenna can be calculated from

$$Z_{\mathrm{in}} = -\int_v [(\mathbf{E}_r^{\mathrm{AR}} + \mathbf{E}_r^D) + \mathbf{E}_r^p] \cdot \mathbf{J}^p \, dv , \tag{6.106}$$

where $(E_r^{\mathrm{AR}} + E_r^D)$ and E_r^p are, respectively, the $\hat{r}$ component of the electric field in the substrate layer due to the patch current and probe current, and $\mathbf{J}^p$ is the probe current given in (5.1). It should also be noted that the expressions of $(E_r^{\mathrm{AR}} + E_r^D)$ and E_r^p are the same as given in (5.21) and (5.24), respectively.

Numerical results for (6.106) have been calculated. Figure 6.33 shows VSWR versus frequency for various sizes of annular-ring patch. Note that the circular patch is excited at the TM_{11} mode and the annular-ring patch is excited at the TM_{12} mode. Also, in the present case ($S = 0.1$ mm) the gap spacing is much less than the substrate thickness ($h = 1$ mm), in order to increase the coupling effect. The patch dimensions are selected such that the resonant frequency of the circular patch at the TM_{11} mode is slightly different from that of the annular-ring patch at the TM_{12} mode. In this case, the input impedance is altered significantly and the antenna bandwidth can be greatly improved. The results for three different annular-ring patches of widths $(r_2 - r_1)$ 1.5, 1.525, and 1.54 cm are presented, and the case for a single circular microstrip antenna is shown for comparison. The antenna bandwidth, determined from VSWR ≤ 2, of a single circular microstrip antenna is seen to be about 150 MHz (about 2.36% with respect to the center frequency at 6.34 GHz). For the case with an annular-ring patch width of 1.525 cm, the antenna bandwidth is seen to be 380 MHz (about 6%), which is about 2.5 times that of a single antenna. The curvature effect on the optimal annular-ring patch size, with the gap spacing fixed, for bandwidth enhancement is also studied. The results are shown in Figure 6.34. The antenna bandwidths for $a = 3$, 5, and 8 cm are found to be about 400, 380, and 360 MHz, respectively. The

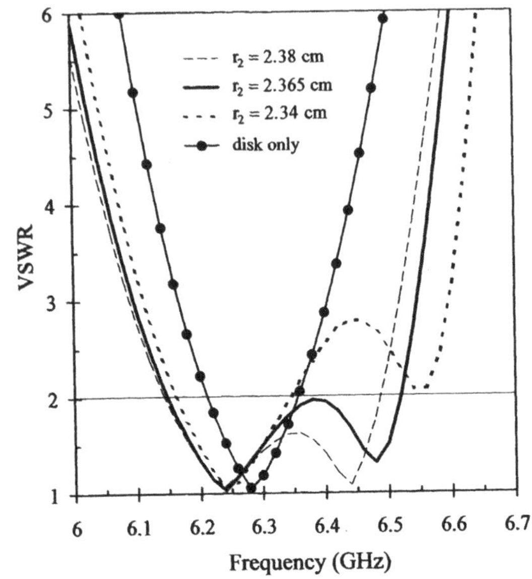

FIGURE 6.33 VSWR versus frequency for a spherical circular microstrip antenna (excited at the TM_{11} mode) coupled to a concentric parasitic annular-ring microstrip patch; $a = 5$ cm, $h = 1$ mm, $\phi_p = 0°$, $\theta_p/\theta_0 = 0.267$, $\epsilon_1 = 2.65$, $r_d\ (= b\theta_0) = 8.3$ mm, $r_1 = 8.4$ mm, $r_2 = 23.65$ mm. (From Ref. [12], © 1997 IEEE, reprinted with permission.)

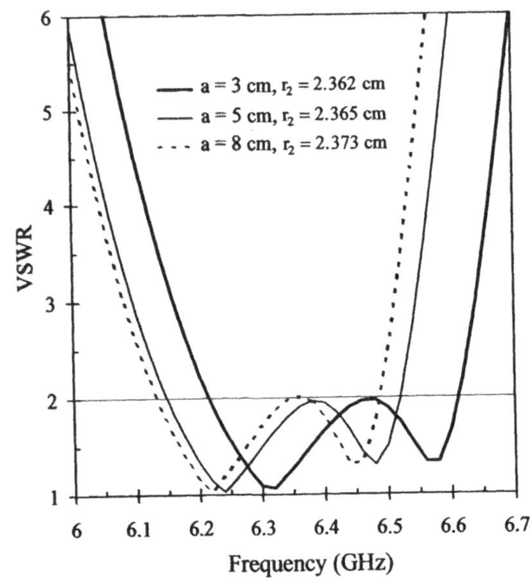

FIGURE 6.34 VSWR versus frequency for the case shown in Figure 6.33 with various sphere radii.

radiation patterns have also been studied. It is observed that the present ARL spherical circular microstrip antenna is still with the broadside radiation characteristic. However, as compared to the single circular patch case, the ARL spherical circular microstrip antenna has larger minor lobes and larger cross-polarized radiation [12].

6.4.2 Circular Patch as a Parasitic Patch

The circular disk as a parasitic patch to the driven annular-ring patch has also been studied [11]. By exciting the annular-ring patch with a probe feed of (5.1) at (θ_p, ϕ_p) and applying the same formulation as in Section 6.4.1, the input impedance of a circular-patch-loaded spherical annular-ring microstrip antenna can be determined. Figure 6.35 shows the VSWR (the feed coax is assumed to be 50 Ω) versus frequency calculated for various sizes of the parasitic disk. The results obtained using three different circular patch sizes are shown. The result of a single annular-ring patch antenna excited at the TM_{12} mode is plotted for comparison. With the resonant frequency of the disk at the TM_{11} mode designed to be slightly higher than that of the annular-ring patch at the TM_{12} mode, the input impedance is affected significantly and the antenna bandwidth can be enhanced. It

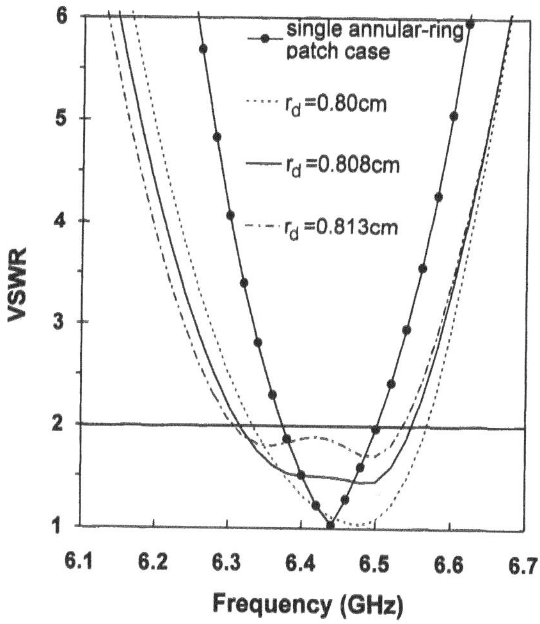

FIGURE 6.35 VSWR versus frequency for a spherical annular-ring microstrip antenna (excited at the TM_{12} mode) coupled to a concentric parasitic circular microstrip patch; $a = 10$ cm, $h = 1$ mm, $\phi_p = 0°$, $\theta_p/\theta_2 = 0.527$, $\epsilon_1 = 2.65$, $r_d = 8$, 8.08, and 8.13 mm, r_1 $(= b\theta_1) = 8.4$ mm, r_2 $(= b\theta_2) = 23.65$ mm. (From Ref. [11], © 1996 John Wiley & Sons, Inc., reprinted with permission.)

can be seen that with $r_d = 0.813$ cm, the antenna bandwidth increases to about 240 MHz, nearly twice that (about 130 MHz) for the single-patch case. For radiation characteristics within the operating bandwidth, behavior similar to that observed for the ARL spherical circular microstrip antenna described in Section 6.4.1 is seen.

REFERENCES

1. S. Y. Ke and K. L. Wong, "Full-wave analysis of mutual coupling between cylindrical-rectangular microstrip antennas," *Microwave Opt. Technol. Lett.*, vol. 7, pp. 419–421, June 20, 1994.
2. K. L. Wong, S. M. Wang, and S. Y. Ke, "Measured input impedance and mutual coupling of rectangular microstrip antennas on a cylindrical surface," *Microwave Opt. Technol. Lett.*, vol. 11, pp. 49–50, Jan. 1996.
3. W. Y. Tam, A. K. Y. Lai, and K. M. Luk, "Mutual coupling between cylindrical rectangular microstrip antennas," *IEEE Trans. Antennas Propagat.*, vol. 43, pp. 897–899, Aug. 1995.
4. S. C. Pan and K. L. Wong, "Mutual coupling between triangular microstrip antennas on a cylindrical body," *Electron. Lett.*, vol. 33, pp. 1005–1006, June 5, 1997.
5. J. S. Chen and K. L. Wong, "Mutual coupling computation of cylindrical-rectangular microstrip antennas using cavity-model theory," *Microwave Opt. Technol. Lett.*, vol. 9, pp. 323–326, Aug. 20, 1995.
6. J. S. Chen and K. L. Wong, "Curvature effect on the mutual coupling of circular microstrip antennas," *Microwave Opt. Technol. Lett.*, vol. 10, pp. 39–41, Sept. 1995.
7. C. Y. Huang and Y. T. Chang, "Curvature effects on the mutual coupling of cylindrical-rectangular microstrip antennas," *Electron. Lett.*, vol. 33, pp. 1108–1109, June 19, 1997.
8. C. Y. Huang and K. L. Wong, "Input impedance and mutual coupling of probe-fed cylindrical-circular microstrip patch antennas," *Microwave Opt. Technol. Lett.*, vol. 11, pp. 260–263, Apr. 5, 1996.
9. S. Y. Ke and K. L. Wong, "Broadband cylindrical-rectangular microstrip antennas using gap-coupled parasitic patches," *Microwave Opt. Technol. Lett.*, vol. 7, pp. 699–701, Oct. 20, 1994.
10. W. Y. Tam, A. K. Y. Lai, and K. M. Luk, "Cylindrical rectangular microstrip antennas with coplanar parasitic patches," *IEE Proc.-Microw. Antennas Propag.*, vol. 142, pp. 300–306, Aug. 1995.
11. H. T. Chen and Y. T. Cheng, "Full-wave analysis of a disk-loaded spherical annular-ring microstrip antenna," *Microwave Opt. Technol. Lett.*, vol. 12, pp. 353–358, Aug. 20, 1996.
12. H. T. Chen, H. D. Chen, and Y. T. Cheng, "Full-wave analysis of the annular-ring loaded spherical-circular microstrip antenna," *IEEE Trans. Antennas Propagat.*, vol. 45, pp. 1581–1583, Nov. 1997.
13. H. T. Chen, J. S. Row, and Y. T. Cheng, "Mutual coupling between spherical annular-ring and circular microstrip antennas," *1997 IEEE AP-S International Symposium Digest*, pp. 1521–1524.

14. R. P. Jedlicka, M. T. Poe, and K. R. Carver, "Measured mutual coupling between microstrip antennas," *IEEE Trans. Antennas Propagat.*, vol. 29, pp. 147–149, Jan. 1981.

15. D. M. Pozar, "Input impedance and mutual coupling of rectangular microstrip antennas," *IEEE Trans. Antennas Propagat.*, vol. 30, pp. 1191–1196, Nov. 1982.

16. E. H. Newman, J. H. Richmond, and B. W. Kwan, "Mutual impedance computation between microstrip antennas," *IEEE Trans. Microwave Theory Tech.*, vol. 31, pp. 941–945, Nov. 1983.

17. R. F. Harrington, *Time-Harmonic Electromagnetic Fields*, McGraw-Hill, New York, 1961, Chap. 3.

18. D. M. Pozar, *Microwave Engineering*, Addison-Wesley Publishing Company, Reading, Mass., 1990, Chap. 5.

19. C. Y. Huang and K. L. Wong, "Mutual coupling computation of probe-fed circular microstrip antennas," *Microwave Opt. Technol. Lett.*, vol. 9, pp. 100–102, June 5, 1995.

20. G. Kumar and K. C. Gupta, "Broad-band microstrip antennas using additional resonators gap-coupled to the radiating edges," *IEEE Trans. Antennas Propagat.*, vol. 32, pp. 1375–1379, Dec. 1984.

21. S. Y. Ke and K. L. Wong, "Input impedance of a probe-fed superstrate-loaded cylindrical-rectangular microstrip antenna," *Microwave Opt. Technol. Lett.*, vol. 7, pp. 232–236, Apr. 5, 1994.

22. Y. K. Cho, G. H. Son, G. S. Chae, L. H. Yun, and J. P. Hong, "Improved analysis method for broadband rectangular microstrip antenna using *E*-plane gap coupling," *Electron. Lett.*, vol. 29, pp. 1907–1909, Oct. 28, 1993.

CHAPTER SEVEN

Conformal Microstrip Arrays

7.1 INTRODUCTION

Microstrip arrays mounted on cylindrical [1–6], spherical [7,8], and conical [9] surfaces, have been reported. In studies of cylindrical microstrip arrays, most of the work emphasizes the design and characterization of N-element wraparound arrays ($1 \times N$ arrays) for omnidirectional radiation in the roll (ϕ) plane of the cylindrical host. Such a radiation pattern can find applications ranging from radio guidance of missiles to mobile-phone base stations. For many applications, $N \times N$ microstrip arrays must be mounted on curved (mainly cylindrical) surfaces, for structural reasons. In this case the radiation patterns will be strongly affected by the curvature of the host [5,6]. To reduce or eliminate this curvature effect on the radiation patterns of the conformal microstrip array, we can introduce an excitation phase difference between columns of the array through design of the feed network to compensate for the propagation path difference between columns of the array. Good compensation can make the directivity of the conformal microstrip array almost unchanged compared to that of the planar array.

Microstrip arrays mounted on a spherical surface have the advantages of wide-angle coverage. A typical geometry is shown in Figure 7.1. Such spherical microstrip arrays are usually designed to have radiation coverage over nearly a full hemisphere, which can find applications in ground station-to-satellite, aircraft-to-satellite, and satellite-to-satellite communication links [7,8]. The conical microstrip array has been used to provide tracking antennas for high-speed missiles [9], where the front end of the missile makes a design using conventional planar microstrip antennas impractical. Other uses are in curved bodies that have conical or nearly conical surfaces.

7.2 CYLINDRICAL MICROSTRIP ARRAYS

The cylindrical wraparound array for achieving an omnidirectional radiation is

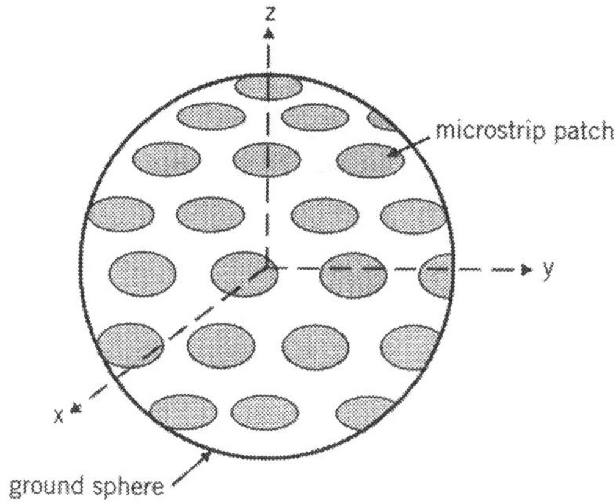

FIGURE 7.1 Geometry of a spherical micrsotrip array.

discussed first. Early work toward realizing this purpose was by wrapping a long microstrip patch around the circumference of the cylindrical host and feeding the patch at a number of points equally spaced along the circumference of the cylinder [1]. The total number (N) of feed points is a multiple of 2 (2^n, $n=1, 2, 3, \ldots$), and spacing between adjacent feed points must be less than one wavelength; that is [1],

$$N > \frac{2\pi a\sqrt{\epsilon_1}}{\lambda_0}, \tag{7.1}$$

where a is the cylinder radius, ϵ_1 the relative permittivity of the substrate, and λ_0 the operating wavelength in air. An improved version [2,3] of the design is to use a number of regular-size microstrip patches (see Figure 7.2), instead of a single microstrip patch of large width (about the circumference of the mounting cylinder) used in [1], employed on the curved surface of the cylindrical host to form a cylindrical wraparound array. This modified design requires fewer feed points than

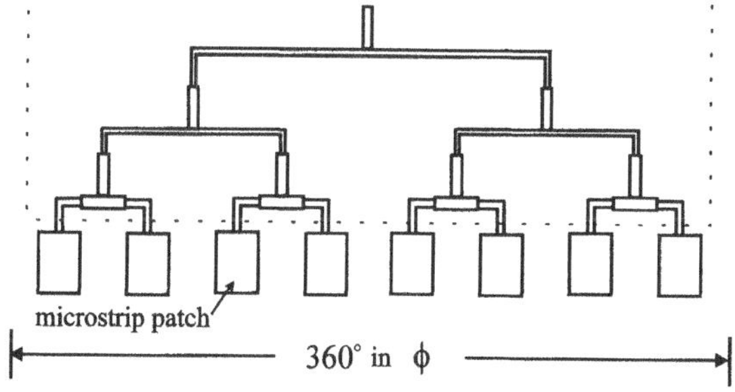

FIGURE 7.2 Geometry of an eight-element cylindrical microstrip wraparound array. The widths of the microstrip lines in the feed network for impedance matching are not to scale.

in the single microstrip patch case, and the feed network is thus simplified. The radiation characteristics of the cylindrical rectangular microstrip wraparound array have been studied using the cavity model and application of the stationary-phase method [3]. By exciting the patches in the axial (z) direction at the TM_{01} mode with equal power and phase, omnidirectional linearly polarized radiation in the roll plane is obtained. By replacing the linearly polarized patches in the wraparound array with circularly polarized patches, an omnidirectional circularly polarized wraparound array has also been designed [4].

For the case of $N \times N$ cylindrical microstrip arrays (see Figure 7.3), investigations have been reported in [5,6]. In the study in [5], the patches in the cylindrical microstrip array are designed to be excited with different excitation phases to compensate for the curvature effect on the propagation path difference between the columns of the microstrip array, with a major effort to demonstrate the design of a cylindrical microstrip array without degradation in the radiation pattern due to the curvature. The propagation path difference between the columns is given as

$$\Delta\psi = k_0 b\left(\cos\frac{1.5S_\phi}{b} - \cos\frac{0.5S_\phi}{b}\right), \tag{7.2}$$

where S_ϕ is the interelement spacing between centers of two adjacent elements in the ϕ direction; $k_0 = \omega\sqrt{\mu_0\epsilon_0}$. This phase difference can be realized by a feed-line length difference given by

$$\Delta l = \frac{\Delta\psi}{k_0\sqrt{\epsilon_{\text{eff}}}}, \tag{7.3}$$

where ϵ_{eff} is the effective relative permittivity in the microstrip line, whose

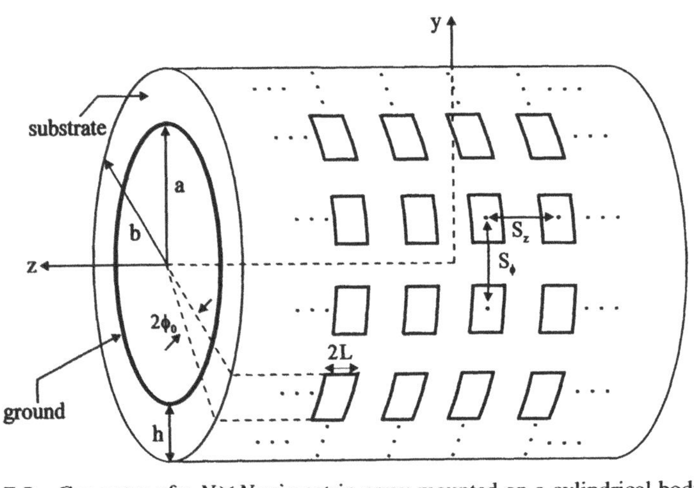

FIGURE 7.3 Geometry of a $N \times N$ microstrip array mounted on a cylindrical body. (From Ref. [6], © 1998 John Wiley & Sons, Inc.)

solution is given in Section 8.2. By considering (7.3) in the feed network design, the curvature effect on the radiation pattern can be compensated.

As for evolution of the radiation pattern due to the curvature variation, results are given in [6], where a full-wave analysis of the radiation patterns of $N \times N$ cylindrical microstrip arrays is presented. The case studied is for the microstrip array operated in the Ku band. The far-zone radiated fields of the cylindrical microstrip array are derived, and the radiation patterns for the array mounted on ground cylinders with various radii are calculated and analyzed. Several Ku-band cylindrical microstrip arrays are also constructed and measured. A comparison of experiment and theory is shown, and variations of the side-lobe level (SLL) with the cylinder radius are also analyzed.

As referred to the geometry shown in Figure 7.3, the cylindrical substrate is of thickness h (= b-a) and relative permittivity ϵ_1. Each element in the array is of the same size and has dimensions of $2L \times 2b\phi_0$, where $2\phi_0$ is the angle subtended by the curved element in the microstrip array. All the elements in the array are assumed to be uniformly excited. The interelement spacing, S (= $S_\phi = S_z$), between the centers of two adjacent elements is also selected to be the same in the ϕ and z directions and is chosen to be in the range 0.7 to $0.9\lambda_0$ to obtain a better array gain [10]. Then, by applying the full-wave approach in Section 4.2, neglecting the patch surface current orthogonal to the excitation direction in each element, and ignoring the mutual coupling between array elements, the far-zone radiated fields of the cylindrical array are derived as

$$\begin{bmatrix} E_\theta \\ E_\phi \end{bmatrix} = \sum_{p=1}^{N} \sum_{q=1}^{N} \begin{bmatrix} E_\theta^{(p,q)} \\ E_\phi^{(p,q)} \end{bmatrix} = \frac{1}{\sin\theta} \begin{bmatrix} -1 & 0 \\ 0 & \eta_0 \end{bmatrix} \begin{bmatrix} E_z^{(0,0)} \\ H_z^{(0,0)} \end{bmatrix}, \tag{7.4}$$

with

$$\begin{bmatrix} E_z^{(0,0)} \\ H_z^{(0,0)} \end{bmatrix} \approx \frac{-je^{jk_0 r}}{\pi r} \sum_{n=-\infty}^{\infty} \frac{(-j)^n e^{jn\phi}}{H_n^{(1)}(k_0 b \sin\theta)} \begin{bmatrix} X_{11} & X_{12} \\ X_{21} & X_{22} \end{bmatrix}^{-1} \begin{bmatrix} 0 \\ \tilde{J}_z^{(0,0)} \end{bmatrix} A_f, \tag{7.5}$$

$$\tilde{J}_z^{(0,0)} = \frac{\sin n\phi_0}{nL} \frac{\sin(\pi/2 - k_0 L \cos\theta)}{(\pi/2L)^2 - (k_0 \cos\theta)^2}, \tag{7.6}$$

$$A_f = \frac{\sin(NnS_\phi/2b)}{\sin(nS_\phi/2b)} \frac{\sin(Nk_0 S_z \cos\theta/2)}{\sin(k_0 S_z \cos\theta/2)}, \tag{7.7}$$

where the elements in the $[X]$ matrix are expressed in (2.35)–(2.38); the superscript (0, 0) denotes an imaginary patch centered in the microstrip array, and A_f is the array factor of the microstrip array; the tilde again represents a Fourier transform, and η_0 in (7.4) is free-space intrinsic impedance.

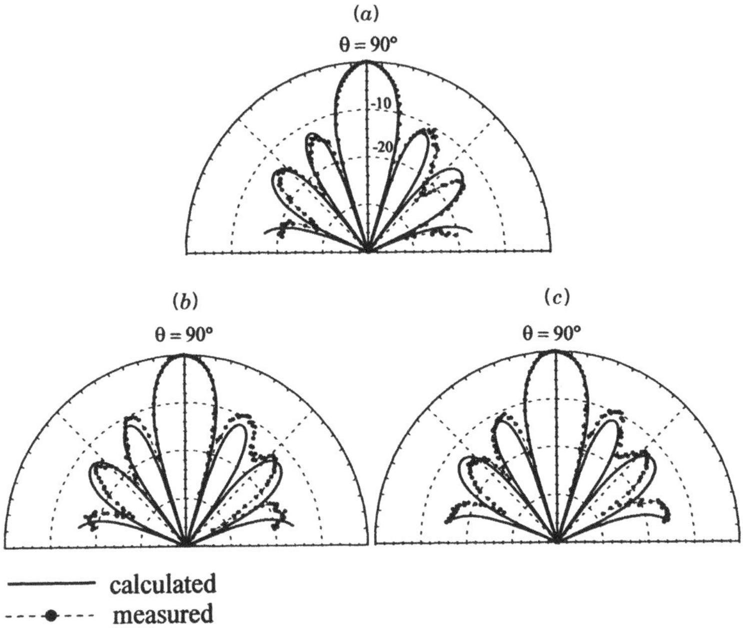

FIGURE 7.4 E-plane patterns calculated and measured for a 4×4 microstrip array at 16.2 GHz; $h = 0.254$ mm, $\epsilon_1 = 2.94$, $S = 0.81\lambda_0$, $2L = 7.2$ mm, $2b\phi_0 = 5.0$ mm. (*a*) planar case; (*b*) $a = 10.6$ cm; (*c*) $a = 7.6$ cm. (From Ref. [6], © 1998 John Wiley & Sons, Inc.)

Typical results of (7.4) calculated for a 4×4 array at 16.2 GHz are presented in Figures 7.4 and 7.5. For the experiment, the microstrip arrays were fabricated using flexible microwave laminates. Good agreement is obtained between the results calculated and experimental data. It is observed that the curvature effects on the E-plane (x–z plane) pattern is very small, while the H-plane (x–y plane) pattern is strongly affected by the cylinder radius variation, and the SLL increases with decreasing cylinder radius. Figure 7.6 presents SLL as a function of cylinder radius for various array sizes and interelement spacings. It is seen that the grating lobe (SLL = 0 dB) in the H-plane pattern occurs at a larger cylinder radius. This suggests that a small curvature variation can have a significant effect on a large array.

7.3 SPHERICAL AND CONICAL MICROSTRIP ARRAYS

Based on the geometry shown in Figure 7.1, a spherical microstrip array consisting of 120 circular microstrip patches have been designed and constructed [8]. This array is designed for satellite communications in the band 2.0 to 2.3 GHz and can produce 1024 beams, to cover the hemisphere with a gain of about 14 dBi within 150° from the vertical axis, and the beam direction is electronically switched by exciting various sets of circular patches in the array. Other spherical microstrip

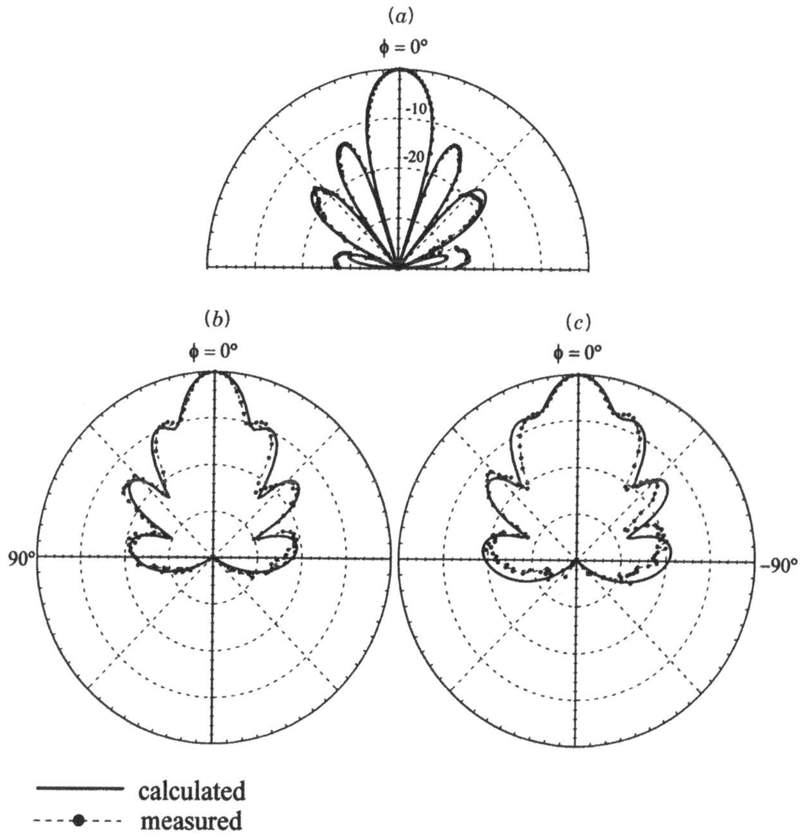

FIGURE 7.5 *H*-plane patterns calculated and measured for the case shown in Figure 7.4: (*a*) planar case; (*b*) $a = 10.6$ cm; (*c*) $a = 7.6$ cm. (From Ref. [6], © 1998 John Wiley & Sons, Inc.)

arrays, such as a six-element spherical array and an 86-element spherical array, have been reported [7]. The six-element array is operated in the L band and radiates a circularly polarized wave with a gain of more than 7 dBi within 60° from the vertical axis, and the 86-element array has a coverage gain of more than 12 dBi within 115° from the vertical axis. As for the conical microstrip array with typical geometry shown in Figure 7.7, reports of related designs are relatively scanty. A typical design reported is a monopulse tracking antenna array, consisting of mounting four radiating microstrip patches on the surface of a cone with a triplate feed network placed below the patches [9]. Such a design is used as a guided-weapon seeker antenna, operated at a center frequency of 10 GHz, for high-speed missiles. Details of the antenna construction and performance are presented in [9]. The case for a single conical microstrip antenna is discussed in Section 5.3. This type of conical microstrip array can be a promising candidate for employment on curved bodies with conical or nearly conical surfaces.

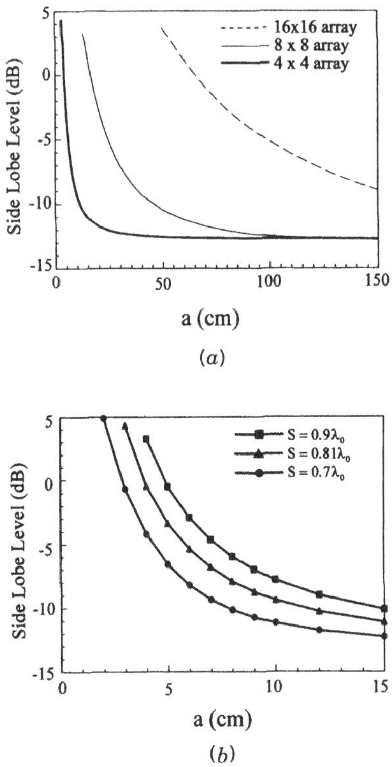

FIGURE 7.6 *H*-plane side-lobe levels calculated as a function of cylinder radius for various (*a*) array sizes and (*b*) interelement spacings of a 4×4 array; the array parameters are as in Figure 7.4. (From Ref. [6], © 1998 John Wiley & Sons, Inc.)

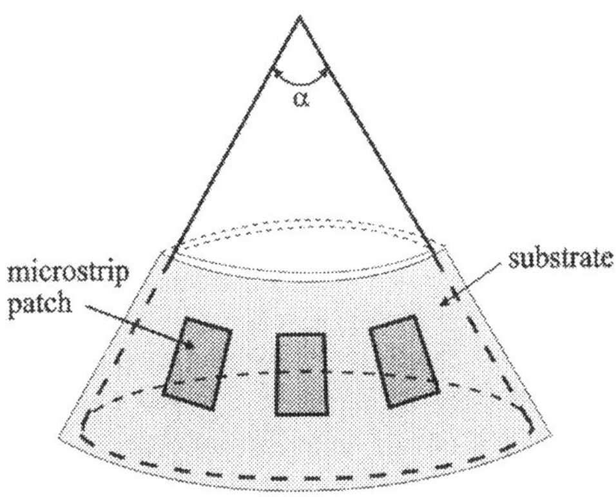

FIGURE 7.7 Geometry of a microstrip array mounted on the surface of a cone; α is the flare angle of the angle.

REFERENCES

1. R. E. Munson, "Conformal microstrip antennas and microstrip phased arrays," *IEEE Trans. Antennas Propagat.*, vol. 22, pp. 74–78, Jan. 1974.
2. I. Jayakumar, R. Garg, B. K. Sarap, and B. Lal, "A conformal cylindrical microstrip array for producing omnidirectional radiation patterns," *IEEE Trans. Antennas Propagat.*, vol. 34, pp. 1258–1261, Oct. 1986.
3. C. M. Silva, F. Lumini, J. C. S. Lacava, and F. P. Richards, "Analysis of cylindrical arrays of microstrip rectangular patches," *Electron. Lett.*, vol. 27, pp. 778–780, Apr. 25, 1991.
4. R. C. Hall and D. I. Wu, "Modeling and design of circularly-polarized cylindrical wraparound microstrip antennas," *1996 IEEE AP-S International Symposium Digest*, pp. 672–675.
5. J. Ashkenazy, S. Shtrikman, and D. Treves, "Conformal microstrip arrays on cylinders," *IEE Proc.*, pt. H, vol. 135, pp. 132–134, Apr. 1988.
6. K. L. Wong and G. B. Hsieh, "Curvature effects on the radiation patterns of microstrip arrays," *Microwave Opt. Technol. Lett.*, vol. 18, June 20, 1998.
7. Fujimoto, T. Hori, S. Nishimura, and K. Hirasawa, in J. R. James and P. S. Halls, eds., *Handbook of Microstrip Antennas*, Peter Peregrinus, London, 1989, pp. 1132–1136.
8. R. Stockton and R. Hockensmith, "Application of spherical arrays—a simple approach," *1977 IEEE AP-S International Symposium Digest*, pp. 202–205.
9. P. Newham and G. Morris, in J. R. James and P. S. Halls, eds., *Handbook of Microstrip Antennas*, Peter Peregrinus, London, 1989, Chap. 20.
10. E. Levine, G. Malamud, S. Shtrikman, and D. Treves, "A study of microstrip array antennas with the feed network," *IEEE Trans. Antennas Propagat.*, vol. 37, pp. 426–434, Apr. 1989.

CHAPTER EIGHT

Cylindrical Microstrip Lines and Coplanar Waveguides

8.1 INTRODUCTION

In this chapter we describe the recent development of some quasistatic models and full-wave approaches for the analysis of cylindrical microstrip lines and coplanar waveguides. Such nonplanar transmission lines can be constructed using flexible substrates mounted on a cylindrical surface for the excitation of cylindrical microstrip antennas and arrays. Two basic quasistatic models for the analysis of cylindrical transmission lines have been employed: one by applying the conformal mapping technique to transform the cylindrical structure into a planar one [1,2] and the other by solving Laplace's equation in cylindrical coordinates [3–6]. As for full-wave approaches, many studies have recently been reported on the analysis of cylindrical microstrip lines [7–12], coupled cylindrical microstrip lines [13,14], slot-coupled double-sided cylindrical microstrip lines [15], cylindrical microstrip discontinuities [16,17], and cylindrical coplanar waveguides [18,19]. Other theoretical models, such as the use of a finite-difference time-domain method for the analysis of cylindrical microstrip lines, have also been demonstrated [20]. In the following sections, the theoretical formulation of a single cylindrical microstrip line and a single coplanar waveguide using the quasistatic model or full-wave approach is described. Also, the coupling and discontinuity characteristics of cylindrical microstrip lines analyzed using the full-wave approach are presented and discussed.

8.2 CYLINDRICAL MICROSTRIP LINES

Two different configurations of inside and outside cylindrical microstrip lines have been studied, as shown in Figure 8.1. Quasistatic solutions obtained by solving

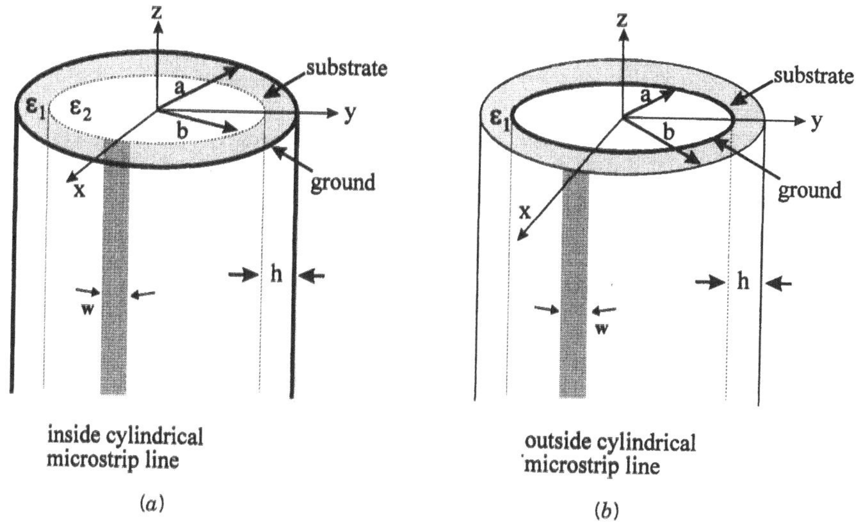

FIGURE 8.1 Configurations of (*a*) an inside cylindrical microstrip line and (*b*) an outside cylindrical microstrip line.

Laplace's equation in cylindrical coordinates, which is useful for analyzing the characteristics of cylindrical microstrip lines at low operating frequencies, are described first. Results obtained with the use of a full-wave formulation and a moment-method calculation are then given. A comparison of full-wave solutions with the measured data is also given, and various characteristics of inside and outside cylindrical micostrip lines are addressed.

8.2.1 Quasistatic Solution

The inside cylindrical microstrip line shown in Figure 8.1*a* is considered first. A cylindrical substrate of thickness h $(=a-b)$ and relative permittivity ϵ_1 is mounted inside a ground cylinder of radius a. The cylindrical microstrip line, with a width of w $(=2b\phi_0)$, is assumed to be infinitely long. The region of $\rho < b$ is assumed to have a relative permittivity of ϵ_2. For a microstrip line operating in the quasi-TEM mode (the dominant mode), the potential function $\Psi(\rho, \phi)$ in all regions satisfies Laplace's equation in cylindrical coordinates [5,21],

$$\nabla^2\Psi = \rho\frac{\partial}{\partial\rho}\left(\frac{\partial\Psi}{\partial\rho}\right) + \frac{\partial^2\Psi}{\partial\phi^2} = 0\,, \tag{8.1}$$

and the charge distribution on the strip line can assume the form

$$\sigma(\rho, \phi) = \sigma(\phi)\delta(\rho - b)\,, \qquad |\phi| \le \phi_0\,, \tag{8.2}$$

with

$$\sigma(\phi)=\begin{cases}\dfrac{1}{b\pi\phi_0\sqrt{1-(\phi/\phi_0)^2}}, & |\phi|\le\phi_0\,,\\ 0, & \text{elsewhere}\,,\end{cases}\tag{8.3}$$

where $\delta(\rho-b)$ is the Dirac delta function and the singularity at the edges of the strip line is considered for the charge distribution [22]. Then, by adopting the following definition of the Fourier transform pair of Ψ:

$$\tilde{\Psi}(\rho,n)=\frac{1}{2\pi}\int_0^{2\pi}\Psi(\rho,\phi)e^{jn\phi}\,d\phi\,,\tag{8.4}$$

and

$$\Psi(\rho,\phi)=\sum_{n=-\infty}^{\infty}\tilde{\Psi}(\rho,n)e^{-jn\phi}\,,\tag{8.5}$$

(8.1) can be transformed into

$$\frac{1}{\rho}\left[\frac{\partial}{\partial\rho}\rho\frac{\partial}{\partial\rho}\tilde{\Psi}(\rho,n)\right]-\frac{n^2}{\rho^2}\tilde{\Psi}(\rho,n)=0\,.\tag{8.6}$$

Using the method of separation of variables, general solutions of (8.6) can be expressed as

$$\tilde{\Psi}_k(\rho,n)=\sum_{n=0}^{\infty}\tilde{\Psi}_{kn}(\rho)\,,\tag{8.7}$$

with

$$\tilde{\Psi}_{kn}(\rho)=\begin{cases}A_{k0}\ln\rho+B_{k0}\,, & n=0\,,\\ A_{kn}\rho^n+B_{kn}\rho^{-n}\,, & n\neq0\,,\end{cases}\tag{8.8}$$

where the subscript $k=1$ denotes region 1 ($a>\rho>b$) and $k=2$ is for region 2 ($\rho<b$), and A_{kn} and B_{kn} are unknown coefficients to be determined. The boundary conditions in the spectral domain give

$$\tilde{\Psi}_{1n}(b,n)=\tilde{\Psi}_{2n}(b,n)\,,\tag{8.9}$$

$$\tilde{\Psi}_{1n}(a,n)=0\,,\tag{8.10}$$

$$\epsilon_0\epsilon_1\frac{\partial}{\partial\rho}\tilde{\Psi}_{1n}(b,n)-\tilde{\sigma}(n)=\epsilon_0\epsilon_2\frac{\partial}{\partial\rho}\tilde{\Psi}_{2n}(b,n)\,,\tag{8.11}$$

where $\tilde{\sigma}(n)$ is the Fourier transform of $\sigma(\phi)$ in (8.3), which can be evaluated to be $J_0(n\phi_0)$ [$=\int_{-\phi_0}^{\phi_0}J_z(\phi)e^{jn\phi}b\,d\phi$], a Bessel function of the first kind of order 0. Upon applying the solutions shown by (8.8) and enforcing the boundary conditions given in (8.9)–(8.11), the potential functions in the spectral domain for each region are obtained as follows: In the substrate region ($a > \rho > b$),

$$\tilde{\Psi}_1(\rho, n) = \begin{cases} \dfrac{b \ln(\rho/a)}{\epsilon_0 \epsilon_1} \tilde{\sigma}(n), & n = 0, \\ \dfrac{b(\rho/b)^n[(1 - (a/\rho)^{2n}]\tilde{\sigma}(n)}{\epsilon_0 \epsilon_2 n[1 - (a/b)^{2n}]\{\epsilon_{12} \coth[n \ln(b/a)] - 1\}}, & n \neq 0, \end{cases} \tag{8.12}$$

and in the region $\rho < b$,

$$\tilde{\Psi}_2(\rho, n) = \begin{cases} \dfrac{b \ln(\rho/a)}{\epsilon_0 \epsilon_1} \tilde{\sigma}(n), & n = 0, \\ \dfrac{b(\rho/b)^n \tilde{\sigma}(n)}{\epsilon_0 \epsilon_2 n\{\epsilon_{12} \coth[n \ln(b/a)] - 1\}}, & n \neq 0, \end{cases} \tag{8.13}$$

where $\epsilon_{12} = \epsilon_1/\epsilon_2$. With the spectral-domain potential functions obtained, the Green's function in the spectral domain due to a unit-amplitude charge placed at $\rho = b$ can be expressed as

$$\tilde{G}_k(\rho, n) = \frac{\tilde{\Psi}_k(\rho, n)}{\tilde{\sigma}(n)}, \qquad k = 1, 2. \tag{8.14}$$

Using the derived spectral-domain Green's function, we can have the capacitance per unit length of the strip line given by the variational expression as [23]

$$\begin{aligned} \frac{1}{C} &= \frac{\left| \int_{-\phi_0}^{\phi_0} \int_{-\phi_0}^{\phi_0} \sigma(b, \phi) G_1(b, \phi; b, \phi') \sigma(b, \phi') b \, d\phi' \, b \, d\phi \right|}{\left[\int_{-\phi_0}^{\phi_0} \sigma(b, \phi') b \, d\phi' \right]^2} \\ &= \frac{1}{2\pi\epsilon_0\epsilon_1} \left\{ \ln \frac{a}{b} + \sum_{n=1}^{\infty} \frac{2\epsilon_{12} J_0^2(n\phi_0)}{n\epsilon_{12} \coth[n \ln(a/b)] + n} \right\}, \end{aligned} \tag{8.15}$$

where $G_1(b, \phi; b, \phi')$ is the Green's function in the space domain obtained via (8.5). Then, for the quasi-TEM mode considered, the propagation constant and characteristic impedance of the microstrip line can be found in terms of the static distributed parameters per unit length. The effective relative permittivity of the microstrip line can be obtained from [22,23]

$$\epsilon_{\text{eff}} = \frac{C}{C_0}$$

$$= \epsilon_1 \frac{\ln(a/b) + \sum_{n=1}^{\infty} \{2J_0^2(n\phi_0)/n \coth[n \ln(a/b)] + n\}}{\ln(a/b) + \sum_{n=1}^{\infty} \{2\epsilon_{12} J_0^2(n\phi_0)/n\epsilon_{12} \coth[n \ln(a/b)] + n\}}, \tag{8.16}$$

where C is the distributed capacitance per unit length of the microstrip line and C_0 is the capacitance of the structure with air ($\epsilon_1 = \epsilon_2 = 1$) as dielectrics. As for the characteristic impedance of the microstrip line, we have [21,22]

$$Z_c = \frac{\sqrt{\epsilon_{\text{eff}}}}{\nu_0 C}$$

$$= \frac{60\sqrt{\epsilon_{\text{eff}}}}{\epsilon_1} \left\{ \ln\frac{a}{b} + \sum_{n=1}^{\infty} \frac{2\epsilon_{12} J_0^2(n\phi_0)}{n\epsilon_{12} \coth[n \ln(a/b)] + n} \right\}, \tag{8.17}$$

where ν_0 is the velocity of light in air. It is also noted that although the quasistatic formulation above is for the structure of an inside cylindrical microstrip line, the corresponding results for an outside cylindrical microstrip line can be obtained from the expressions above by interchanging radii a and b in the expressions.

From (8.16)–(8.17), the low-frequency characteristics of the inside cylindrical microstrip line are evaluated. The effective relative permittivity as a function of w/h for various curvilinear coefficients is first calculated, with the curvilinear coefficient R_c defined as $(a-h)/a$. Typical results are shown in Figure 8.2, with the region $\rho < b$ chosen to be air. From the results it is seen that the effective

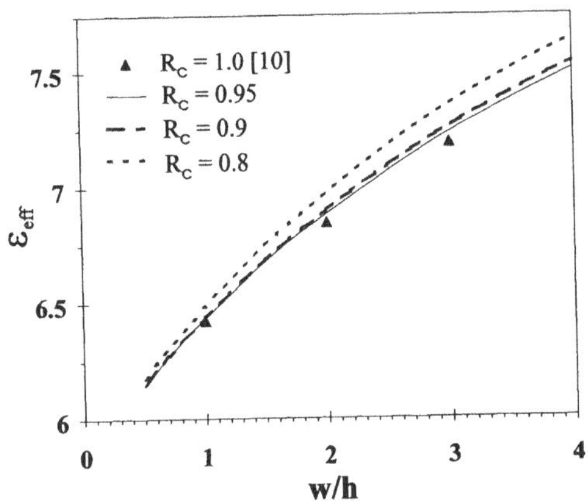

FIGURE 8.2 Effective relative permittivity ϵ_{eff} as a function of w/h for various curvilinear coefficients; $\epsilon_1 = 9.6$, $\epsilon_2 = 1.0$, $h = 3.04$ mm. (From Ref. [3], © 1995 John Wiley & Sons, Inc.)

relative permittivity decreases with increasing values of R_c (or decreasing curvature), and the results for higher values of R_c are also seen to approach the results for a corresponding planar microstrip line ($R_c \to 1$). The results for the characteristic impedance are shown in Figure 8.3. It is found that the characteristic impedance is larger when R_c increases. By comparing the quasistatic behavior of ϵ_{eff} and Z_c for an inside cylindrical microstrip line with various curvatures obtained here to that [10] of the outside cylindrical microstrip line shown by Figure 8.1*b*, different curvature effects between a cylindrical microstrip line mounted inside and outside a ground cylinder are observed. This is probably due to the different bending directions (one is concave and the other is convex) of the microstrip line with respect to a planar ground surface.

8.2.2 Full-Wave Solution

From the quasistatic results shown in Section 8.2.1 it is seen that the effective relative permittivity and characteristic impedance of a cylindrical microstrip line vary greatly with the curvilinear coefficient, especially for a wider microstrip line. Accurate design of the microstrip line mounted on a cylindrical surface is thus important. For this purpose, rigorous full-wave solutions have been reported. In this section, the case for an inside cylindrical microstrip line with a narrow width solved using a full-wave approach is first described, and then the approach for an outside cylindrical microstrip line with a wide width is discussed.

A. Inside Cylindrical Microstrip Line Given the geometry in Figure 8.1*a* with a narrow microstrip line (i.e., $w \ll 2\pi/\beta$, where β is the effective wavenumber), the $\hat{\phi}$-directed surface current density J_ϕ on the conducting strip can be neglected,

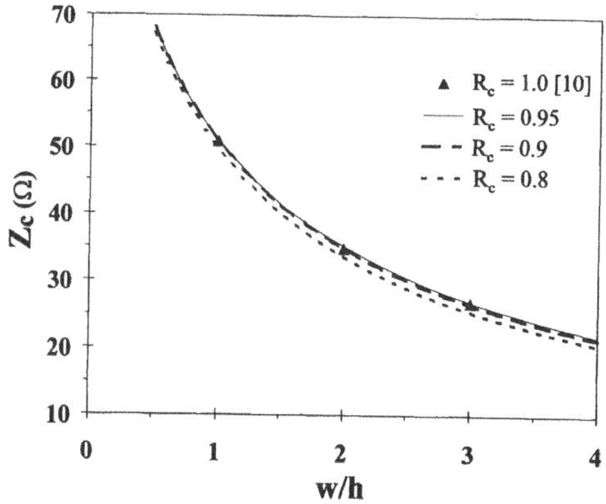

FIGURE 8.3 Characteristic impedance Z_c as a function of w/h for various curvilinear coefficients; $\epsilon_1 = 9.6$, $\epsilon_2 = 1.0$, h = 3.04 mm. (From Ref. [3], © 1995 John Wiley & Sons, Inc.)

and the $\hat{z}$-directed current density J_z can be assumed to have a traveling-wave current of the form $e^{j\beta z}$ given by

$$\mathbf{J}(\phi, z) = \hat{z}J_z(\phi)e^{j\beta z}\,, \tag{8.18}$$

with

$$J_z(\phi) = \begin{cases} \dfrac{1}{b\pi\phi_0\sqrt{1-(\phi/\phi_0)^2}}\,, & |\phi| \le \phi_0\,, \\ 0\,, & \text{elsewhere}\,, \end{cases} \tag{8.19}$$

where an edge-singularity condition is assumed in the expression. Then, by using the spectral-domain approach to solve Maxwell's equation in cylindrical coordinates, the $\hat{z}$-directed electric and magnetic fields satisfy the following wave equations: In region 1 (the substrate layer),

$$\left[\frac{1}{\rho}\frac{d}{d\rho}\left(\rho\frac{d}{d\rho}\right) + (\epsilon_1 k_0^2 - k_z^2)\right]\tilde{\Psi}_z(\rho, n, k_z) = 0\,, \tag{8.20}$$

and in region 2 (the inner region),

$$\left[\frac{1}{\rho}\frac{d}{d\rho}\left(\rho\frac{d}{d\rho}\right) + (\epsilon_2 k_0^2 - k_z^2)\right]\tilde{\Psi}_z(\rho, n, k_z) = 0\,, \tag{8.21}$$

where $\tilde{\Psi}_z$ denotes $\tilde{E}_z$ or $\tilde{H}_z$; $k_0 = \omega\sqrt{\mu_0\epsilon_0}$. The spectral-domain field amplitude is defined as

$$\tilde{\Psi}(\rho, n, k_z) = \frac{1}{2\pi}\int_{-\pi}^{\pi}\int_{-\infty}^{\infty} \Psi(\rho, \phi, z)e^{-jn\phi}e^{-jk_z z}\,dz\,d\phi\,. \tag{8.22}$$

By solving the foregoing spectral-domain wave equations, exact Green's functions of the cylindrical structure that are crucial to the solutions can be derived. In the cylindrical structure considered here, the $\hat{z}$-directed electric field at $\rho = b$ can be derived as related to the surface current density J_z by the following relationship:

$$E_z(b, \phi, z) = \frac{1}{2\pi}\int_{-\infty}^{\infty}\sum_{n=-\infty}^{\infty} \tilde{G}_{zz}^{EJ}(b, n, k_z)\tilde{J}_z(n, k_z)e^{jn\phi}e^{jk_z z}\,dk_z\,, \tag{8.23}$$

where $\tilde{G}_{zz}^{EJ}$ is the Green's function denoting the $\hat{z}$-directed electric field at $\rho = b$ due to a unit-amplitude $\hat{z}$-directed electric current element at $\rho = b$, whose full expression can be found in [7] and is similar to that derived in Section 2.2.1. Then, to apply Galerkin's moment method to yield the result for the effective relative permittivity, one can multiply (8.23) using $J_z(\phi)$ as the weighting function, integrate over the conducting strip area, and impose the boundary condition that the surface current density and the electric field are complementary to each other. We have

$$\int_{-\phi_0}^{\phi_0} E_z(b, \phi, z) J_z(\phi) b \, d\phi = \frac{1}{2\pi} \sum_{n=-\infty}^{\infty} \tilde{G}_{zz}^{EJ}(b, n, \beta) \tilde{J}_0^2(n\phi_0) = 0 \, . \tag{8.24}$$

Thus, by seeking the root of the characteristic equation

$$\sum_{n=-\infty}^{\infty} \tilde{G}_{zz}^{EJ}(b, n, \beta) \tilde{J}_0^2(n\phi_0) = 0 \, , \tag{8.25}$$

the effective propagation constant β and, in turn, the effective relative permittivity can be obtained from

$$\epsilon_{\text{eff}} = \left(\frac{\beta}{k_0} \right)^2 . \tag{8.26}$$

As for calculation of the characteristic impedance of a cylindrical microstrip line, we adopt the power-current definition [24]:

$$Z_c = \frac{2P}{I_z^2} \, , \tag{8.27}$$

where I_z is the total $\hat{z}$-directed surface current; P is the time-average Poynting power flowing along the z axis inside the region of $0 < \rho < a$ and can be calculated from

$$P = \tfrac{1}{2} \, \text{Re} \left[\int_0^a \int_0^{2\pi} (\mathbf{E} \times \mathbf{H}^*) \cdot \hat{z} \rho \, d\phi \, d\rho \right] . \tag{8.28}$$

The asterisk denotes a complex conjugate. By further applying Parseval's theorem, the time-average power P of (8.28) can be expressed in the spectral domain as

$$P = \frac{1}{4\pi} \sum_{n=-\infty}^{\infty} (\tilde{P}_1 + \tilde{P}_2) \, , \tag{8.29}$$

with

$$\begin{aligned} \tilde{P}_i &= \int_0^a [\tilde{E}_\rho^{(i)} \tilde{H}_\phi^{(i)*} - \tilde{E}_\phi^{(i)} \tilde{H}_\rho^{(i)*}] \rho \, d\rho \\ &= \int_0^a [\tilde{G}_{\rho z}^{EJ^{(i)}} \tilde{G}_{\phi z}^{HJ^{(i)}*} - \tilde{G}_{\phi z}^{EJ^{(i)}} \tilde{G}_{\rho z}^{HJ^{(i)}*}] \tilde{J}_n^2(n) \rho \, d\rho \, , \qquad i = 1, 2 \, , \end{aligned} \tag{8.30}$$

where the spectral-domain Green's functions $\tilde{G}_{\rho z}^{EJ^{(i)}}$, $\tilde{G}_{\phi z}^{HJ^{(i)}}$, $\tilde{G}_{\phi z}^{EJ^{(i)}}$, and $\tilde{G}_{\rho z}^{HJ^{(i)}}$ for region i (= 1, 2) have been derived in [7].

Numerical results have been calculated. Figure 8.4 presents the effective relative permittivity as a function of $k_0 a$ for various curvilinear coefficients. Two different stripline widths of $w/h = 1.0$ and 2.0 are shown. It is seen that the effective relative permittivity obtained is very sensitive to the values of R_c and

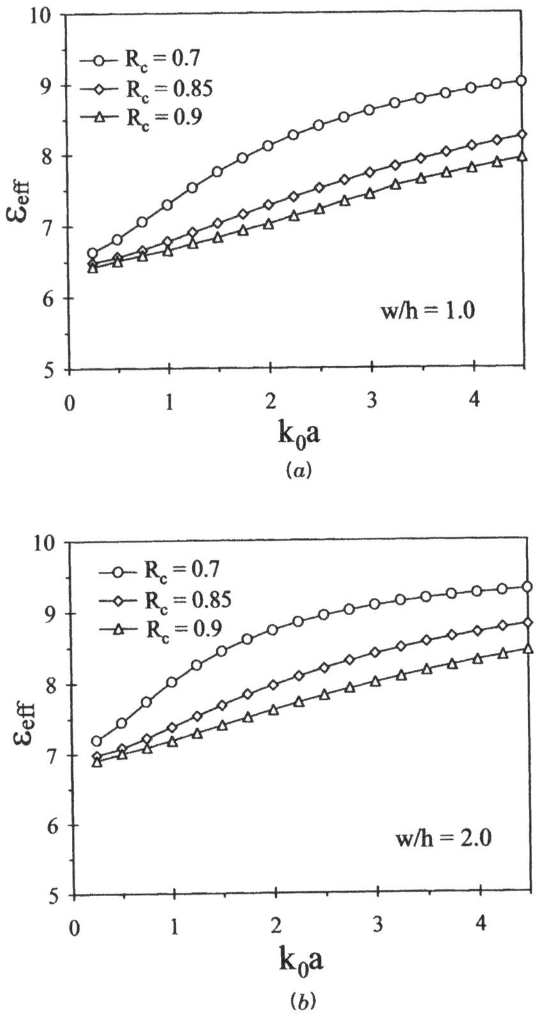

FIGURE 8.4 Effective relative permittivity ϵ_{eff} as a function of k_0a for various curvilinear coefficients; $\epsilon_1 = 9.6$, $\epsilon_2 = 1.0$, $h = 3.04$ mm. (*a*) $w/h = 1.0$; (*b*) $w/h = 2.0$. (From Ref. [7], © 1995 IEEE, reprinted with permission.)

k_0a. Another interesting result of the dependence of ϵ_{eff} on the stripline width is presented in Figure 8.5. It is found that ϵ_{eff} approaches the value of ϵ_1 for a larger stripline width. For the limiting case that the stripline width is $2\pi b$ (i.e., the structure becomes a coaxial line), the value of ϵ_{eff} is found to be the same as ϵ_1. Figure 8.6 presents the results of characteristic impedance as a function of k_0a. It is observed that the variation of characteristic impedance with k_0a is more sensitive for smaller values of R_c, and the variation is also more significant for a narrower microstrip line.

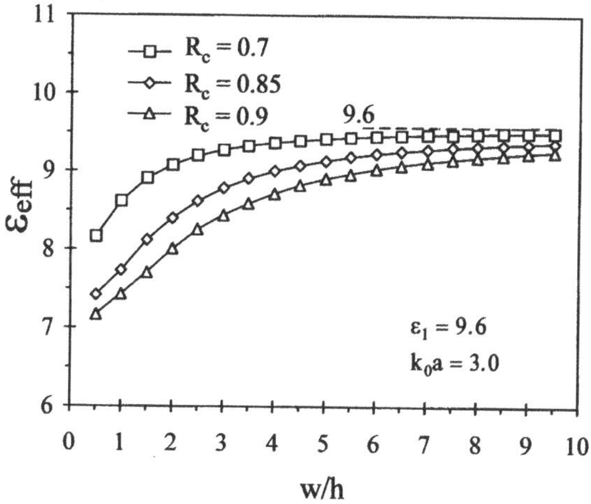

FIGURE 8.5 Effective relative permittivity ϵ_{eff} as a function of w/h for various curvilinear coefficients; $\epsilon_1 = 9.6$, $\epsilon_2 = 1.0$, $h = 3.04$ mm, $k_0a = 3.0$. (From Ref. [7], © 1995 IEEE, reprinted with permission.)

B. Outside Cylindrical Microstrip Line In this section, both the longitudinal and transverse components of the surface current density are considered for computation of the effective relative permittivity and characteristic impedance of the cylindrical microstrip line. Experiments are also conducted to verify the results calculated, and the contribution of the transverse current component to the characteristics of a wide cylindrical microstrip line is discussed.

Referring to the geometry in Figure 8.1*b*, the outside cylindrical microstrip line is assumed to be infinitely long and has a width of w. The surface current density of (8.18) is modified to be

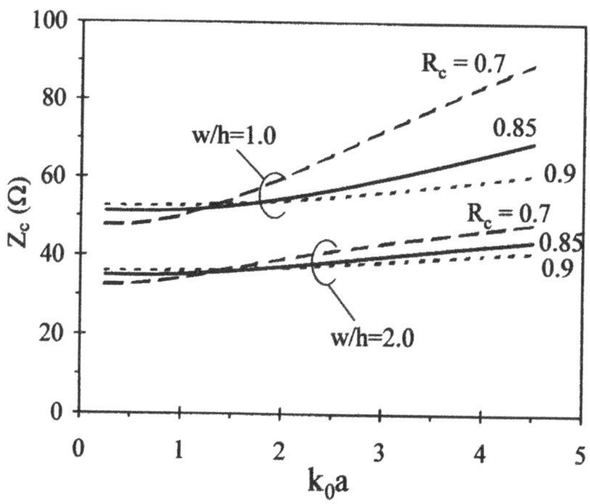

FIGURE 8.6 Characteristic impedance Z_c as a function of k_0a; $\epsilon_1 = 9.6$, $\epsilon_2 = 1.0$, $h = 3.04$ mm. (From Ref. [7], © 1995 IEEE, reprinted with permission.)

$$\mathbf{J}(\phi, z) = e^{-j\beta z}\left[\hat{\phi} \sum_{n=1}^{N} I_{\phi n} J_{\phi n}(\phi) + \hat{z} \sum_{m=1}^{M} I_{zm} J_{zm}(\phi)\right], \tag{8.31}$$

with

$$J_{\phi n}(\phi) = \frac{\sin[k_e(D_t/2 - |\phi - \phi_n|)]}{\sin(k_e D_t/2)}, \qquad |\phi - \phi_n| < \frac{D_t}{2}, \qquad k_e = k_0\sqrt{\frac{1+\epsilon_1}{2}}, \tag{8.32}$$

$$J_{zm} = 1, \qquad |\phi - \phi_m| < \frac{D_z}{2}, \tag{8.33}$$

where $J_{\phi n}$ are piecewise sinusoidal basis functions for the transverse current and J_{zm} are pulse basis functions for the longitudinal current; $I_{\phi n}$ and I_{zm} are the unknown coefficients of the basis functions $J_{\phi n}$ and J_{zm}, respectively; and D_t and D_z are defined as $2w/(N+1)$ and w/M, respectively. By following a similar dyadic Green's function formulation in Section 2.2, the tangential electric field on the microstrip line can be related to the surface current density as follows:

$$\begin{bmatrix} \tilde{E}_\phi \\ \tilde{E}_z \end{bmatrix} = \tilde{\bar{\bar{G}}}\begin{bmatrix} \sum_{n=1}^{N} I_{\phi n} \tilde{J}_{\phi n} & \sum_{m=1}^{M} I_{zm} \tilde{J}_{zm} \end{bmatrix}^{\mathrm{T}}, \tag{8.34}$$

where $\tilde{\bar{\bar{G}}} = \hat{\phi}\tilde{G}_{\phi\phi}\hat{\phi} + \hat{\phi}\tilde{G}_{\phi z}\hat{z} + \hat{z}\tilde{G}_{z\phi}\hat{\phi} + \hat{z}\tilde{G}_{zz}\hat{z}$ is the spectral-domain dyadic Green's function interpreted and expressed in Section 2.2.1. Then, by applying the boundary condition that the surface current density and the electric field are complementary to each other at $\rho = b$ and imposing the Galerkin's moment method, a matrix equation similar to (2.72) is obtained:

$$\begin{bmatrix} (Z_{ln}^{\phi\phi})_{N\times N} & (Z_{lm}^{\phi z})_{N\times M} \\ (Z_{kn}^{z\phi})_{M\times N} & (Z_{km}^{zz})_{M\times M} \end{bmatrix}\begin{bmatrix} (I_{\phi n})_{N\times 1} \\ (I_{zm})_{M\times 1} \end{bmatrix} = \begin{bmatrix} 0 \\ 0 \end{bmatrix}, \tag{8.35}$$

where

$$Z_{ln}^{\phi\phi} = \frac{1}{4\pi^2} \sum_{p=-\infty}^{\infty} \tilde{G}_{\phi\phi}(b, p, \beta) A_p^2 \exp[-jp(\phi_l - \phi_n)], \tag{8.36}$$

$$Z_{lm}^{\phi z} = \frac{1}{2p\pi^2} \sum_{p=-\infty}^{\infty} \tilde{G}_{\phi z}(b, p, \beta) A_p \sin\frac{pD_z}{2} \exp[-jp(\phi_l - \phi_m)], \tag{8.37}$$

$$Z_{kn}^{z\phi} = \frac{1}{2p\pi^2} \sum_{p=-\infty}^{\infty} \tilde{G}_{z\phi}(b, p, \beta) A_p \sin\frac{pD_z}{2} \exp[-jp(\phi_k - \phi_n)], \tag{8.38}$$

$$Z_{km}^{zz} = \frac{1}{p^2\pi^2} \sum_{p=-\infty}^{\infty} \tilde{G}_{zz}(b, p, \beta) \sin^2 \frac{pD_z}{2} \exp[-jp(\phi_k - \phi_m)] , \tag{8.39}$$

$$A_p = \frac{2k_e[\cos(pD_t/2) - \cos(k_e D_t/2)]}{(k_e^2 - p^2) \sin(k_e D_t/2)} , \tag{8.40}$$

$$k, m = 1, 2, \ldots, M , \qquad l, n = 1, 2, \ldots, N .$$

To have nontrivial solutions for $I_{\phi n}$ and I_{zm} in (8.35), the determinant of (8.35) must vanish; that is,

$$\det \begin{bmatrix} (Z_{ln}^{\phi\phi})_{N\times N} & (Z_{lm}^{\phi z})_{N\times M} \\ (Z_{kn}^{z\phi})_{M\times N} & (Z_{km}^{zz})_{M\times M} \end{bmatrix} = 0 . \tag{8.41}$$

The solution to the characteristic equation above gives the propagation constant β of the cylindrical microstrip line, and the effective relative permittivity can be calculated from (8.26).

For the calculation of characteristic impedance, we adopt the voltage–current definition [10]:

$$Z_c = \frac{V_0}{I_z} , \tag{8.42}$$

with

$$V_0 = \frac{2\pi b}{|I_z|^2} \sum_{p=-\infty}^{\infty} \tilde{J}_z(-p)\tilde{V}_p \tilde{J}_z(p)\delta(k_z - \beta) , \tag{8.43}$$

$$\tilde{V}_p = \int_b^a \tilde{G}_{\rho z}(\rho, p, k_z)\, d\rho , \tag{8.44}$$

where V_0 is the potential difference of the conducting strip to the ground cylinder, I_z the total surface current in the $\hat{z}$ direction, and $\tilde{G}_{\rho z}$ the spectral-domain Green's function for the $\hat{\rho}$-directed electric field in the substrate due to a unit-amplitude $\hat{z}$ -directed current element at $\rho = b$. By a straightforward manipulation, (8.42) can be rewritten as

$$Z_c = \frac{2}{\pi b p^2 D_z^2} \frac{\sum_{p=-\infty}^{\infty} \tilde{V}_p \sin^2(pD_z/2) \sum_{k=1}^{M} I_{zk} e^{jp\phi_{zk}} \sum_{m=1}^{M} I_{zm} e^{-jp\phi_{zm}}}{\left(\sum_{m=1}^{M} I_{zm} \right)^2} . \tag{8.45}$$

To obtain good convergence in the numerical computation, a total of six basis functions ($N = M = 3$) for expanding the surface current density on the strip line is used. When more basis functions are used, the results obtained differ very slightly from those obtained using $N = M = 3$. It is also noted that $N = 0$ represents omission of the transverse current component. Cylindrical microstrip lines with various widths are also constructed using flexible copper-clad laminates, and the effective relative permittivity and characteristic impedance are measured using a vector network analyzer, based on the measurement method described in [25]. The theoretical results both with and without considering the transverse current

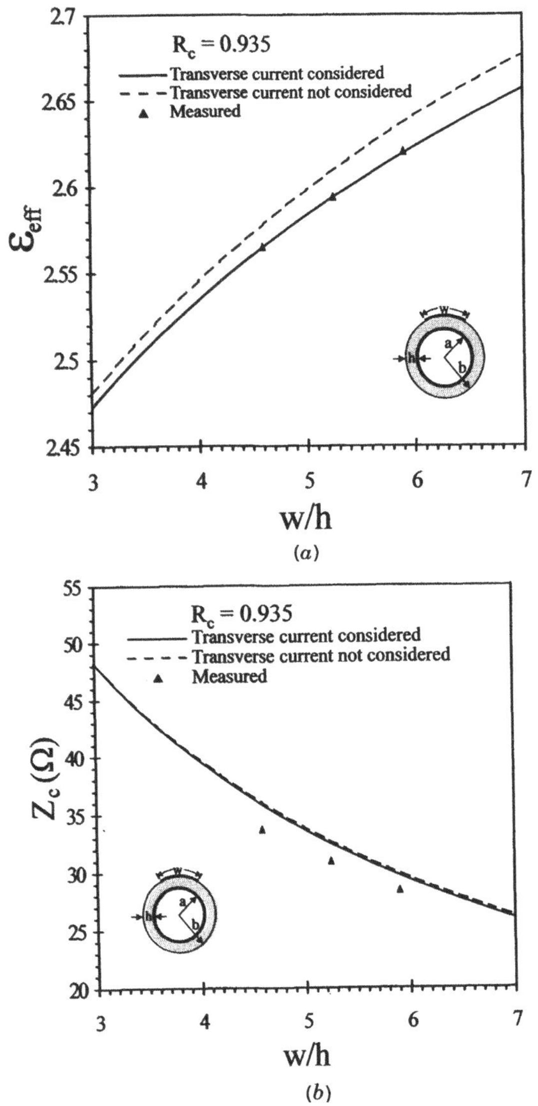

FIGURE 8.7 (*a*) Effective relative permittivity and (*b*) characteristic impedance as a function of w/h; $\epsilon_1 = 3.0$, $h = 0.762$ mm, $R_c = 0.935$, f = 5 GHz. (From Ref. [8], © 1996 John Wiley & Sons, Inc.)

component are calculated, and typical results versus w/h for a cylindrical microstrip line with $R_c = 0.935$ are shown in Figure 8.7. The data measured are shown in the figure for comparison. It is seen that, as expected, results with the transverse current component agree better with the data measured. The theoretical and measured results are plotted versus frequency in Figure 8.8. The results calculated with the transverse current component again agree better with the data

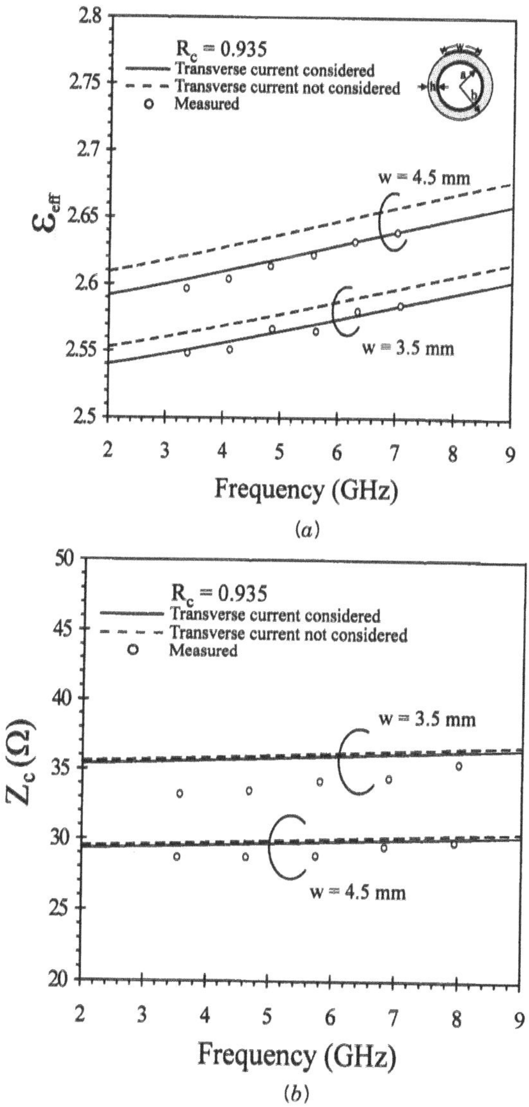

FIGURE 8.8 (*a*) Effective relative permittivity and (*b*) characteristic impedance as a function of frequency; $\epsilon_r = 3.0$, $h = 0.762$ mm, $R_c = 0.935$. (From Ref. [8], © 1996 John Wiley & Sons, Inc.)

measured, and the transverse current contribution is more significant for wider cylindrical microstrip lines. As expected, the contribution of the transverse current component becomes important and should be considered when the width of the microstrip line increases.

8.3 COUPLED CYLINDRICAL MICROSTRIP LINES

Coupled cylindrical microstrip lines have applications for directional couplers, filters, and other important microstrip circuits on (outside or inside) cylindrical surfaces. The coupled microstrip lines can support two propagation modes (even and odd modes), and the propagation constant and characteristic impedance of these two modes are in general unequal. To study the characteristics of coupled microstrip lines accurately, the longitudinal and transverse components of the surface current on the coupled microstrip lines must both be considered in the formulation. The even- and odd-mode cases are both considered. Typical results showing the effects of curvature variation and line separation on the characteristics of coupled microstrip lines are presented.

In the following formulation, coupled microstrip lines mounted inside a ground cylinder as shown in Figure 8.9*a* are studied. The line separation of the coupled microstrip lines is denoted as *S*. Other parameters have the same meanings as in

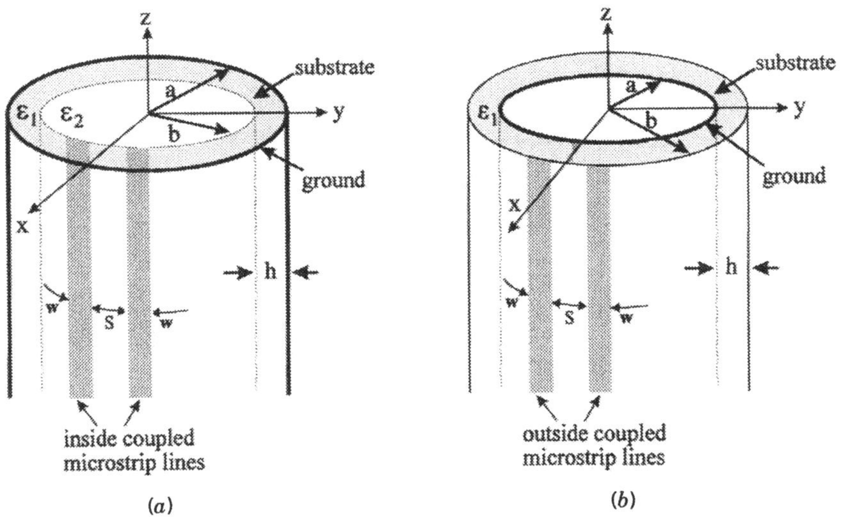

FIGURE 8.9 Configurations of coupled cylindrical microstrip lines mounted (*a*) inside and (*b*) outside a ground cylinder.

Figure 8.1. By following a theoretical formulation similar to that for the wide cylindrical microstrip line discussed in Section 8.2.2B, the current density on the coupled lines is first expanded in terms of linear combinations of known basis functions as given by (8.31)–(8.33). By further consideration of the even- and odd-mode operations of the coupled lines, we can place a magnetic or electric wall at the center ($\phi = 0°$ plane) of the coupled lines. In this case the even- and odd-mode basis functions are linear combinations of (8.32) and (8.33) and can be, respectively, written as follows (see Figure 8.10): For the even mode,

$$J_{\phi n}^{(e)} = \frac{J_{\phi n}^{(+)} - J_{\phi n}^{(-)}}{2}, \tag{8.46}$$

$$J_{zm}^{(e)} = \frac{J_{zm}^{(+)} + J_{zm}^{(-)}}{2}, \tag{8.47}$$

and for the odd mode,

$$J_{\phi n}^{(o)} = \frac{J_{\phi n}^{(+)} + J_{\phi n}^{(-)}}{2}, \tag{8.48}$$

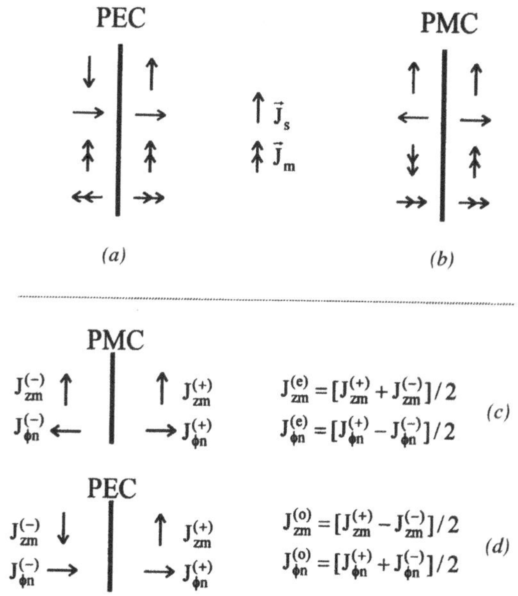

FIGURE 8.10 Images for the cases of (*a*) electric and (*b*) magnetic conducting planes; (*c*) even-mode and (*d*) odd-mode basis functions that are linear combinations of $J_{\phi n}^{(\pm)}$ and $J_{zm}^{(\pm)}$.

$$J_{zm}^{(o)} = \frac{J_{zm}^{(+)} - J_{zm}^{(-)}}{2} . \tag{8.49}$$

Then, taking the Fourier transform of (8.46)–(8.49), we have

$$\tilde{J}_{\phi n}^{(e)} = j \frac{1}{2\pi} A_p \sin p\phi_n , \tag{8.50}$$

$$\tilde{J}_{zm}^{(e)} = \frac{1}{p\pi} \sin \frac{pD_z}{2} \cos p\phi_m , \tag{8.51}$$

and

$$\tilde{J}_{\phi n}^{(o)} = \frac{1}{2\pi} A_p \cos p\phi_n , \tag{8.52}$$

$$\tilde{J}_{zm}^{(o)} = \frac{j}{p\pi} \sin \frac{pD_z}{2} \sin p\phi_m , \tag{8.53}$$

where A_p is given in (8.40). Based on the basis functions above, the expressions of the elements in the [Z] matrix given in (8.36)–(8.39) can be replaced by

$$Z_{ln}^{\phi\phi} = \frac{1}{4\pi^2} \sum_{p=-\infty}^{\infty} \tilde{G}_{\phi\phi}(b, p, \beta) A_p^2 \begin{matrix}\sin\\ \cos\end{matrix}(p\phi_l) \begin{matrix}\sin\\ \cos\end{matrix}(p\phi_n) , \tag{8.54}$$

$$Z_{lm}^{\phi z} = \pm \frac{j}{2p\pi^2} \sum_{p=-\infty}^{\infty} \tilde{G}_{\phi z}(b, p, \beta) A_p \sin \frac{pD_z}{2} \begin{matrix}\sin\\ \cos\end{matrix}(p\phi_l) \begin{matrix}\cos\\ \sin\end{matrix}(p\phi_m) , \tag{8.55}$$

$$Z_{kn}^{z\phi} = \mp \frac{j}{2p\pi^2} \sum_{p=-\infty}^{\infty} \tilde{G}_{z\phi}(b, p, \beta) A_p \sin \frac{pD_z}{2} \begin{matrix}\cos\\ \sin\end{matrix}(p\phi_k) \begin{matrix}\sin\\ \cos\end{matrix}(p\phi_n) , \tag{8.56}$$

$$Z_{km}^{zz} = \frac{1}{2p^2\pi^2} \sum_{p=-\infty}^{\infty} \tilde{G}_{zz}(b, p, \beta) A_p \sin \frac{pD_z}{2} \begin{matrix}\cos\\ \sin\end{matrix}(p\phi_k) \begin{matrix}\cos\\ \sin\end{matrix}(p\phi_m) , \tag{8.57}$$

$$k, m = 1, 2, \ldots, M , \qquad l, n = 1, 2, \ldots, N ,$$

where uppercase expressions denote the even mode and lower case, the odd mode. By applying (8.54)–(8.57) to the characteristic matrix equation of (8.41), the effective relative permittivity of the coupled lines can be evaluated from (8.26).

For calculation of the characteristic impedance of the coupled lines, the variational expression given in (8.42)–(8.44) can also be adopted. By a straightforward manipulation, the corresponding characteristic impedance for the even and odd modes can be given, respectively, by

$$Z_c^{(e)} = \frac{2}{\pi b p^2 D_z^2} \frac{\left\{ \sum_{p=-\infty}^{\infty} \tilde{V}_p \sin^2(pD_z/2) \sum_{k=1}^{M} I_{zk} \cos(p\phi_{zk}) \sum_{m=1}^{M} I_{zm} \cos(p\phi_{zm}) \right\}}{\left(\sum_{m=1}^{M} I_{zm} \right)^2}, \tag{8.58}$$

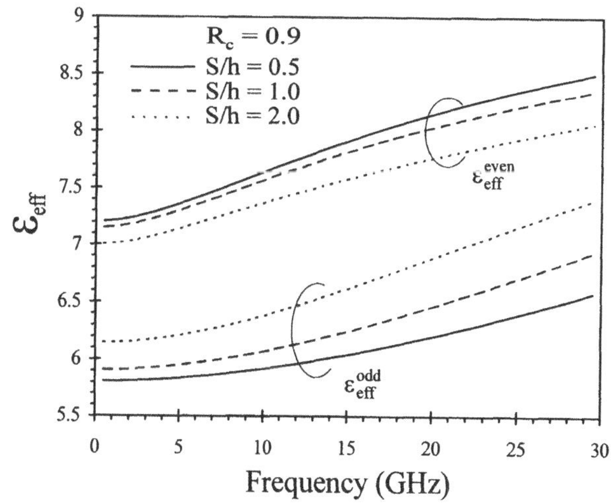

(*a*)

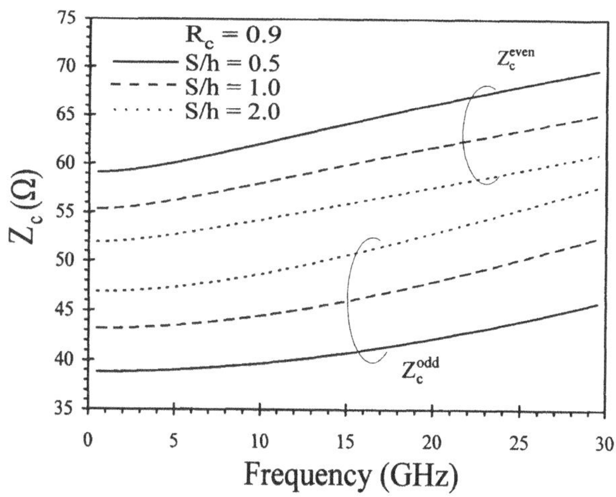

(*b*)

FIGURE 8.11 (*a*) Effective relative permittivity and (*b*) characteristic impedance as a function of frequency for various line spacings; $R_c = 0.9$, $\epsilon_1 = 9.8$, $h = 0.635$ mm, $w/h = 1.0$. (From Ref. [14], © 1995 John Wiley & Sons, Inc.)

$$Z_c^{(o)} = \frac{2}{\pi b p^2 D_z^2} \frac{\left\{ \sum_{p=-\infty}^{\infty} \tilde{V}_p \sin^2(pD_z/2) \sum_{k=1}^{M} I_{zk} \sin(p\phi_{zk}) \sum_{m=1}^{M} I_{zm} \sin(p\phi_{zm}) \right\}}{\left(\sum_{m=1}^{M} I_{zm} \right)^2} . \tag{8.59}$$

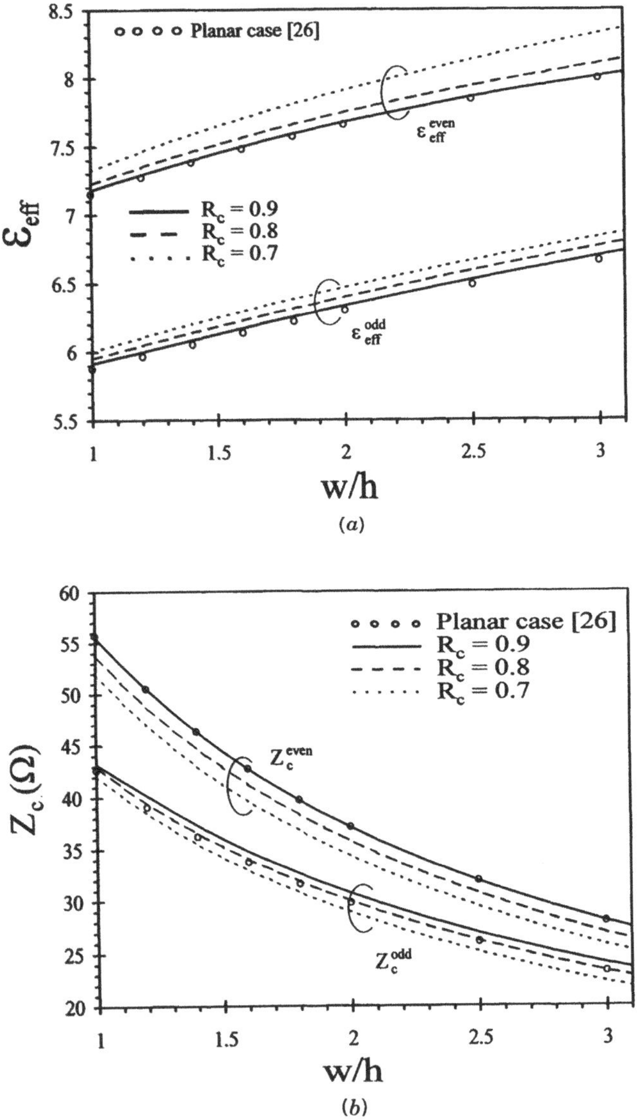

FIGURE 8.12 (*a*) Effective relative permittivity and (*b*) characteristic impedance as a function of w/h for various curvilinear coefficients; $\epsilon_1 = 9.8$, $h = 0.635$ mm, $f = 2$ GHz, $S/h = 1.0$. (From Ref. [14], © 1995 John Wiley & Sons, Inc.)

A total of six basis functions ($N = M = 3$) for modeling the surface current density on the coupled microstrip lines is selected in the numerical computation, which shows good numerical convergence. Figure 8.11 shows typical results of ϵ_{eff} and Z_c as a function of frequency for various line separations. The coupled lines are chosen to have a curvilinear coefficient of $R_c = 0.9$. It can be seen that the

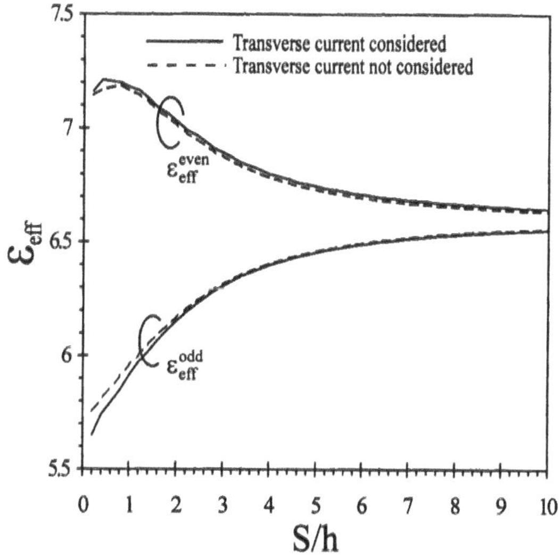

(*a*)

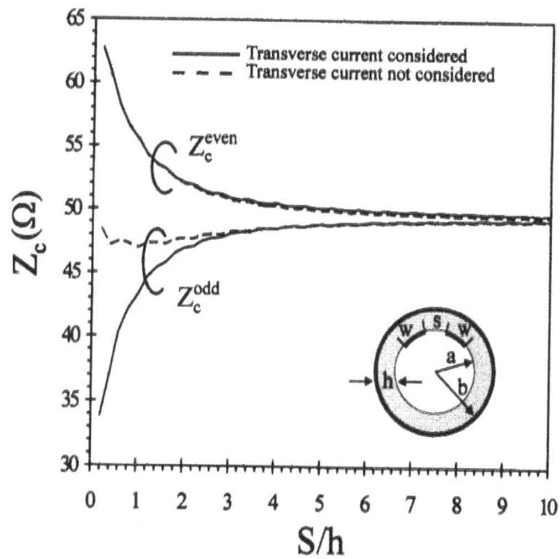

(*b*)

FIGURE 8.13 (*a*) Effective relative permittivity and (*b*) characteristic impedance as a function of S/h for the cases with and without considering the transverse current component; $\epsilon_1 = 9.8$, $h = 0.635$ mm, $f = 2$ GHz, $w/h = 1.0$, $R_c = 0.9$.

coupled microstrip lines have different even- and odd-mode ϵ_{eff} and Z_c. This is due to different field distributions for the even and odd modes. The even mode has less fringing field in the air region than the odd mode; that is, the electric field energy is more concentrated in the substrate region for the even mode, which leads to the

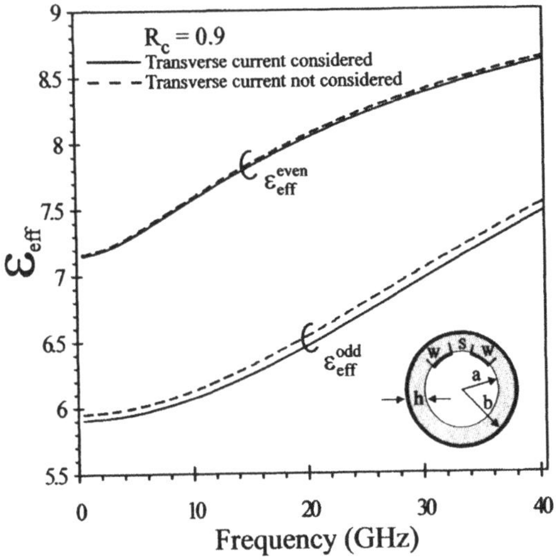

(a)

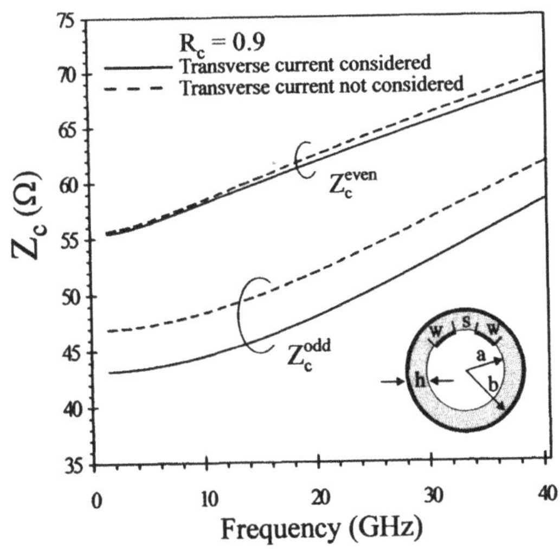

(b)

FIGURE 8.14 (*a*) Effective relative permittivity and (*b*) characteristic impedance as a function of frequency with and without the transverse current component; $\epsilon_1 = 9.8$, $h = 0.635$ mm, $S/h = 1.0$, $w/h = 1.0$, $R_c = 0.9$.

results that the even-mode ϵ_{eff} and Z_c are greater than those of the odd mode, as shown in Figure 8.11.

Figure 8.12 shows ϵ_{eff} and Z_c as a function of w/h for various curvilinear coefficients. The corresponding results of the planar case [26] are also plotted for comparison. From the results, it is observed that except for the even-mode ϵ_{eff}, the variations of ϵ_{eff} and Z_c with the curvature are in contrast to the case of coupled cylindrical microstrip lines mounted outside a ground cylinder [13]. This is again probably due to the different bending directions of the coupled microstrip lines.

Figures 8.13 presents ϵ_{eff} and Z_c of the coupled microstrip lines as a function of line separation. It can be seen that when the separation increases, the differences in the characteristics of even- and odd-mode propagation become small. Also, in such a case, the transverse current contribution to the coupled-line propagation can be ignored. Values of ϵ_{eff} and Z_c as a function of frequency are also calculated and shown in Figure 8.14. It is seen that results obtained without considering the transverse current component differ slightly from those obtained considering the transverse current component for even-mode propagation. However, for odd-mode propagation, a large discrepancy is observed when the transverse current component is not considered. In this case the transverse current contribution should be taken into consideration in the theoretical formulation.

8.4 SLOT-COUPLED DOUBLE-SIDED CYLINDRICAL MICROSTRIP LINES

When multilayer structures are considered, surface-to-surface transitions between printed transmission lines on different layers are required for the transfer of power between two adjacent layers. Among these transitions, slot-coupled microstrip lines have found applications in multilayer microstrip antennas as feed networks or in microwave circuits as directional couplers and filters, because slot coupling offers several well-known advantages, such as no direct connection and no radiation from the feed network interfering with the main radiation pattern.

In this section, slot-coupled double-sided (SCDS) cylindrical microstrip lines are investigated theoretically by using the reciprocity theorem described in Section 4.5 and the exact Green's functions incorporating a moment-method calculation. In this case the coupling slot in the common ground plane and the coupled line can be viewed as an equivalent series load at the slot position seen by the feed line. Expressions for the amplitudes of reflected and transmitted waves on the slot-coupled cylindrical microstrip lines are derived first. The solution is based on the moment-method calculation, and the slot field is represented in terms of a set of piecewise sinusoidal (PWS) mode with unknown expansion coefficients. Then, S-parameter expressions for the SCDS cylindrical microstrip lines as a coupler are shown, and numerical results are calculated and analyzed. In the study, a four-port coupler using SCDS cylindrical microstrip lines was constructed. Experiments are also conducted to compare with theoretical results. The curvature effects of SCDS cylindrical microstrip lines are discussed. The coupling coefficient is also presented as a function of the coupling slot length.

Figure 8.15 shows the geometry of SCDS cylindrical microstrip lines. The inside cylindrical microstrip line is treated as a feed line, and the outside cylindrical microstrip line is the coupled line. These two lines are assumed to be infinitely long and are coupled through a rectangular narrow slot with a length of L_s $(=2a\phi_s)$ and a width of W_s centered at $z=0$ in the common cylindrical ground cylinder of radius a. The widths of the feed and coupled lines are W_f $(=2b_f\phi_f)$ and W_c $(=2b_c\phi_c)$, respectively. The feed substrate has a thickness h_f and a relative permittivity ϵ_f; the coupled substrate has a thickness h_c and a relative permittivity ϵ_c. The inner $(\rho < b_f)$ and outer $(\rho > b_c)$ regions are assumed to be air. To begin with, the SCDS microstrip lines are considered as a four-port coupler and we assume that the input power at port 1 is 1 watt and that the microstrip lines are propagating quasi-TEM waves. By the narrow slot assumption $(L_s >> W_s)$, the electric field in the slot can be approximated as follows, similar to (4.217):

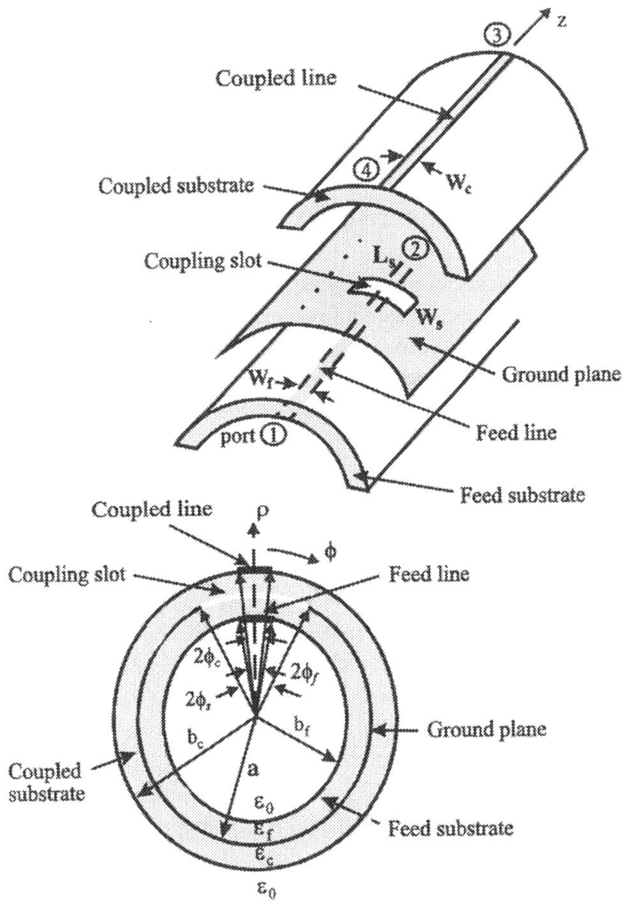

FIGURE 8.15 Geometry of slot-coupled double-sided (SCDS) cylindrical microstrip lines. (From Ref. [15], © 1996 IEEE, reprinted with permission.)

$$\mathbf{E}_s = \hat{z} \sum_{q=1}^{N} V_q E_q^s(\phi, z)$$

$$= \hat{z} \sum_{q=1}^{N} V_q \frac{1}{W_s} f_p^s(\phi - \phi_q), \qquad |z| < \frac{W_s}{2}, \quad |\phi| < \phi_s, \tag{8.60}$$

with

$$f_p^s(\phi) = \begin{cases} \dfrac{\sin k_e(a\phi_h - a|\phi|)}{\sin k_e a\phi_h}, & |\phi| < \phi_h, \\ 0, & |\phi| > \phi_h, \end{cases} \tag{8.61}$$

$$\phi_q = \left(\frac{2q}{N+1} - 1\right)\phi_s, \qquad \phi_h = \frac{2\phi_s}{N+1}, \qquad \phi_s = \frac{L_s}{2a}, \tag{8.62}$$

$$k_e = k_0\sqrt{\frac{\epsilon_f + \epsilon_c}{2}}, \qquad k_0 = \omega\sqrt{\mu_0\epsilon_0}, \tag{8.63}$$

where $f_p^s(\phi)$ is a PWS basis function for the slot field, ϕ_q the center point of the qth expansion mode, $a\phi_h$ the half-length of the PWS function, and V_q the unknown coefficient of the qth expansion mode.

To solve the unknown coefficient V_q, two boundary conditions are assumed: continuity of the tangential magnetic field at the slot position, and zero tangential electric field on the coupled line. For the first boundary condition, we have

$$H_\phi^f + H_\phi^{sf} = H_\phi^c + H_\phi^{sc}, \tag{8.64}$$

where H_ϕ^f and H_ϕ^c are, respectively, the tangential magnetic fields at $z = 0$, $\rho = a^-$ and $z = 0$, $\rho = a^+$ contributed from the feed line and the coupled line in the absence of the coupling slot; H_ϕ^{sf} and H_ϕ^{sc} are the tangential magnetic fields at $\rho = a^-$ and $\rho = a^+$ contributed from the slot field, respectively. Knowing that $1 - R = T$ on the feed line [see (4.205)], we have

$$(1 - R)h_\phi^f + H_\phi^{sf} = H_\phi^c + H_\phi^{sc}, \tag{8.65}$$

where R $(= S_{11})$ is the reflection coefficient on the feed line and h_ϕ^f is the normalized transverse magnetic field at the slot position caused by the electric surface current on the feed line. By imposing the equivalence principle, the coupling slot can be closed off and then replaced by an equivalent magnetic surface current $\mathbf{M}^{sf}$ $[= -\hat{\phi}M_\phi^{sf} = \mathbf{E}^s \times (-\hat{\rho})]$ inside the ground cylinder at $\rho = a^-$ and $\mathbf{M}^{sc}$ $(= -\mathbf{M}^{sf})$ outside the ground cylinder at $\rho = a^+$, where the negative sign ensures that the tangential component of the electric field is continuous across the slot. Then, by deriving the appropriate Green's functions, such as $G_{\phi\phi}^{HMf}$, $G_{\phi\phi}^{HMc}$, $G_{\phi z}^{HJf}$, and $G_{\phi z}^{HJc}$ for the cylindrical structure studied here to account for the $\hat{\phi}$ component of the magnetic fields at $\rho = a^-$ and $\rho = a^+$ due to a unit-amplitude

$\hat{\phi}$-directed equivalent magnetic current at the slot position and the $\hat{z}$-directed electric surface currents on the feed line and coupled line, the magnetic fields in (8.65) can be expressed as

$$H_\phi^{sf} = \iint_{S_a} G_{\phi\phi}^{HMf}(a^-, \phi, z) M_\phi^{sf}\, dS\,, \tag{8.66}$$

$$H_\phi^{sc} = \iint_{S_a} G_{\phi\phi}^{HMc}(a^+, \phi, z) M_\phi^{sc}\, dS\,, \tag{8.67}$$

$$H_\phi^{c} = \int_{-\infty}^{\infty} \int_{-W_c/2}^{W_c/2} G_{\phi z}^{HJc}(a^+, \phi, z) J_z^c\, dS\,, \tag{8.68}$$

where J_z^c is the current density on the coupled line, S_a is the slot area, and the related Green's functions have expressions similar to those given in [27]. Multiplying (8.65) by the expansion function of the slot electric field and integrating over the slot area, we can obtain

$$([Y^{sf}] + [Y^{sc}])[V] = -(1-R)[\Delta v^f][\Delta v^c]\,, \tag{8.69}$$

where $[Y^{sf}]$ and $[Y^{sc}]$ are the admittance matrices for the slot admittance looking at $\rho = a^-$ and $\rho = a^+$, respectively; $[V]$ is the unknown expansion coefficient matrix for V_q, $[\Delta v^f]$ is a matrix representing the voltage discontinuity across the slot, and $[\Delta v^c]$ denotes the reaction between the slot field and the current on the coupled line. The expressions of the elements Y_{mq}^{sf} in $[Y^{sf}]$, Y_{mq}^{sc} in $[Y^{sc}]$, Δv_q^f in $[\Delta v^f]$, and Δv_q^c in $[\Delta v^c]$ are derived as

$$Y_{mq}^{sf} = \frac{2}{W_s^2 \pi} \sum_{n=-\infty}^{\infty} \int_{-\infty}^{\infty} \frac{1}{k_z^2} \tilde{G}_{\phi\phi}^{HMf}(a, n, k_z) \sin^2 \frac{k_z W_s}{2} \tilde{f}_p^{s\,2}(n) \cos[n(\phi_m - \phi_q)]\, dk_z\,, \tag{8.70}$$

$$Y_{mq}^{sc} = \frac{2}{W_s^2 \pi} \sum_{n=-\infty}^{\infty} \int_{-\infty}^{\infty} \frac{1}{k_z^2} \tilde{G}_{\phi\phi}^{HMc}(a, n, k_z) \sin^2 \frac{k_z W_s}{2} \tilde{f}_p^{s\,2}(n) \cos[n(\phi_m - \phi_q)]\, dk_z\,, \tag{8.71}$$

$$\Delta v_q^f = \frac{-2}{\phi_f \beta_f W_s \sqrt{Z_c^f}} \sum_{n=-\infty}^{\infty} \frac{1}{n} \tilde{G}_{\phi z}^{HJf}(a, n, -\beta_f) \sin n\phi_f \sin^2\!\left(-\frac{\beta_f W_s}{2}\right) \tilde{f}_p^s(n) \cos n\phi_q\,, \tag{8.72}$$

$$\Delta v_q^c = \frac{-1}{W_s \pi} \sum_{n=-\infty}^{\infty} \int_{-\infty}^{\infty} \frac{1}{k_z} \tilde{G}_{\phi z}^{HJc}(a, n, k_z) \tilde{J}_z^c \sin \frac{k_z W_s}{2} \tilde{f}_p^s(n) \cos n\phi_q\, dk_z\,, \tag{8.73}$$

with

$$\tilde{f}^{s}_{p}(n) = \frac{2k_e(\cos n\phi_h - \cos k_e a\phi_h)}{[k_e^2 - (n/a)^2]\sin k_e a\phi_h}, \tag{8.74}$$

where Z^f_c and β_f are, respectively, the characteristic impedance and propagation constant of the feed line, whose expressions have been described in Section 8.2.2.

As for applying the second boundary condition that the electric field on the coupled line must vanish, we can have

$$\int_{-\infty}^{\infty}\int_{-W_c/2}^{W_c/2} G^{EJc}_{zz}(b_c, \phi, z)J^c_z\, dS + \iint_{S_a} G^{EMc}_{z\phi}(b_c, \phi, z)M^{sc}_{\phi}\, dS = 0, \tag{8.75}$$

where G^{EJc}_{zz} and $G^{EMc}_{z\phi}$ are Green's functions showing the tangential electric field on the coupled line due to a unit-amplitude $\hat{z}$-directed electric surface current on the coupled line and a unit-amplitude $\hat{\phi}$-directed equivalent magnetic current at the slot, respectively. By assuming that the coupled line also propagates a quasi-TEM wave in the z direction as the feed line, J^c_z can be expressed as

$$J^c_z(\phi, z) = \begin{cases} \dfrac{I_0}{W_c} e^{j\beta_c z}, & z > 0, \\ \dfrac{I_0}{W_c} e^{-j\beta_c z}, & z < 0, \end{cases} \tag{8.76}$$

where β_c is the propagation constant of the coupled line, I_0 the unknown current amplitude coupled from the feed line, and a pulse function $(1/W_c)$ is assumed for the ϕ dependence. Then, by following a similar theoretical treatment described in [28] for (8.75), a relationship between I_0 and the unknown expansion coefficient V_q for the slot electric field can be obtained; that is,

$$ZI_0 + [V]^{\mathrm{T}}[N^s] = 0, \tag{8.77}$$

with

$$Z = -\frac{1}{2\pi}\sum_{n=-\infty}^{\infty} \tilde{G}^{EJc}_{zz}(b_c, n, -\beta_c)\,\mathrm{sinc}^2(n\phi_c)\tilde{f}^{c}_{p}(-\beta_c), \tag{8.78}$$

$$N^s_q = \frac{1}{W_s\phi_c\pi}\sum_{n=-\infty}^{\infty}\int_{-\infty}^{\infty}\frac{1}{k_z n}\tilde{G}^{EMc}_{z\phi}(b_c, n, k_z)\sin\frac{k_z W_s}{2}\sin n\phi_c \tilde{f}^{s}_{p}(n)\tilde{f}^{c}_{p}(k_z)\cos n\phi_q\, dk_z, \tag{8.79}$$

$$\tilde{f}^{c}_{p}(k_z) = \frac{2k_e(\cos k_z d - \cos k_e d)}{(k_e^2 - k_z^2)\sin k_e d}, \tag{8.80}$$

where $[V]^{\mathrm{T}}$ is the transpose of $[V]$, N_q^s the element in $[N^s]$, and $\tilde{f}_p^c(k_z)$ the Fourier transform of a PWS function in (8.61) with a half-length of d.

By further substituting the Fourier transform of (8.76) into (8.73) and applying the relationship of $\tilde{G}_{\phi z}^{HJc} = -(b_c/a)\tilde{G}_{z\phi}^{EMc}$ [27], we have the element Δv_q^c expressed as

$$\Delta v_q^c = I_0 N_q^c , \tag{8.81}$$

with

$$N_q^c = \frac{b_c}{W_s \beta_c \phi_c \pi a} \sum_{n=-\infty}^{\infty} \frac{1}{n} \tilde{G}_{z\phi}^{EMc}(b_c, n, -\beta_c) \sin\left(-\frac{\beta_c W_s}{2}\right) \sin n\phi_c \tilde{f}_p^s(n) \cos n\phi_q . \tag{8.82}$$

By rewriting (8.81) into a matrix form and substituting it into (8.69), we have

$$([Y^{sf}] + [Y^{sc}] + [Y^c])[V] = -(1-R)[\Delta v^f] , \tag{8.83}$$

with

$$[Y^c] = \frac{[N^s][N^c]^{\mathrm{T}}}{Z} . \tag{8.84}$$

With the reflection coefficient R expressed as [(4.220)]

$$R = \frac{-1}{2} [V]^{\mathrm{T}} [\Delta v^f] , \tag{8.85}$$

the unknown coefficient matrix $[V]$ for the slot electric field can readily be calculated from (8.83) and written as

$$[V] = -\{[Y^{sf}] + [Y^{sc}] + [Y^c] + \tfrac{1}{2}[\Delta v^f][\Delta v^f]^{\mathrm{T}}\}^{-1}[\Delta v^f] . \tag{8.86}$$

With $[V]$ determined, the coupling coefficient S_{31}, defined as the power ratio between ports 3 and 1, can be written as

$$\begin{aligned} S_{31} &= I_0\sqrt{Z_c^c} \\ &= -\frac{[V]^{\mathrm{T}}[N^s]\sqrt{Z_c^c}}{Z} , \end{aligned} \tag{8.87}$$

where Z_c^c is the characteristic impedance of the coupled line, whose expression is given by (8.45) in Section 8.2.2. Typical results for the S parameters [S_{11} $(=R)$, the reflection coefficient; S_{21} $(=1-R)$, the transmission coefficient; and S_{31} the coupling coefficient] are calculated and compared with the data measured. Figure 8.16 compares the results calculated and data measured. Three ($N=3$) PWS basis functions for the expansion of the slot electric field are used, which ensures good

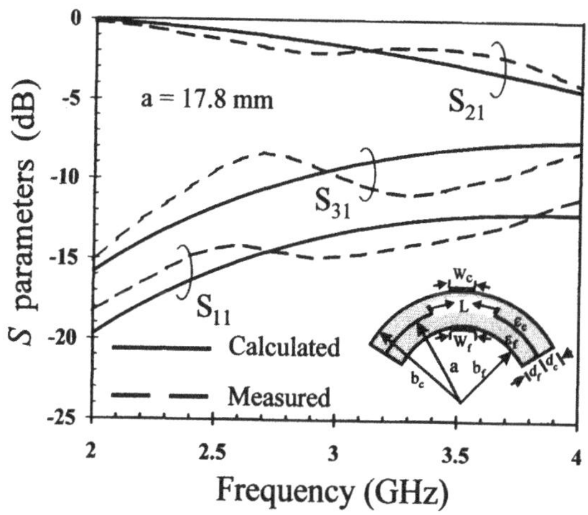

FIGURE 8.16 S parameters calculated and measured for SCDS cylindrical microstrip lines; $a = 17.8$ mm, $\epsilon_f = \epsilon_c = 3.0$, $h_f = h_c = 0.762$ mm, $L_s = 15$ mm, $W_s = 1.0$ mm, $W_f = W_c = 1.9$ mm. (From Ref. [15], © 1996 IEEE, reprinted with permission.)

numerical convergence. For the experiment, several SCDS cylindrical microstrip lines were constructed and measured. The characteristic impedances of the feed line and coupled line were both designed to be 50 Ω in the planar geometry, and 50-Ω terminators were connected at ports 3 and 4 for the measurement of S_{11} and S_{21}. For measuring S_{31}, ports 2 and 4 were both connected to 50-Ω terminators. It is observed that the results calculated in general agree with the data measured except for some ripples appearing in the measured data. The ripples are probably due to shifting of the characteristic impedance (Z_c^f and Z_c^c) of the inside and outside cylindrical microstrip lines away from 50 Ω, due to the curvature variation, which results in a mismatch of the microstrip lines to the 50-Ω terminator.

S parameters calculated versus frequency for various cylinder radii are presented in Figure 8.17. It is seen that S_{11} increases with increasing cylinder radius, while S_{21} and S_{31} decrease with increasing cylinder radius. Figure 8.18 shows variation in S parameters versus coupling-slot length for various cylinder radii. Similar dependence of the S parameters on cylinder radius as seen in Figure 8.17 is observed. Finally, S parameters calculated versus normalized coupling-slot length are presented in Figure 8.19. It is found that as the slot length nears $0.5\lambda_s$ [$\lambda_s = \lambda_0/\sqrt{(\epsilon_f + \epsilon_c)/2}$], the coupling coefficient S_{31} has a maximum value. This implies that when the coupling slot is at resonance, we can have optimal coupling between the slot-coupled cylindrical microstrip lines. However, it should be noted that the power loss (radiation and surface-wave loss) of the slot-coupled structure also increases with increasing coupling-slot length [29]. When the slot length is greater than $0.2\lambda_s$, the power loss of the structure is usually greater than 1%, which should be considered in practical designs.

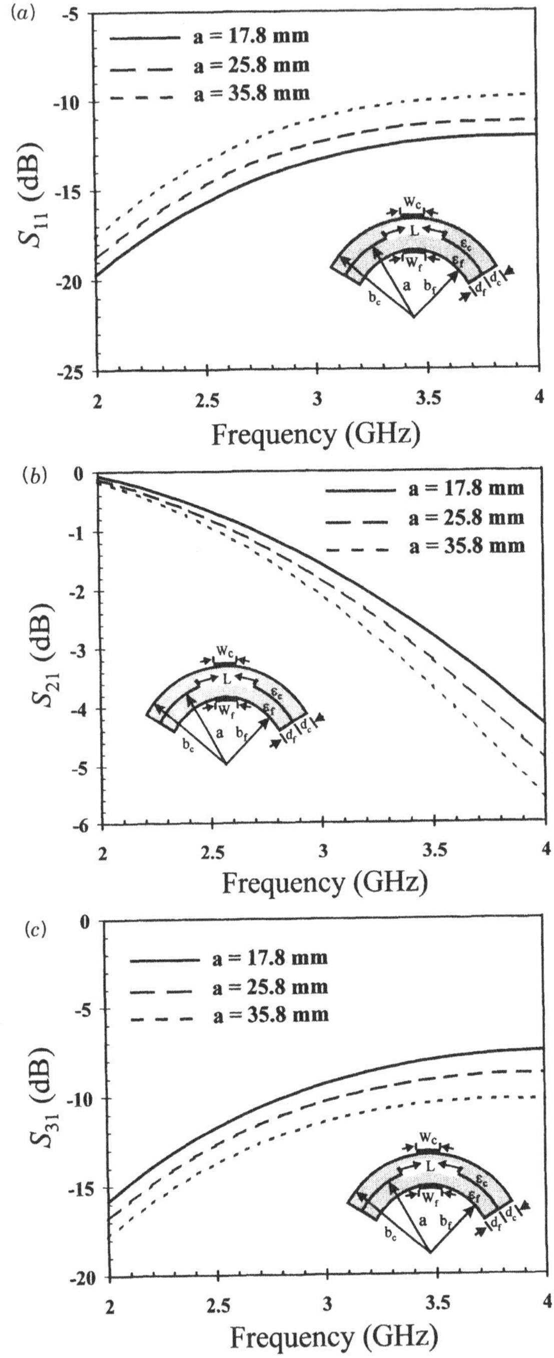

FIGURE 8.17 *S* parameters calculated versus frequency for SCDS cylindrical microstrip lines with various cylinder radii. Parameters of the microstrip lines are as given in Figure 8.16. (*a*) S_{11}; (*b*) S_{21}; (*c*) S_{31}.

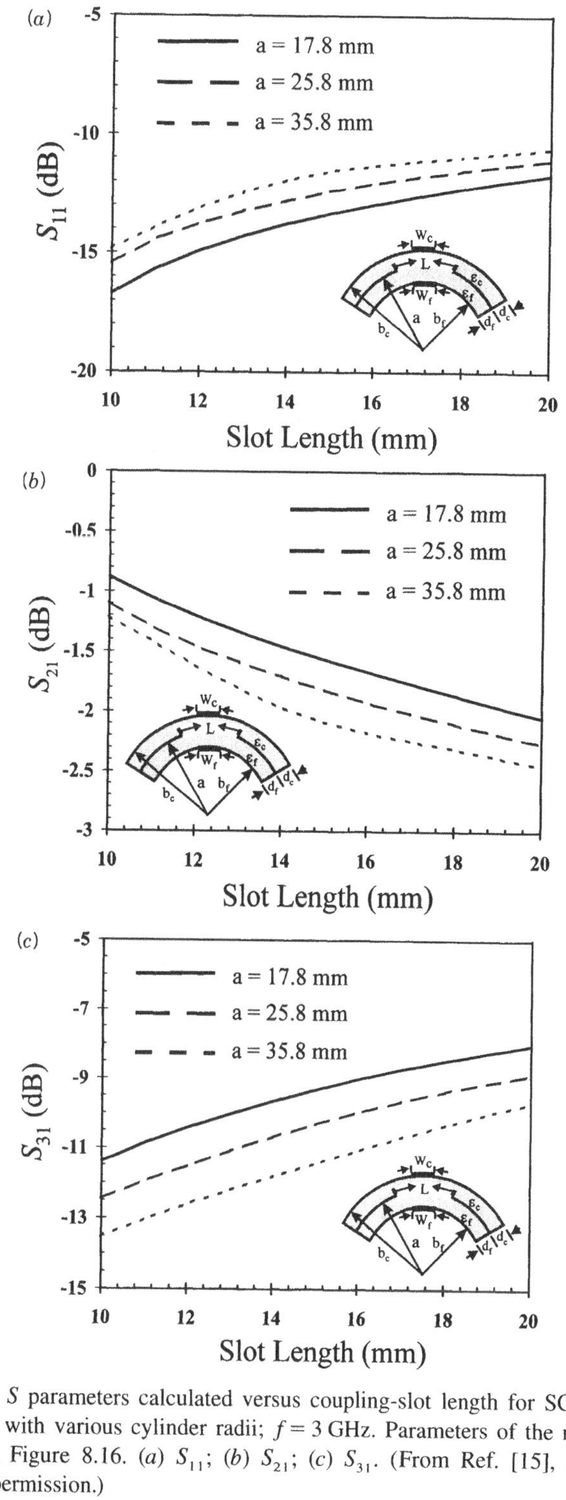

FIGURE 8.18 *S* parameters calculated versus coupling-slot length for SCDS cylindrical microstrip lines with various cylinder radii; $f = 3$ GHz. Parameters of the microstrip lines are as given in Figure 8.16. (*a*) S_{11}; (*b*) S_{21}; (*c*) S_{31}. (From Ref. [15], © 1996 IEEE, reprinted with permission.)

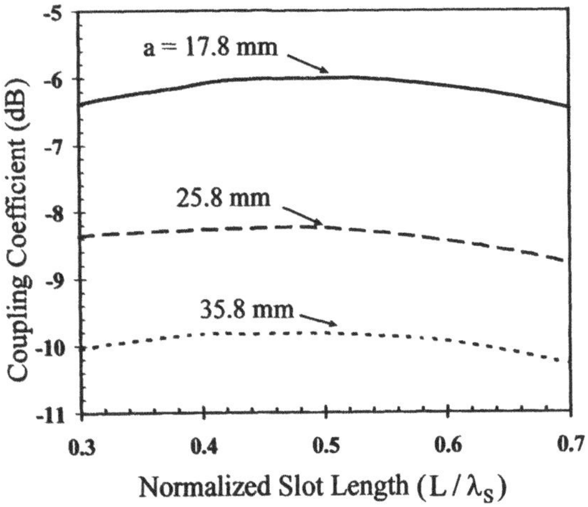

FIGURE 8.19 Coupling coefficient calculated versus normalized coupling slot length (L/λ_s, where λ_s is the wavelength in the slot) for SCDS cylindrical microstrip lines with various cylinder radii; $f = 3.5$ GHz. Parameters of the microstrip lines are as given in Figure 8.16.

8.5 CYLINDRICAL MICROSTRIP DISCONTINUITIES

Discontinuities in microstrip lines are caused by abrupt changes in the geometry of the strip conductor. Typical microstrip discontinuities include open-end and gap discontinuities, which can be used not only in the design of matching stubs and coupled filters for applications in microwave integrated circuits but also in the microstrip circuitry that forms the microstrip antenna or array excitation network. Since such microstrip discontinuities may generate radiating and surface waves, accurate characterization of the discontinuity characteristics of microstrip lines is important. In this section we give a full-wave solution for the characteristics of cylindrical microstrip open-end discontinuities. Numerical results are obtained using exact Green's functions in a moment-method calculation. Details of the formulation and results are given below.

8.5.1 Microstrip Open-End Discontinuity

Figure 8.20 shows the geometry of a cylindrical microstrip open-end discontinuity, which can be treated as the special case of gap discontinuity shown in Figure 8.21. A microstrip line with width w has a discontinuity at $z = 0$. For simplicity, only the $\hat{z}$-directed electric current is assumed to flow on the microstrip line, which is a good approximation when narrow lines are considered (see Section 8.2.2). To begin, the current density on the microstrip line is modeled. For the current far away from the open end, a traveling-wave propagating mode is assumed; that is,

$$\mathbf{J}(\phi, z) = \hat{z}f(\phi)\exp(j\beta z), \tag{8.88}$$

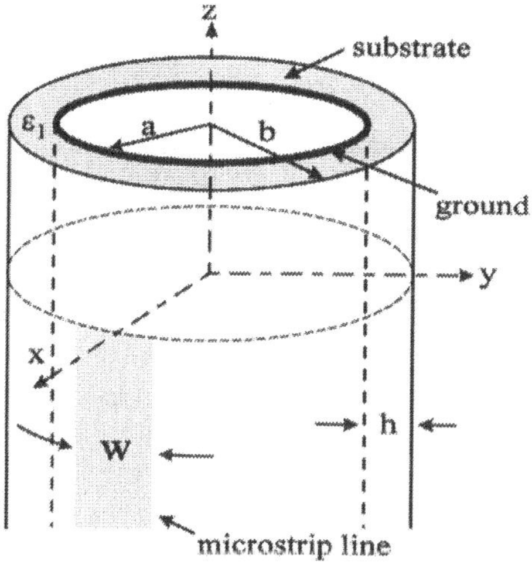

FIGURE 8.20 Geometry of a cylindrical microstrip open-end discontinuity.

where β is the effective propagation constant of the microstrip line with an infinite length and $f(\phi)$ is chosen to be uniform. The traveling-wave mode corresponds to the fundamental mode of the microstrip line, and the propagation constant β can be obtained by solving the characteristic equation (8.25). Next, the current density near the open end is modeled as a combination of the semi-infinite traveling-wave mode and the local subdomain mode. The subdomain modes are used in the vicinity of the discontinuity to account for higher-order-mode effects. Figure 8.22 shows how the various propagating modes are arranged near the open-end and gap discontinuities. The current density near the open end is modeled as

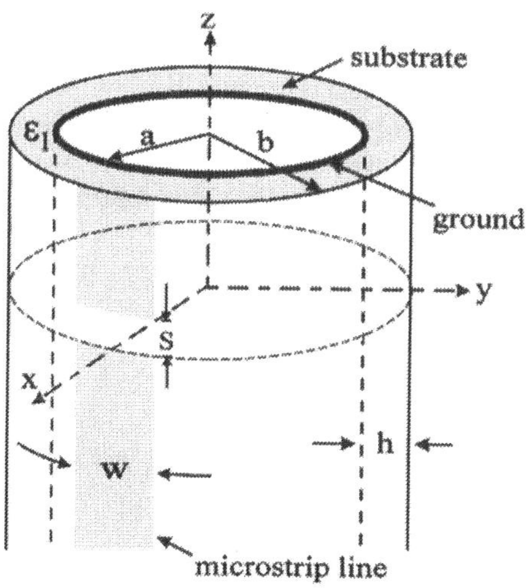

FIGURE 8.21 Geometry of a cylindrical microstrip gap discontinuity.

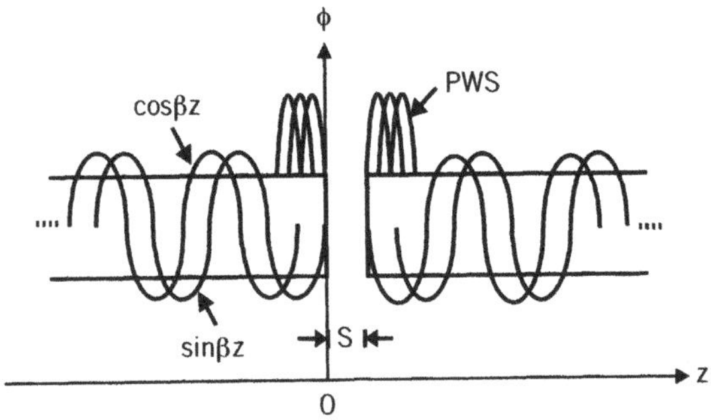

FIGURE 8.22 Expansion modes of the current density near the gap discontinuity shown in Figure 8.21.

$$\mathbf{J}(\phi, z) = \hat{z}f(\phi)g(z)\,, \tag{8.89}$$

with

$$g(z) = (1-R)f_s\left(\beta z + \frac{\pi}{2}\right) + j(1+R)f_s(\beta z) + \sum_{n=1}^{N} I_n g_n(z)\,, \qquad z \le 0\,, \tag{8.90}$$

$$g_n(z) = \frac{\sin \beta(d - |z - z_n|)}{\sin \beta d}\,, \qquad |z - z_n| < d, \quad z_n = -nd\,, \quad n = 1, 2, 3, \ldots, \tag{8.91}$$

$$f_s(u) = \begin{cases} \sin u\,, & 0 > u > -m\pi\,, \\ 0\,, & \text{elsewhere}\,, \end{cases} \tag{8.92}$$

where R is the reflection coefficient from the discontinuity at $z = 0$; $g(z)$ is a PWS basis function chosen to represent currents that are higher-order propagating modes; I_n are the unknown expansion coefficients for the PWS basis functions with a half-length of d. The sinusoidal functions of (8.92), chosen to be several $(m/2)$ cycles in length, represent the incident and reflected traveling waves of the fundamental propagation mode.

To solve for the unknown expansion coefficient, I_n, the boundary condition that the electric field on the microstrip line must be zero is again applied, which yields [similar to (8.24)]

$$\int_{-\infty}^{0} \int_{-w/2b}^{w/2b} G_{zz}(\phi, z)J(\phi, z)\,d\phi\,dz = \frac{1}{2\pi} \sum_{p=-\infty}^{\infty} \tilde{G}_{zz}(b, p, k_z)\tilde{J}(p, k_z)e^{j(p\phi + k_z z)}\,dk_z = 0\,, \tag{8.93}$$

with

$$\tilde{J}(p, k_z) = \tilde{f}(p)\left\{[\tilde{f}_c(k_z) + j\tilde{f}_s(k_z)] - R[\tilde{f}_c(k_z) - j\tilde{f}_s(k_z)] + \sum_{n=1}^{N} I_n \tilde{g}_n(k_z)\right\}. \tag{8.94}$$

In the expressions above, $\tilde{g}_n$, $\tilde{f}_s$, and $\tilde{f}_c$ are Fourier transforms of g_n, $f_s(\beta z)$, and $f_s(\beta z + \pi/2)$, respectively, and are given as

$$\tilde{g}_n(k_z) = \frac{2\beta(\cos k_z d - \cos \beta d)}{(\beta^2 - k_z^2)\sin \beta d} e^{-jk_z z_n}, \tag{8.95}$$

$$\tilde{f}_s(k_z) = \frac{\beta}{k_z^2 - \beta^2}[1 - (-1)^m e^{jm\pi k_z/\beta}], \tag{8.96}$$

$$\tilde{f}_c(k_z) = e^{jk_z \pi/2\beta} \tilde{f}_s(k_z). \tag{8.97}$$

Then, by following Galerkin's procedure, the integral equation in (8.93) can be converted into a matrix equation,

$$[[Z_{mn}]_{(N+1)\times N} \quad [-(Z_{mc} - jZ_{ms})]_{(N+1)\times 1}]\begin{bmatrix}[I_n]_{N\times 1} \\ R\end{bmatrix} = [-(Z_{mc} + jZ_{ms})]_{(N+1)\times 1}, \tag{8.98}$$

with

$$Z_{mn} = \sum_{p=-\infty}^{\infty} \int_{-\infty}^{\infty} \tilde{G}_{zz}(b, p, k_z)\tilde{f}^2(p)\tilde{g}_m(-k_z)\tilde{g}_n(k_z)\, dk_z, \tag{8.99}$$

$$Z_{ms} = \sum_{p=-\infty}^{\infty} \int_{-\infty}^{\infty} \tilde{G}_{zz}(b, p, k_z)\tilde{f}^2(p)\tilde{g}_m(-k_z)\tilde{f}_s(k_z)\, dk_z, \tag{8.100}$$

$$Z_{mc} = \sum_{p=-\infty}^{\infty} \int_{-\infty}^{\infty} \tilde{G}_{zz}(b, p, k_z)\tilde{f}^2(p)\tilde{g}_m(-k_z)\tilde{f}_c(k_z)\, dk_z. \tag{8.101}$$

By solving (8.98), the reflection coefficient R and coefficient I_n for the PWS

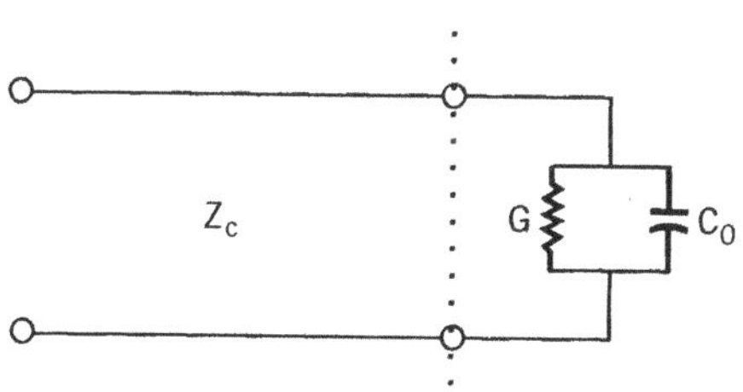

FIGURE 8.23 Equivalent circuit of a cylindrical microstrip open-end discontinuity.

expansion functions can be obtained. Once R is obtained, an open-end admittance Y can be defined as

$$Y = \frac{1 - R}{Z_c(1 + R)} = G + j\omega C_0, \tag{8.102}$$

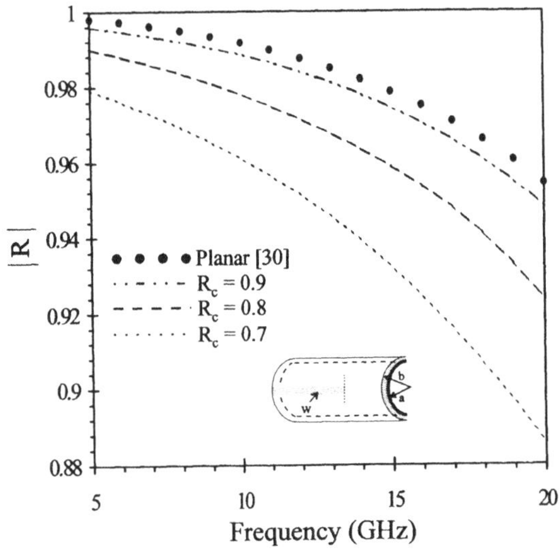

(a)

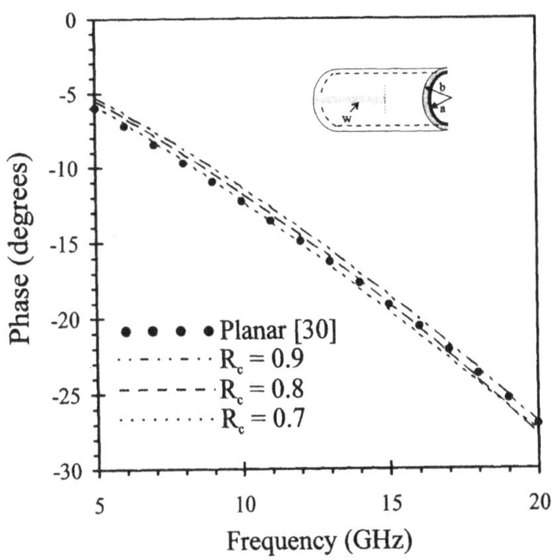

(b)

FIGURE 8.24 (a) Magnitude and (b) phase of the reflection coefficient for a cylindrical microstrip open-end discontinuity; $\epsilon_1 = 9.9$, $h = 0.635$ mm, $w = 0.635$ mm.

where Z_c is the characteristic impedance defined in (8.27) or (8.42), G the open-end conductance, and C_0 the open-end capacitance. That is, the open-end discontinuity can be characterized by an equivalent circuit with a terminal conductance and a terminal capacitance, as shown in Figure 8.23. The conductance accounts for the radiation and surface wave losses, and the capacitance is due to fringing electric field at the open end.

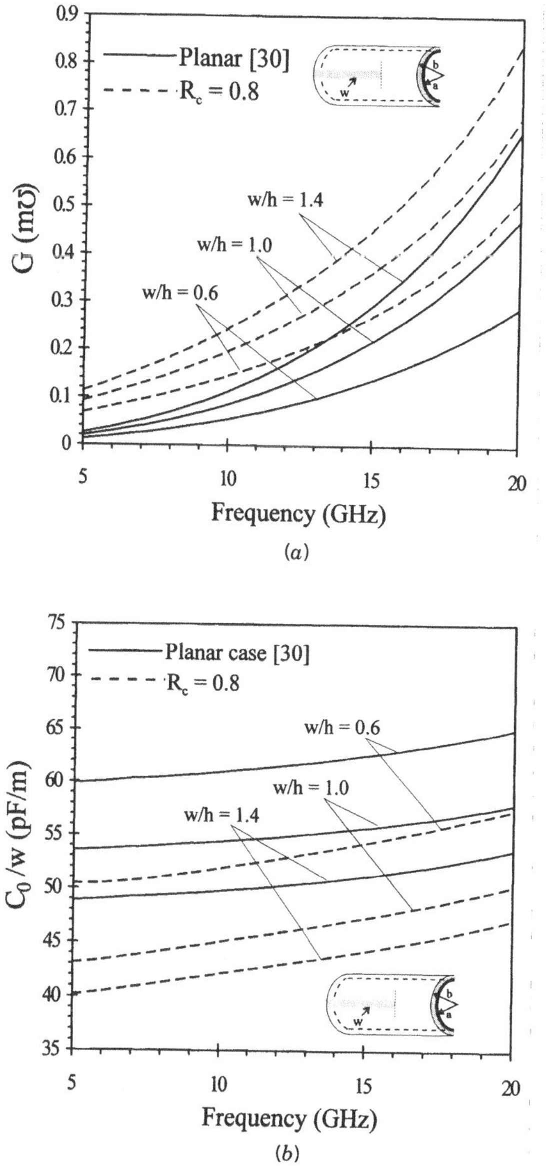

FIGURE 8.25 (*a*) Equivalent terminal conductance and (*b*) capacitance for a cylindrical microstrip open-end discontinuity; $R_c = 0.8$, $\epsilon_1 = 9.9$, $h = 0.635$ mm, $w = 0.635$ mm.

Figure 8.24 shows the calculated reflection coefficient for a cylindrical microstrip open-end discontinuity. In the calculation, four PWS functions ($N = 4$) in (8.91) and six cycles of the sinusoidal functions ($m = 12$) in (8.92) are used for expansion of the current on the microstrip line, which results in good convergent solutions. From the results it is found that the curvature effect on the reflection coefficient magnitude is significant. As for the phase of the reflection coefficient, however, the results obtained show very small variations for different curvatures. Figure 8.25 presents the results of the equivalent terminal conductance and capacitance for an open-end microstrip discontinuity with $R_c = 0.8$. It is seen that the terminal conductance is larger for a cylindrical open-end discontinuity then for the planar case [30], which suggests that the radiation and surface-wave losses at a microstrip discontinuity increases when the curvature increases (i.e., the value of R_c decreases). On the other hand, it is seen that the curvature decreases the terminal capacitance at a microstrip open-end discontinuity.

8.5.2 Microstrip Gap Discontinuity

In the geometry shown in Figure 8.21, the gap spacing of the microstrip line is denoted as S. Similar to the formulation in Section 8.5.1, the currents (incident, reflected, and transmitted currents) on the microstrip line far away from the discontinuity are treated as a traveling-wave propagating mode of (8.88). As for the currents near the gap (see Figure 8.22), we have

$$\mathbf{J}(\phi, z) = \hat{z} f(\phi) g(z) , \tag{8.103}$$

with

$$g(z) = \begin{cases} e^{j\beta z} - Re^{-j\beta z} + \sum_{n=1}^{N} I_n^a g_n^a(z) , & z \leq 0 , \\ Te^{j\beta(z-S)} + \sum_{n=1}^{N} I_n^b g_n^b(z) , & z \geq S , \end{cases} \tag{8.104}$$

$$g_n^a(z) = \frac{\sin \beta(d - |z - z_n^a|)}{\sin \beta d} , \qquad |z - z_n^a| < d , \quad z_n^a = -nd , \qquad n = 1, 2, 3, \ldots , \tag{8.105}$$

$$g_n^b(z) = \frac{\sin \beta(d - |z - z_n^b|)}{\sin \beta d} , \qquad |z - z_n^b| < d , \quad z_n^b = nd + S , \quad n = 1, 2, 3 \ldots , \tag{8.106}$$

where $g_n^a(z)$ and $g_n^b(z)$ are PWS basis functions with a half-length of d chosen to represent currents that are higher-order propagation modes; R and T are, respectively, the reflection and transmission coefficients from the gap discontinuity; I_n^a and I_n^b are unknown expansion coefficients for the PWS functions. To deal with real expansion modes only and eliminate current discontinuities, $g(z)$ is modified to be

$$g(z)=\begin{cases}(1-R)f_s^a\left(\beta z+\dfrac{\pi}{2}\right)+j(1+R)f_s^a(\beta z)+\displaystyle\sum_{n=1}^{N} I_n^a g_n^a(z)\,, & z\le 0\,,\\ -Tf_s^b\left[\beta(z-S)+\dfrac{\pi}{2}\right]+jTf_s^b[\beta(z-S)]+\displaystyle\sum_{n=1}^{N} I_n^b g_n^b(z)\,, & z\ge S\,,\end{cases}$$

(8.107)

with

$$f_s^a(u)=\begin{cases}\sin u\,, & 0>u>-m\pi\,,\\ 0\,, & \text{elsewhere}\,,\end{cases} \tag{8.108}$$

$$f_s^b(u)=\begin{cases}\sin u\,, & 0<u<m\pi\,,\\ 0\,, & \text{elsewhere}\,.\end{cases} \tag{8.109}$$

Then a matrix equation similar to (8.98) can be derived as follows:

$$\begin{bmatrix}[Z_{mn}^{aa}]_{(N+1)\times N} & [-(Z_{mc}^{aa}-jZ_{ms}^{aa})]_{(N+1)\times 1} & [Z_{mn}^{ab}]_{(N+1)\times N} & [-(Z_{mc}^{ab}-jZ_{ms}^{ab})]_{(N+1)\times 1}\\ [Z_{mn}^{ba}]_{(N+1)\times N} & [-(Z_{mc}^{ba}-jZ_{ms}^{ba})]_{(N+1)\times 1} & [Z_{mn}^{bb}]_{(N+1)\times N} & [-(Z_{mc}^{bb}-jZ_{ms}^{bb})]_{(N+1)\times 1}\end{bmatrix}$$

$$\times\begin{bmatrix}[I_n^a]_{N\times 1}\\ R\\ [I_n^b]_{N\times 1}\\ T\end{bmatrix}=\begin{bmatrix}[-(Z_{mc}^{aa}+jZ_{ms}^{aa})]_{(N+1)\times 1}\\ [-(Z_{mc}^{ba}+jZ_{ms}^{ba})]_{(N+1)\times 1}\end{bmatrix}, \tag{8.110}$$

with

$$Z_{mn}^{ij}=\sum_{p=-\infty}^{\infty}\int_{-\infty}^{\infty}\tilde{G}_{zz}(b,p,k_z)\tilde{f}^2(p)\tilde{g}_m^i(-k_z)\tilde{g}_n^j(k_z)\,dk_z\,, \tag{8.111}$$

$$Z_{ms}^{ij}=\sum_{p=-\infty}^{\infty}\int_{-\infty}^{\infty}\tilde{G}_{zz}(b,p,k_z)\tilde{f}^2(p)\tilde{g}_m^i(-k_z)\tilde{f}_s^j(k_z)\,dk_z\,, \tag{8.112}$$

$$Z_{mc}^{ij}=\sum_{p=-\infty}^{\infty}\int_{-\infty}^{\infty}\tilde{G}_{zz}(b,p,k_z)\tilde{f}^2(p)\tilde{g}_m^i(-k_z)\tilde{f}_c^j(k_z)\,dk_z\,, \tag{8.113}$$

$$i=a,\,b;\quad j=a,\,b.$$

Solving (8.110), the reflection coefficient R, transmission coefficient T, and unknown coefficients I_n^a and I_n^b can be obtained. Once R and T are evaluated, an equivalent circuit shown in Figure 8.26 for the gap discontinuity can be obtained, where $G_p+j\omega C_p$ and $G_g+j\omega C_g$ are the parallel and series admittances. These equivalent-circuit elements can be expressed as

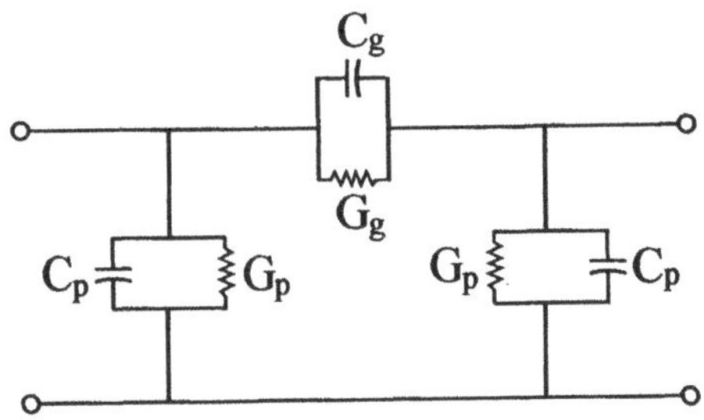

FIGURE 8.26 Equivalent circuit of a cylindrical microstrip gap discontinuity. (From Ref. [16], © 1995 John Wiley & Sons, Inc.)

$$G_p + j\omega C_p = \frac{1 - R - T}{Z_c(1 + R + T)}, \tag{8.114}$$

$$G_g + j\omega C_g = \frac{2T}{Z_c(1 + R + T)(1 + R - T)}. \tag{8.115}$$

The inclusion of G_p and G_g in the equivalent circuit accounts for the radiation and surface-wave losses, and C_p and C_g are due mainly to the fringing electric field at the gap discontinuity.

Numerical results are calculated using four PWS basis functions ($N = 4$) and six cycles of the sinusoidal basis functions ($m = 12$), similar to that for the open-end case. Figure 8.27 shows the reflection coefficients versus frequency for various curvilinear coefficients. It is seen that the reflection coefficient increases

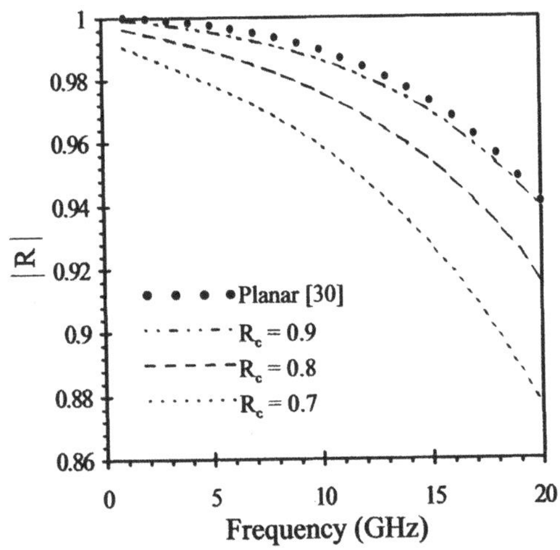

FIGURE 8.27 Reflection coefficient from a microstrip gap discontinuity; $\epsilon_1 = 9.9$, $h = 0.635$ mm, $w = 0.635$ mm, $S = 0.8h$. (From Ref. [16], © 1995 John Wiley & Sons, Inc.)

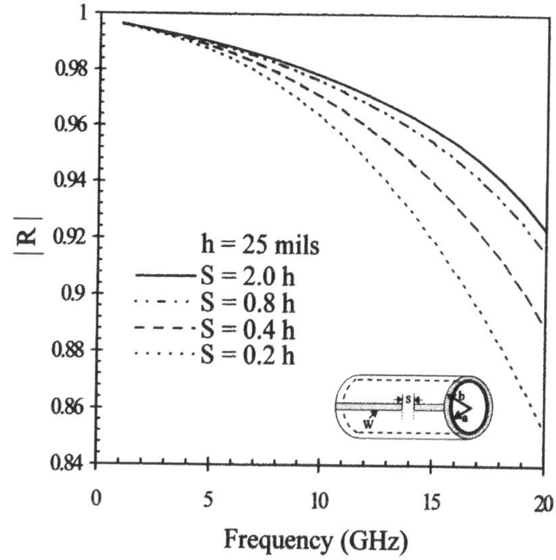

(*a*)

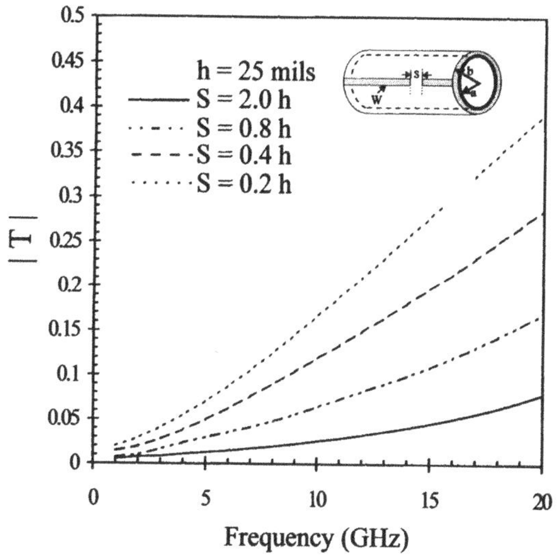

(*b*)

FIGURE 8.28 (*a*) Reflection and (*b*) transmission coefficients versus frequency for a microstrip gap discontinuity; $\epsilon_1 = 9.9$, $h = 0.635$ mm, $w = 0.635$ mm, $R_c = 0.8$. (From Ref. [16], © 1995 John Wiley & Sons, Inc.)

with increasing curvilinear coefficient. In Figure 8.28, the reflection and transmission coefficients versus frequency for various gap spacings with $R_c = 0.8$ are shown. It is observed that for the case with a large gap spacing, the gap discontinuity behaves like an open-end discontinuity. The gap capacitances and conductances versus gap spacing for various values of R_c are presented in Figure 8.29. The planar results of gap capacitance obtained using the quasistatic method [31] are shown in Figure 8.29*a* for comparison. A large curvature effect on the gap capacitance is seen. It is also observed that the values of C_p and C_g for a large

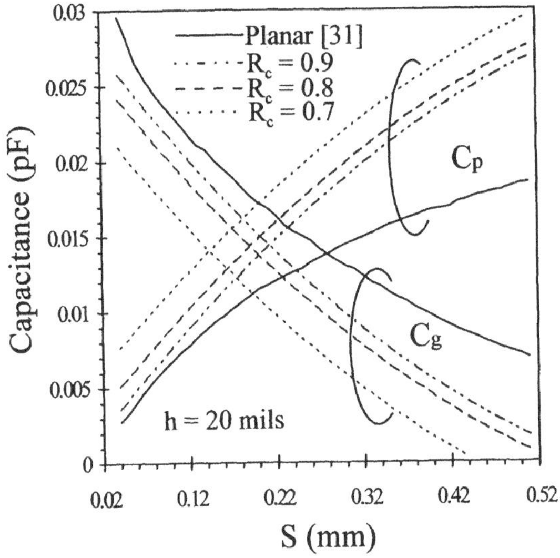

(*a*)

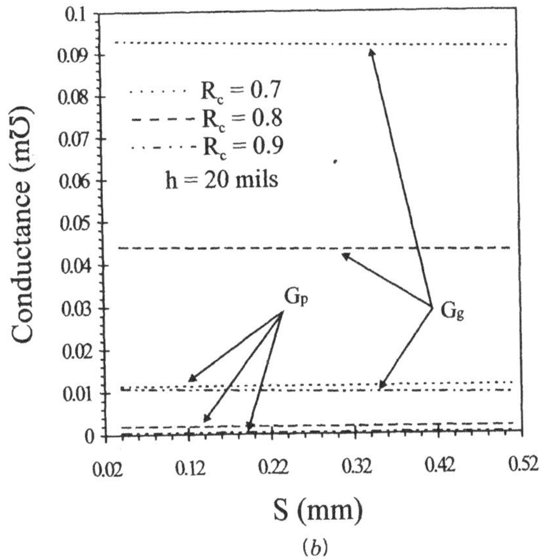

(*b*)

FIGURE 8.29 (*a*) Gap capacitance and (*b*) gap conductance versus gap spacing for a microstrip gap discontinuity; $\epsilon_r = 8.875$, $h = w = 0.508$ mm, $f = 1.3$ GHz. (From Ref. [16], © 1995 John Wiley & Sons, Inc.)

value of R_c in general agree with the quasistatic solutions for the planar case, except for the case of large gap spacing. The discrepancy at large values of S is probably due to the radiation and surface-wave losses that are not included in the quasistatic approach [31]. As for the gap conductance in Figure 8.29*b*, it is found that both the open-end conductance G_p and gap conductance G_g are insensitive to the variation in gap spacing, but are affected strongly by the curvature variation.

It should be noted that the microstrip discontinuities discussed above are for microstrip lines mounted on (outside) a ground cylinder. The discontinuity characteristics of lines mounted inside a ground cylinder have been reported in [17]. It is observed that curvature effects on the characteristics of inside cylindrical open-end and gap discontinuities are in contrast to those for outside cylindrical microstrip discontinuities discussed in this section.

8.6 CYLINDRICAL COPLANAR WAVEGUIDES

Recently, coplanar waveguides (CPWs) have been used widely in microwave and millimeter-wave integrated circuits as an alternative to microstrip lines. The principal advantage of a CPW is that the location of signal grounds are on the same substrate surface as the signal line. This eliminates the need for via holes and thus simplifies the fabrication process. CPWs also permit easy connection with both series and shunt components. They are often used in combination with microstrip and slot lines to expand microstrip circuit applications: for example, in designing power dividers, balanced mixers, couplers, and filters. Because of its attractive features, CPWs have been employed in many practical RF circuit designs and also as a feed for excitation of microstrip antennas [32–34].

Although CPWs exhibit attractive advantages, it is noted that studies of conformal CPWs reported in the open literature are very scant. In this section, details of the characteristics of a CPW mounted on a cylindrical surface are presented. Two different numerical techniques, using conformal mapping [2] and full-wave formulation [18,19] for the analysis of a CPW printed on a cylindrical surface, are described. Both outside and inside cylindrical CPWs (see Figure 8.30) are investigated. Using the conformal mapping technique [35], which is a theoretical approach based on a quasistatic approximation, the effective relative permittivity and characteristic impedance of cylindrical CPWs for various curvatures can easily be calculated. The quasistatic solutions obtained can give reasonably accurate results in the frequency range of several GHz or even in a much higher frequency range, when the characteristics of some particular CPW structures are not frequency sensitive [35]. On the other hand, the full-wave formulation incorporating the dyadic Green's functions and Galerkin's moment-method calculation is applied to analyze the frequency-dependent performance of the cylindrical CPWs. In addition to the cases of outside and inside cylindrical CPWs, the cylindrical CPW in a substrate–superstrate geometry is also studied. Numerical results of the frequency-dependent effective relative permittivity and

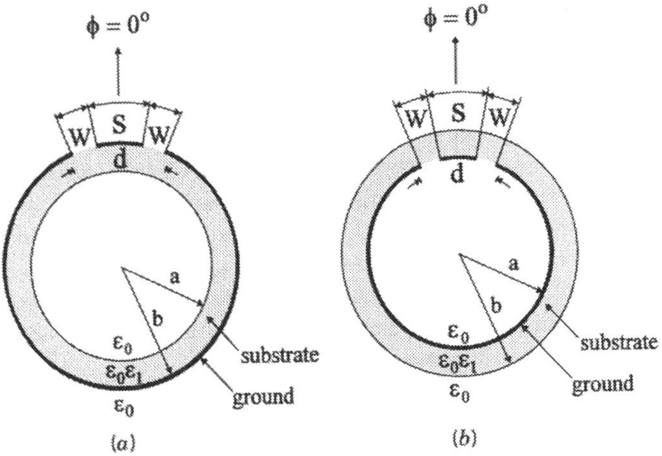

FIGURE 8.30 Geometry of cylindrical CPWs: (*a*) outside cylindrical CPW; (*b*) inside cylindrical CPW.

characteristic impedance of the odd mode are calculated and analyzed. Measured data are also presented for comparison with the numerical results.

8.6.1 Quasistatic Solution

Figure 8.30 shows the structures of outside and inside cylindrical CPWs; a and b are, respectively, inner and outer radii of the cylindrical substrate of thickness h and relative permittivity ϵ_1. In this analysis the metallic conductors are assumed to be infinitely thin and perfectly conducting, and the substrate layer is considered to be lossless. The signal strip (center conductor), of width S, is centered in the $\phi = 0°$ plane. The distance between the signal strip and the coplanar ground is W. Ground-to-ground spacing is denoted as d. The inner ($\rho < a$) and outer ($\rho > b$) regions are assumed to be air. It is also assumed that the air–substrate interfaces between the signal strip and coplanar ground can be modeled as perfect magnetic walls, which ensures that no electric field lines emanating into the air from the signal strip will enter the air–substrate interfaces. This assumption can be justified when W is small. It is also noted that since the ground planes on both sides of the signal strip are in contact and thus at the same potential, excitation of the parasitic mode (even mode) [36] is suppressed. Thus only odd-mode CPW propagation is considered.

The outside cylindrical CPW is considered first. Figures 8.31 to 8.33 show the sequence of conformal mapping to transform the original structure of Figure 8.30*a* into a plane-parallel capacitor. At first, we transform the outside cylindrical CPW into a planar CPW with a finite ground plane (z-plane shown in Figure 8.31*b*) through the mapping function

$$z = -j \ln \frac{b}{r} + \phi\,. \tag{8.116}$$

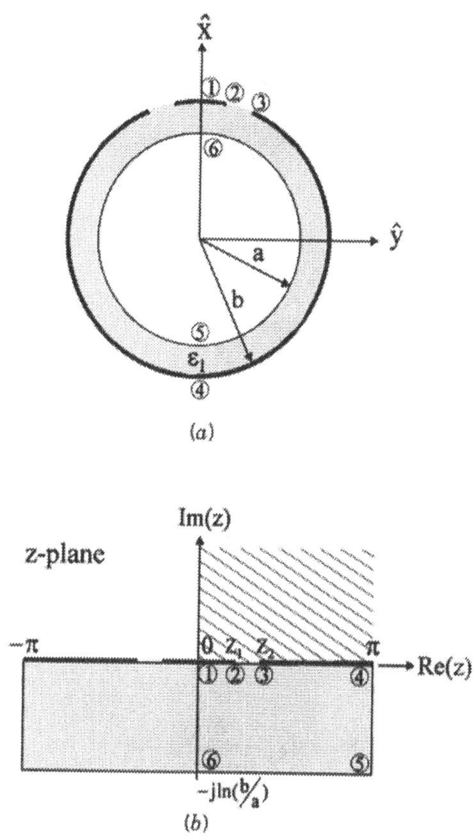

FIGURE 8.31 Conformal mapping for an outside cylindrical CPW: (*a*) original problem; (*b*) intermediate transformed plane mapped into the planar structure. (From Ref. [2], © 1997 John Wiley & Sons, Inc.)

For calculating the free-space capacitance, the first quadrant of Figure 8.31*b* is further transformed into the upper t half-plane of Figure 8.32*a* through the mapping

$$t = z^2, \tag{8.117}$$

and then into the free-space rectangular region (plane-parallel capacitor) shown in Figure 8.32*b* through the mapping formula

$$w = \int_{t_0}^{t} \frac{dt}{\sqrt{t(t-t_1)(t-t_2)(t-t_3)}}, \tag{8.118}$$

where t_0 is an arbitrary beginning point and t is the ending point. In this case the

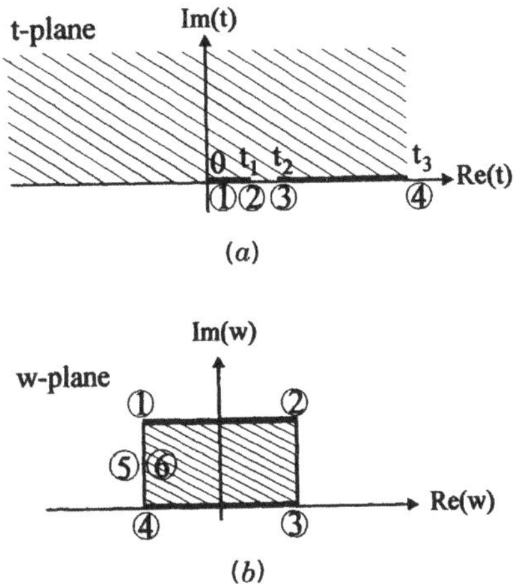

FIGURE 8.32 Conformal mapping for an outside cylindrical CPW. (*a*) Intermediate transformed plane for the dashed region in Figure 8.31*b*; (*b*) final mapping into a plane-parallel capacitor. (From Ref. [2], © 1997 John Wiley & Sons, Inc.)

free-space capacitance per unit length of the structure, considering all four quadrants of the free-space region in Figure 8.31*b*, is determined by

$$C_a = 4\epsilon_0 \frac{K(k_1)}{K(\sqrt{1-k_1^2})}, \tag{8.119}$$

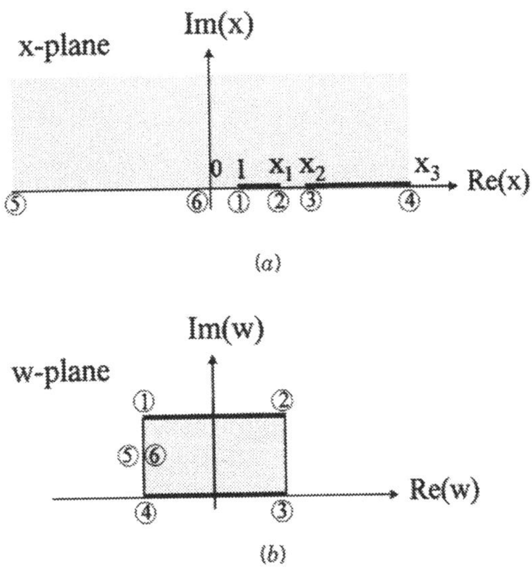

FIGURE 8.33 Conformal mapping for an outside cylindrical CPW: (*a*) intermediate transformed plane for the shaded region in Figure 8.31*b*. (*b*) final mapping into a plane-parallel capacitor. (From Ref. [2], © 1997 John Wiley & Sons, Inc.)

with

$$k_1 = \frac{S}{S + 2W}\sqrt{\frac{1 - (S + 2W)^2/4b^2\pi^2}{1 - S^2/4b^2\pi^2}}, \tag{8.120}$$

where K is a complete elliptic integral of the first kind.

For determination of the second capacitance contribution, the air–substrate interfaces are replaced by magnetic walls [35]. Then, through mapping using

$$x = \cosh^2 \frac{\pi z}{2\ln(b/a)}, \tag{8.121}$$

and

$$w = \int_{x_0}^{x} \frac{dx}{\sqrt{(x-1)(x-x_1)(x-x_2)(x-x_3)}}, \tag{8.122}$$

the right-half-side of the shaded region in Figure 8.31*b* is first transformed into the upper x half-plane (Figure 8.33*a*) and finally, into the rectangular region (plane-parallel capacitor) of relative permittivity $\epsilon_1 - 1$ shown by Figure 8.33*b*. Thus the capacitance for the total shaded region in Figure 8.31*b* can be calculated to be

$$C_s = 2\epsilon_0(\epsilon_1 - 1)\frac{K(k_2)}{K(\sqrt{1-k_2^2})}, \tag{8.123}$$

with

$$k_2 = \frac{\sinh(AS)}{\sinh[A(S+2W)]}\sqrt{\frac{1 - \sinh^2[A(S+2W)]/\sinh^2(2Ab\pi)}{1 - \sinh^2(AS)/\sinh^2(2Ab\pi)}}, \tag{8.124}$$

$$A = \frac{\pi}{4b\ln(b/a)}. \tag{8.125}$$

From (8.119) and (8.123), we can obtain the overall capacitance of the original structure (Figure 8.31*a*) expressed as $C = C_a + C_s$. Thus the effective relative permittivity is determined from

$$\epsilon_{\text{eff}} = \frac{C}{C_a} = 1 + \frac{C_s}{C_a}$$

$$= 1 + \frac{\epsilon_1 - 1}{2}\frac{K(\sqrt{1-k_1^2})}{K(k_1)}\frac{K(k_2)}{K(\sqrt{1-k_2^2})}. \tag{8.126}$$

The characteristic impedance can be calculated from

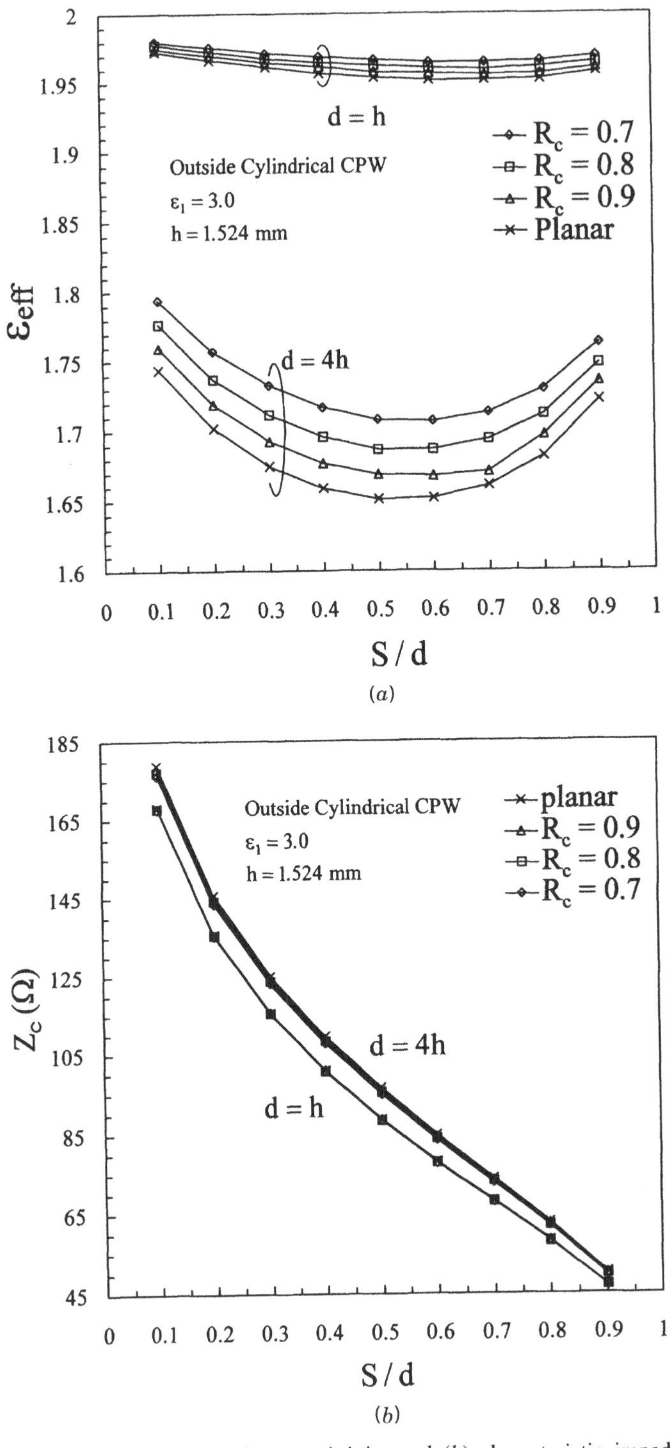

FIGURE 8.34 (*a*) Effective relative permittivity and (*b*) characteristic impedance of an outside cylindrical CPW; $\epsilon_1 = 3.0$, $h = 1.524$ mm. (From Ref. [2], © 1997 John Wiley & Sons, Inc.)

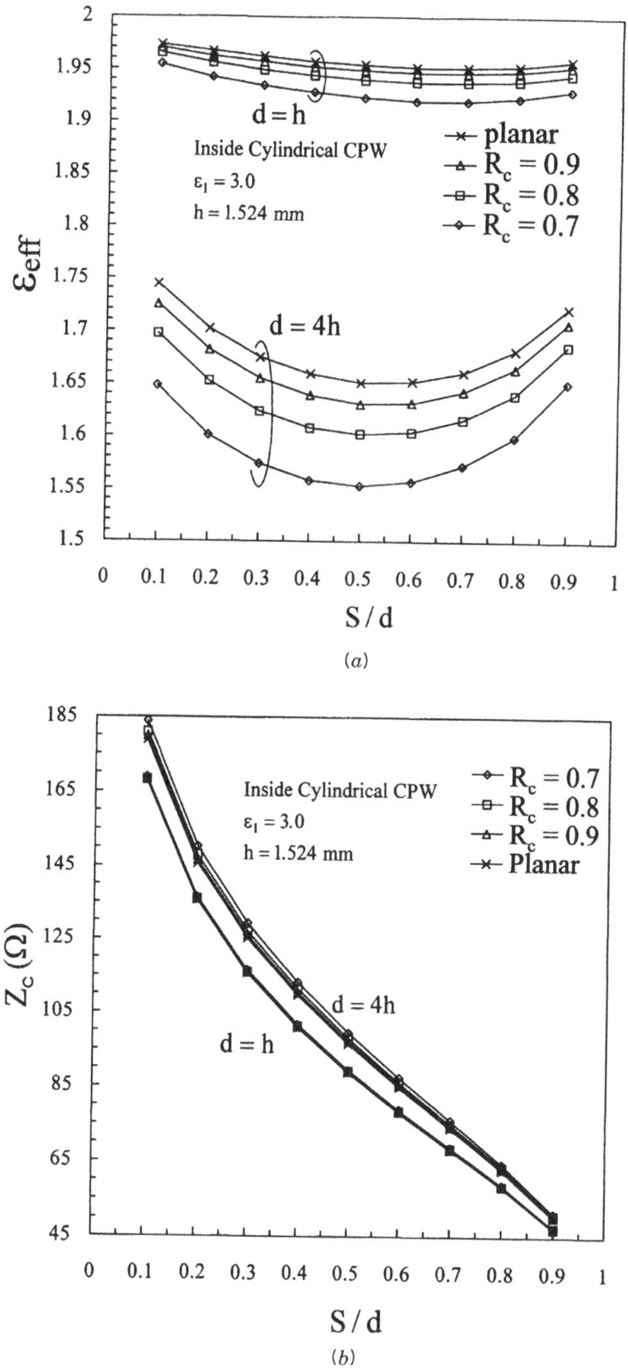

FIGURE 8.35 (*a*) Effective relative permittivity and (*b*) characteristic impedance of an inside cylindrical CPW; $\epsilon_1 = 3.0$, $h = 1.524$ mm. (From Ref. [2], © 1997 John Wiley & Sons, Inc.)

$$Z_c = \frac{\sqrt{\epsilon_{\mathrm{eff}}}}{Cv_0} = \frac{120\pi\epsilon_0}{C_a\sqrt{\epsilon_{\mathrm{eff}}}}$$

$$= \frac{30\pi}{\sqrt{\epsilon_{\mathrm{eff}}}} \frac{K(\sqrt{1-k_1^2})}{K(k_1)}, \tag{8.127}$$

where v_0 is the velocity of light in air. By following the mapping procedure above, we can also derive the quasistatic parameters of an inside cylindrical CPW. The effective relative permittivity and characteristic impedance obtained are of the same form as expressed in (8.126) and (8.127), with interchange of a and b in the expressions.

Typical quasi-static results for cylindrical CPWs are calculated. Figure 8.34 shows the effective relative permittivity and characteristic impedance of an outside cylindrical CPW calculated as a function of S/d. The corresponding results for an inside cylindrical CPW are presented in Figure 8.35. The curvilinear coefficient R_c shown in the figures is again defined as $a/(a+h)$. For both cases the results are calculated for planar CPWs, using the analytical formulas derived in [35] and shown for comparison. From the results it is seen that the characteristic impedance of CPWs is relatively insensitive to curvature variation but that the effective relative permittivity depends strongly on the curvature of cylindrical CPWs. It is also observed that with increasing curvature (R_c decreases), the effective relative permittivity increases for outside cylindrical CPWs and decreases inside cylindrical CPWs. This phenomenon is the same as that observed in Section 8.2 for cylindrical microstrip lines.

8.6.2 Full-Wave Solution

In this section, the structures of outside and inside cylindrical CPWs and the cylindrical CPW in a substrate–superstrate geometry are investigated using a full-wave approach, and numerical results for frequency-dependent effective relative permittivity and characteristic impedance are calculated and analyzed. Measured data are also presented for comparison with the obtained full-wave solutions, and the curvature effects on the characteristics of a cylindrical CPW are discussed.

A. Outside Cylindrical CPWs The geometry shown in Figure 8.36 is studied. The inner region ($\rho < a$) is assumed to have a relative permittivity ϵ_2 and the CPW substrate a relative permittivity ϵ_1. To begin with, the spectral-domain Helmholtz equations in each region of the structure are solved, which gives cylindrical dyadic Green's functions and expressions of electric and magnetic fields in each region of the cylindrical structure. Then, by applying the equivalence principle, the slot region between the signal strip and the ground can be closed off and replaced by an equivalent magnetic surface current density $\mathbf{M}_s$ $(= M_\phi \hat{\phi} +$

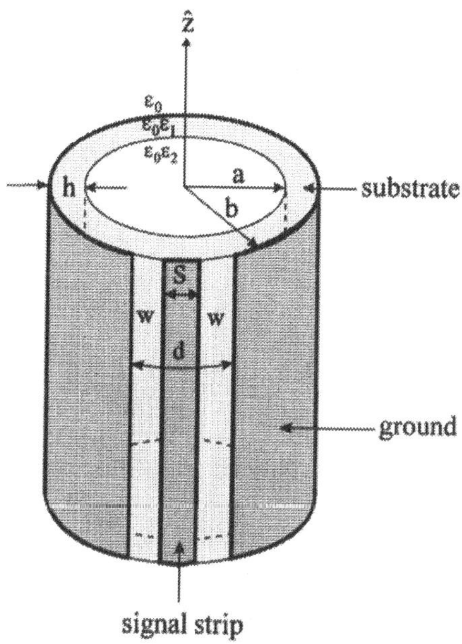

FIGURE 8.36 Geometry of a cylindrical CPW printed on a ground cylinder.

$M_z\hat{z}$) at (b^-, ϕ, z) and $-\mathbf{M}_s$ at (b^+, ϕ, z). When imposing boundary conditions on the structure and manipulating the derived field components, we can have

$$\begin{bmatrix} \Delta\tilde{H}_\phi \\ \Delta\tilde{H}_z \end{bmatrix} = (\tilde{\bar{\bar{G}}}^{(s)} + \tilde{\bar{\bar{G}}}^{(a)}) \begin{bmatrix} \tilde{M}_\phi \\ \tilde{M}_z \end{bmatrix} = \begin{bmatrix} 0 \\ 0 \end{bmatrix}, \tag{8.128}$$

where $\tilde{\bar{\bar{G}}}^{(s,a)} = \hat{\phi}\tilde{G}_{\phi\phi}^{HM(s,a)}\hat{\phi} + \hat{\phi}\tilde{G}_{\phi z}^{HM(s,a)}\hat{z} + \hat{z}\tilde{G}_{z\phi}^{HM(s,a)}\hat{\phi} + \hat{z}\tilde{G}_{zz}^{HM(s,a)}\hat{z}$ is the cylindrical dyadic Green's function [18] showing $\hat{\phi}$- or $\hat{z}$-directed magnetic fields on the substrate (s) or air (a) sides of the slot region due to a unit-amplitude M_ϕ or M_z at the slot region; ΔH denotes the difference between the tangential magnetic fields at $\rho = b^-$ and at $\rho = b^+$, which must be zero to satisfy the boundary condition that continuity of the tangential magnetic fields at the slot region must hold.

With the assumption that a cylindrical CPW is infinitely long, the magnetic surface current density at the slot region can be described as

$$\mathbf{M}(\phi, z) = \mathbf{M}(\phi)e^{j\beta z}, \tag{8.129}$$

where β is the effective propagation constant of the cylindrical CPW to be determined and $e^{j\beta z}$ is the traveling-wave form. For a moment-method calculation, $\mathbf{M}(\phi)$ is expanded in terms of a linear combination of known basis functions; that is,

$$\mathbf{M}(\phi) = \hat{\phi} \sum_{n=1}^{N} I_{\phi n} M_{\phi n}(\phi) + \hat{z} \sum_{m=1}^{M} I_{zm} M_{zm}(\phi), \tag{8.130}$$

where $I_{\phi n}$ and I_{zm} are unknown coefficients of the basis functions $M_{\phi n}$ and M_{zm}, respectively. We choose N rooftop basis functions of the form

$$M_{\phi n} = 1 - \frac{2|\phi - \phi_{\phi n}|}{D_t}, \qquad |\phi - \phi_{\phi n}| < \frac{D_t}{2}, \tag{8.131}$$

with

$$\phi_{\phi n} = \frac{S}{2b} + \frac{nD_t}{2}, \qquad D_t = \frac{d - S}{2b(N+1)}, \tag{8.132}$$

to expand $M_\phi(\phi)$, and M pulse basis functions of the form

$$M_{zm} = 1, \qquad |\phi - \phi_{zm}| < \frac{D_z}{2}, \tag{8.133}$$

with

$$\phi_{zm} = \frac{S}{2b} + \frac{(2m-1)D_z}{2}, \qquad D_z = \frac{d - S}{2bM}, \tag{8.134}$$

to expand $M_z(\phi)$. When considering odd-mode propagation, a magnetic wall at the center of the two slot regions can be assumed, which results in odd-mode basis functions in the spectral domain written as

$$\tilde{M}^{(o)}_{\phi n} = \frac{4}{\pi D_t p^2} \sin^2 \frac{pD_t}{4} \cos p\phi_{\phi n}, \tag{8.135}$$

$$\tilde{M}^{(o)}_{zm} = \frac{j}{p\pi} \sin \frac{pD_z}{2} \sin p\phi_{zm}. \tag{8.136}$$

Then, using the basis functions above as testing functions and applying Galerkin's method to (8.128), a homogeneous matrix equation can be obtained:

$$\begin{bmatrix} (Y^{\phi\phi}_{ln})_{N\times N} & (Y^{\phi z}_{lm})_{N\times M} \\ (Y^{z\phi}_{kn})_{M\times N} & (Y^{zz}_{km})_{M\times M} \end{bmatrix} \begin{bmatrix} (I_{\phi n})_{N\times 1} \\ (I_{zm})_{M\times 1} \end{bmatrix} = \begin{bmatrix} 0 \\ 0 \end{bmatrix}, \tag{8.137}$$

where

$$Y^{\phi\phi}_{ln} = \sum_{p=-\infty}^{\infty} \tilde{M}^{(o)}_{\phi l}(-p)[\tilde{G}^{HM^{(s)}}_{\phi\phi}(b, p, \beta) - \tilde{G}^{HM^{(a)}}_{\phi\phi}(b, p, \beta)]\tilde{M}^{(o)}_{\phi n}(p), \tag{8.138}$$

$$Y^{\phi z}_{lm} = \sum_{p=-\infty}^{\infty} \tilde{M}^{(o)}_{\phi l}(-p)[\tilde{G}^{HM^{(s)}}_{\phi z}(b, p, \beta) - \tilde{G}^{HM^{(a)}}_{\phi z}(b, p, \beta)]\tilde{M}^{(o)}_{zm}(p), \tag{8.139}$$

$$Y_{kn}^{z\phi} = \sum_{p=-\infty}^{\infty} \tilde{M}_{zk}^{(o)}(-p)[\tilde{G}_{z\phi}^{HM^{(s)}}(b, p, \beta) - \tilde{G}_{z\phi}^{HM^{(a)}}(b, p, \beta)]\tilde{M}_{\phi n}^{(o)}(p)\,, \tag{8.140}$$

$$Y_{km}^{zz} = \sum_{p=-\infty}^{\infty} \tilde{M}_{zk}^{(o)}(-p)[\tilde{G}_{zz}^{HM^{(s)}}(b, p, \beta) - \tilde{G}_{zz}^{HM^{(a)}}(b, p, \beta)]\tilde{M}_{zm}^{(o)}(p)\,, \tag{8.141}$$

$$l, n = 1, 2, \ldots, N\,, \qquad k, m = 1, 2, \ldots, M\,.$$

To have nontrivial solutions for $I_{\phi n}$ and I_{zm}, the determinant of (8.137) must vanish; that is,

$$\det\begin{bmatrix} (Y_{ln}^{\phi\phi})_{N\times N} & (Y_{lm}^{\phi z})_{N\times M} \\ (Y_{kn}^{z\phi})_{M\times N} & (Y_{km}^{zz})_{M\times M} \end{bmatrix} = 0\,. \tag{8.142}$$

Solving (8.142), the effective propagation constant β of cylindrical CPWs is obtained, and the effective relative permittivity ϵ_{eff} can be calculated from $(\beta/k_0)^2$.

For computation of the characteristic impedance, the voltage–current definition is adopted [36]; that is,

$$Z_c = \frac{V_0}{I_z}\,, \tag{8.143}$$

where V_0 is the potential difference of the signal strip to the ground, and I_z is the total surface current in the z direction on the signal strip and can be evaluated from

$$I_z = \int_{-S/2b}^{S/2b} \Delta H_\phi b \, d\phi\,. \tag{8.144}$$

Typical numerical results are computed. Figure 8.37 shows the effective relative permittivity measured and calculated versus frequency for a curvilinear coefficient of 0.97. The region $\rho < a$ is here assumed to be air. The theoretical results for planar CPWs are calculated using a full-wave approach described in [36]. From the results it is seen that the theory is in good agreement with experiment. Results also indicate that cylindrical CPWs have a larger effective relative permittivity than planar CPWs. The characteristic impedance is presented in Figure 8.38. Good agreement between theory and experiment is also seen. Figures 8.39 and 8.40 present, respectively, ϵ_{eff} and Z_c for various substrate thicknesses and curvilinear coefficients. Results show that the curvature effect is greater for a smaller substrate thickness. In Figures 8.41 and 8.42 we present, respectively, ϵ_{eff} and Z_c versus normalized CPW size. It is found that the curvature effect becomes more significant with increased CPW size.

B. Inside Cylindrical CPWs By considering the inside cylindrical CPW shown in Figure 8.43 and applying the similar formulation described above, the effective relative permittivity and characteristic impedance can also be evaluated from

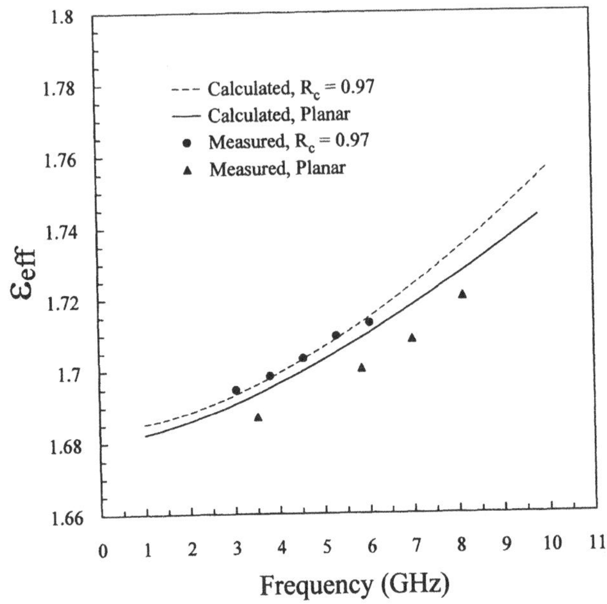

FIGURE 8.37 Effective relative permittivity versus frequency for an outside cylindrical CPW; $\epsilon_1 = 3.0$, $h = 1.524$ mm, $\epsilon_2 = 1.0$, $d = 7$ mm, $S/d = 0.572$. (From Ref. [18], © 1996 IEEE, reprinted with permission.)

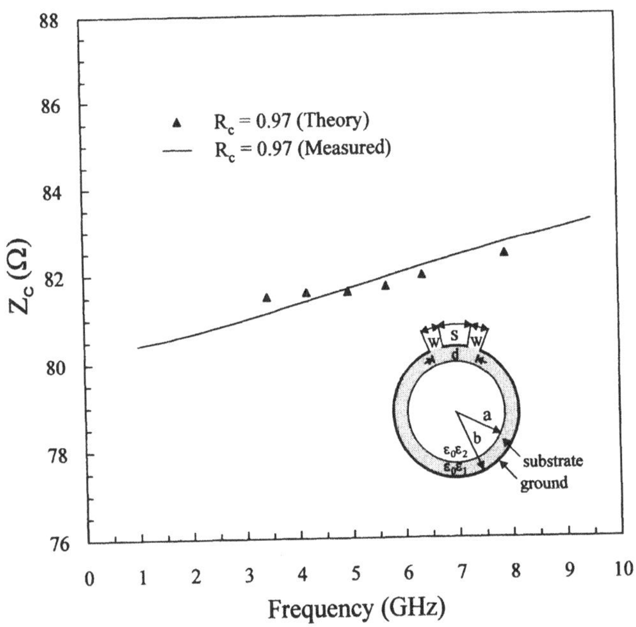

FIGURE 8.38 Characteristic impedance versus frequency for an outside cylindrical CPW; $\epsilon_1 = 3.0$, $h = 1.524$ mm, $\epsilon_2 = 1.0$, $d = 7$ mm, $S/d = 0.572$.

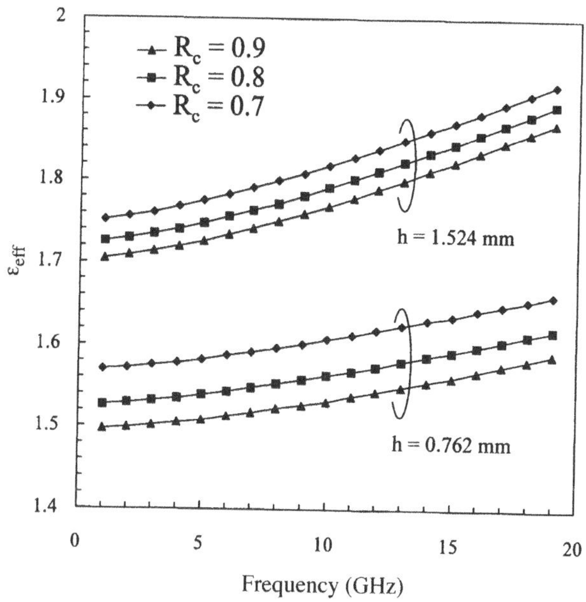

FIGURE 8.39 Effective relative permittivity versus frequency for an outside cylindrical CPW; $\epsilon_1 = 3.0$, $\epsilon_2 = 1.0$, $d = 7$ mm, $S/d = 0.572$. (From Ref. [18], © 1996 IEEE, reprinted with permission.)

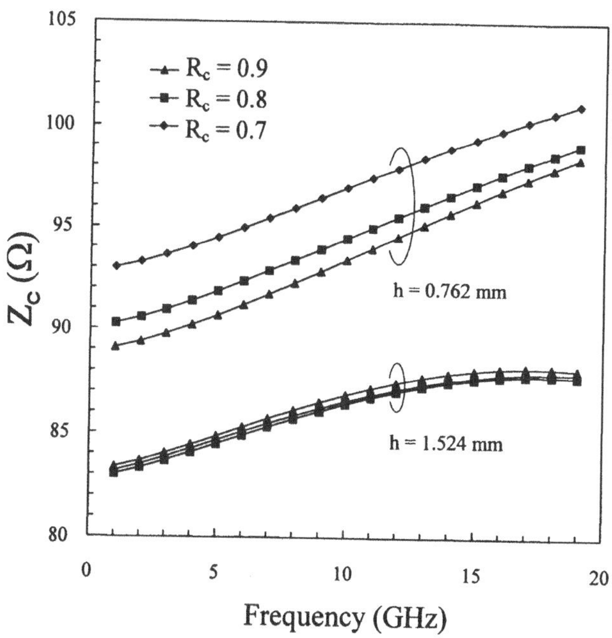

FIGURE 8.40 Characteristic impedance versus frequency for an outside cylindrical CPW; $\epsilon_1 = 3.0$, $\epsilon_2 = 1.0$, $d = 7$ mm, $S/d = 0.572$.

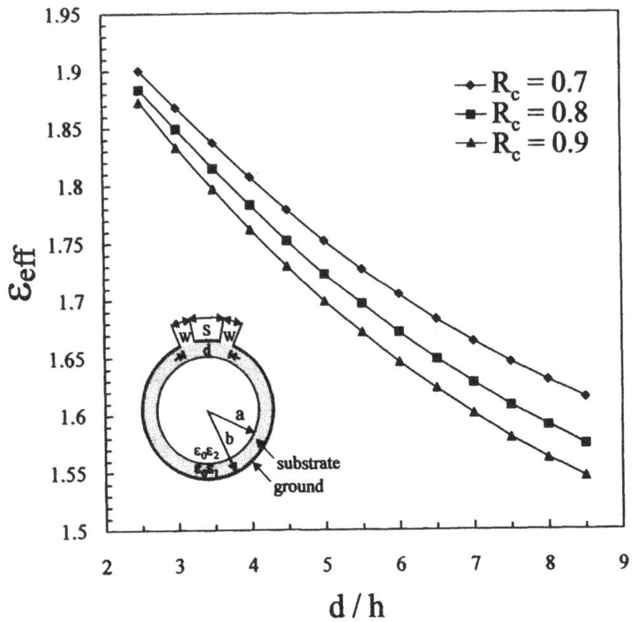

FIGURE 8.41 Effective relative permittivity versus normalized CPW size (ground-to-ground spacing) for an outside cylindrical CPW; $\epsilon_1 = 3.0$, $h = 1.524$ mm, $\epsilon_2 = 1.0$, $S = 2W$, $f = 10$ GHz. (From Ref. [18], © 1996 IEEE, reprinted with permission.)

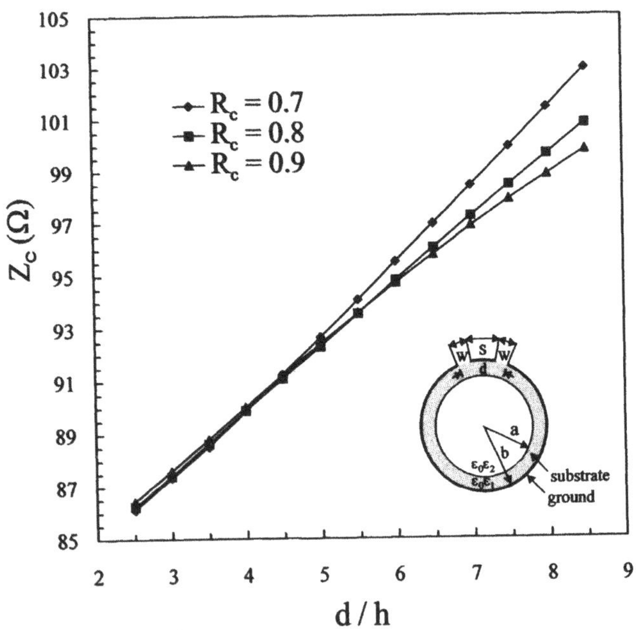

FIGURE 8.42 Characteristic impedance versus normalized CPW size (ground-to-ground spacing) for an outside cylindrical CPW; $\epsilon_1 = 3.0$, $h = 1.524$ mm, $\epsilon_2 = 1.0$, $S = 2W$, $f = 10$ GHz.

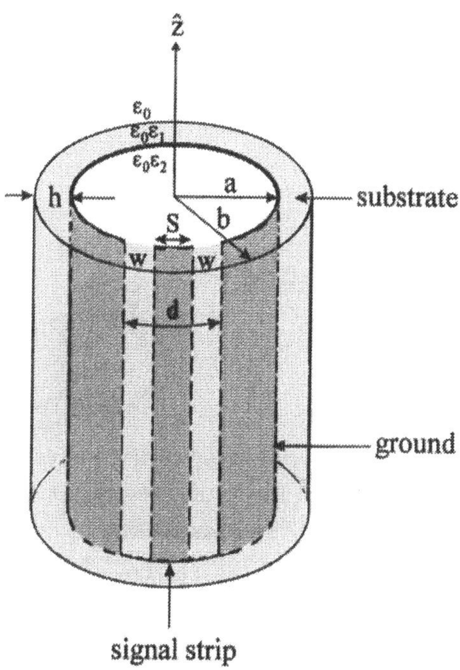

FIGURE 8.43 Geometry of a cylindrical CPW printed inside a ground cylinder.

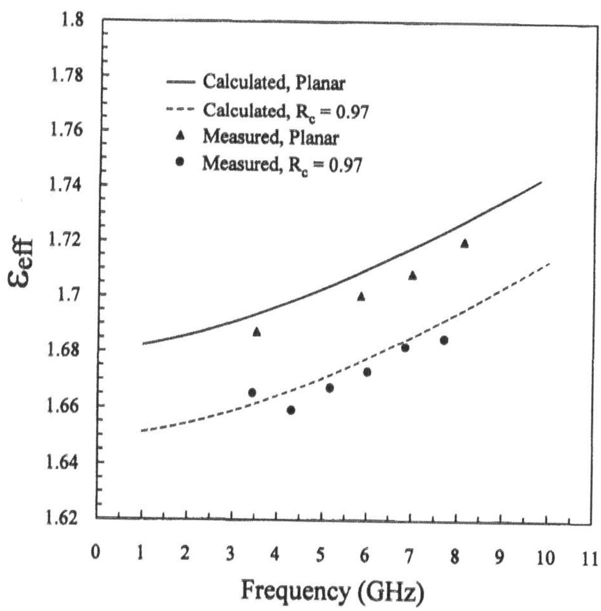

FIGURE 8.44 Effective relative permittivity versus frequency for an inside cylindrical CPW; $\epsilon_1 = 3.0$, $h = 1.524$ mm, $\epsilon_2 = 1.0$, $d = 7$ mm, $S/d = 0.572$. (From Ref. [19], © 1996 John Wiley & Sons, Inc.)

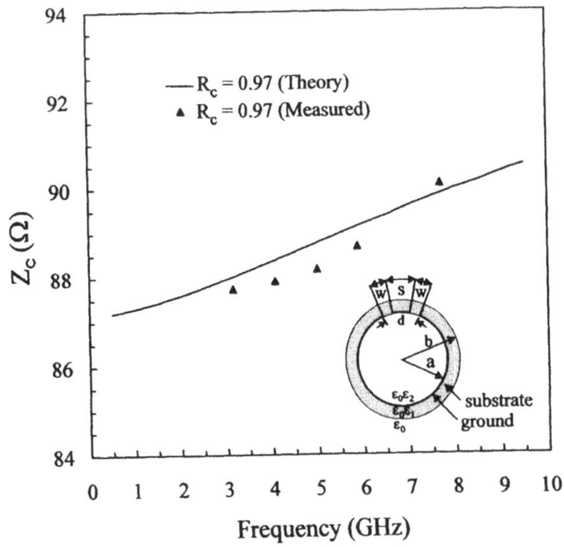

FIGURE 8.45 Characteristic impedance versus frequency for an inside cylindrical CPW; $\epsilon_1 = 3.0$, $h = 1.524$ mm, $\epsilon_2 = 1.0$, $d = 7$ mm, $S/d = 0.572$.

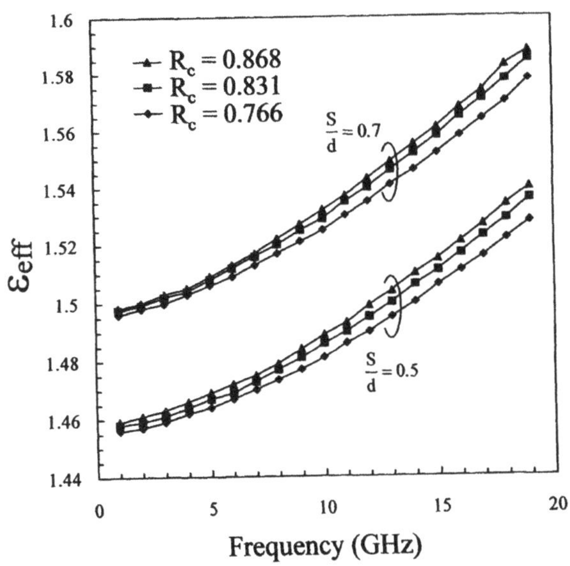

FIGURE 8.46 Effective relative permittivity versus frequency for an inside cylindrical CPW; $\epsilon_1 = 3.0$, $h = 0.762$ mm, $\epsilon_2 = 1.0$, $d = 7$ mm, $S/d = 0.5, 0.7$. (From Ref. [19], © 1996 John Wiley & Sons, Inc.)

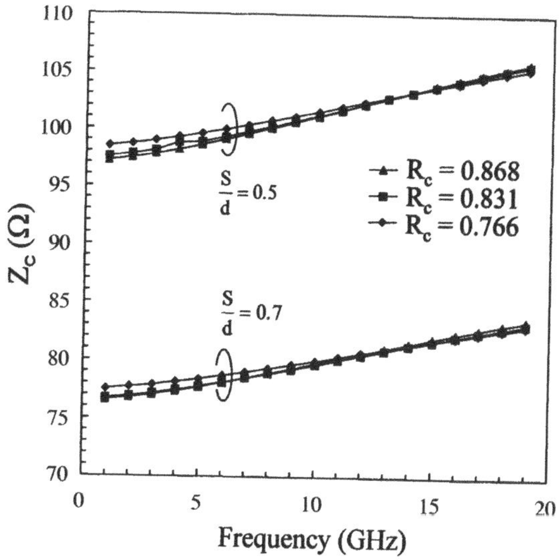

FIGURE 8.47 Characteristic impedance versus frequency for an inside cylindrical CPW; $\epsilon_1 = 3.0$, $h = 0.762$ mm, $\epsilon_2 = 1.0$, $d = 7$ mm, $S/d = 0.5$, 0.7.

(8.142) and (8.143). Experiments were conducted to verify the numerical results. Figure 8.44 shows the results for effective relative permittivity, and the characteristic impedance results are presented in Figure 8.45. The results calculated are seen to agree with the data measured. More theoretical results are shown in Figures 8.46 and 8.47, where curvature effects differ from those observed for the outside cylindrical case.

C. Cylindrical CPWs in a Substrate–Superstrate Structure Figure 8.48 shows the geometry of a cylindrical CPW in a substrate–superstrate structure. The region

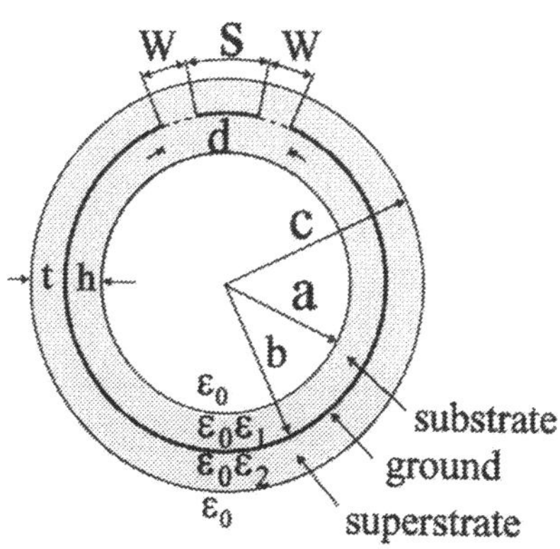

FIGURE 8.48 Geometry of a cylindrical CPW in a substrate–superstrate structure.

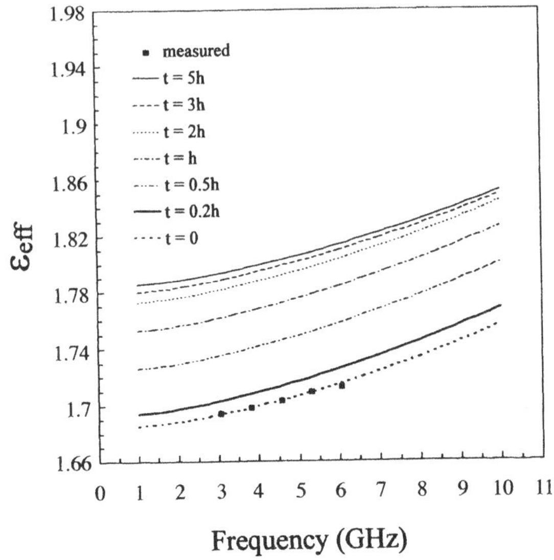

FIGURE 8.49 Effective relative permittivity versus frequency for various superstrate thicknesses; $\epsilon_1 = 3.0$, $h = 1.524$ mm, $\epsilon_2 = 1.2$, $d = 7$ mm, $S/d = 0.572$, $R_c = 0.97$.

$a < \rho < b$ is treated as the substrate for the CPW, and the region $b < \rho < c$ is superstrate loading for the CPW. Other regions are considered to be free space. Also, by using a theoretical formulation similar to that described in Section 8.6.2A, the effective relative permittivity and characteristic impedance of the

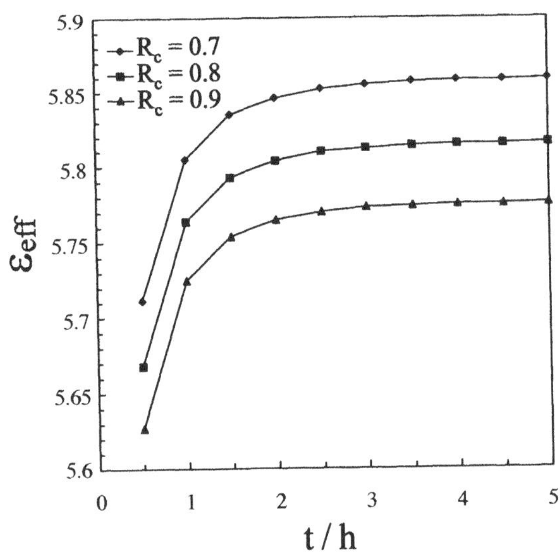

FIGURE 8.50 Effective relative permittivity versus normalized superstrate thickness; $\epsilon_1 = 9.8$, $h = 0.635$ mm, $\epsilon_2 = 2.4$, $d = 2.2h$, $W = h$, $S = 0.2h$, $f = 10$ GHz.

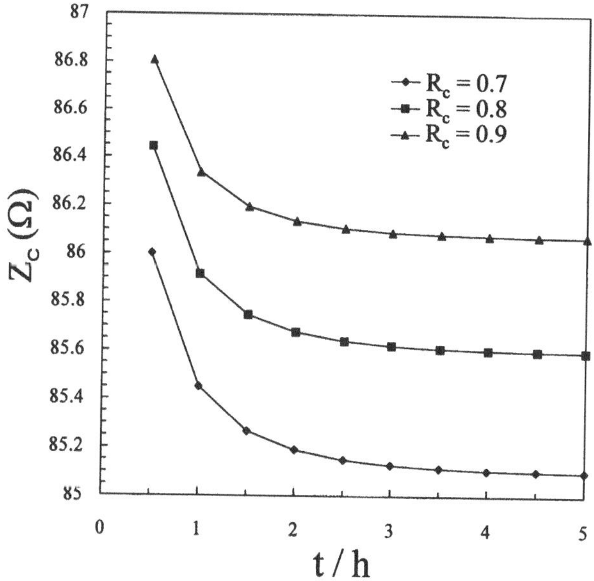

FIGURE 8.51 Characteristic impedance versus normalized superstrate thickness; $\epsilon_1 = 9.8$, $h = 0.635$ mm, $\epsilon_2 = 2.4$, $d = 2.2h$, $W = h$, $S = 0.2h$, $f = 10$ GHz.

superstrate-loaded cylindrical CPW can be evaluated. Figure 8.49 shows ϵ_{eff} calculated versus frequency for cylindrical CPWs with various superstrate thicknesses. Data measured when no superstrate is present are also shown in the figure for comparison. From the results it is seen that the effective relative permittivity increases quickly with addition of the superstrate layer; however, the variation becomes smaller when the superstrate thickness is greater than about three times the substrate thickness. Figures 8.50 and 8.51 show, respectively, the results of ϵ_{eff} and Z_c versus normalized superstrate thickness for various curvilinear coefficients. The effective relative permittivity for various curvatures reaches a steady value when the normalized superstrate thickness (t/h) is greater than 3. The characteristic impedance shows similar behavior.

REFERENCES

1. L. R. Zeng and Y. Wang, "Accurate solutions of elliptical and cylindrical striplines and microstrip lines," *IEEE Trans. Microwave Theory Tech.*, vol. 34, pp. 259–265, Feb. 1986.
2. H. C. Su and K. L. Wong, "Quasistatic solutions of cylindrical coplanar waveguides," *Microwave Opt. Technol. Lett.*, vol. 14, pp. 347–351, Apr. 20, 1997.
3. R. B. Tsai and K. L. Wong, "Quasistatic solution of a cylindrical microstrip line mounted inside a ground cylinder," *Microwave Opt. Technol. Lett.*, vol. 8, pp. 136–138, Feb. 20, 1995.

4. C. J. Reddy and M. D. Deshpande, "Analysis of cylindrical stripline with multilayer dielectrics," *IEEE Trans. Microwave Theory Tech.*, vol. 34, pp. 701–706, June 1986.

5. C. H. Chan and R. Mittra, "Analysis of a class of cylindrical multiconductor transmission lines using an iterative approach," *IEEE Trans. Microwave Theory Tech.*, vol. 35, pp. 415–423, Apr. 1987.

6. Y. Wang, "Cylindrical and cylindrically wraped strip and microstriplines," *IEEE Trans. Microwave Theory Tech.*, vol. 26, pp. 20–23, Jan. 1978.

7. R. B. Tsai and K. L. Wong, "Characteristics of cylindrical microstriplines mounted inside a ground cylindrical surface," *IEEE Trans. Microwave Theory Tech.*, vol. 43, pp. 1607–1610, July 1995.

8. H. M. Chen and K. L. Wong, "A study of the transverse current contribution to the characteristics of a wide cylindrical microstrip line," *Microwave Opt. Technol. Lett.*, vol. 11, pp. 339–342, Apr. 20, 1996.

9. F. C. Silva, S. B. A. Fonseca, A. J. M. Soares, A. J. Giarola, "Effect of a dielectric overlay in a microstripline on a circular cylindrical surface," *IEEE Microwave Guided Wave Lett.*, vol. 2, pp. 359–360, Sept. 1992.

10. N. G. Alexopoulos and A. Nakatani, "Cylindrical substrate microstrip line characterization," *IEEE Trans. Microwave Theory Tech.*, vol. 35, pp. 843–849, Sept. 1987.

11. A. Nakatani and N. G. Alexopoulos, "Microstrip elements on cylindrical substrates–general algorithm and numerical results," *Electromagnetics*, vol. 9, pp. 405–426, 1989.

12. W. Y. Tam, "The characteristic impedance of a cylindrical strip line and a microstrip line," *Microwave Opt. Technol. Lett.*, vol. 12, pp. 372–375, Aug. 20, 1996.

13. A. Nakatani and N. Alexopoulos, "Coupled microstrip lines on a cylindrical substrate," *IEEE Trans. Microwave Theory Tech.*, vol. 35, pp. 1392–1398, Dec. 1987.

14. H. M. Chen and K. L. Wong, "Characterization of coupled cylindrical microstrip lines mounted inside a ground cylinder," *Microwave Opt. Technol. Lett.*, vol. 10, pp. 330–333, Dec. 20, 1995.

15. J. H. Lu and K. L. Wong, "Analysis of slot-coupled double-sided cylindrical microstrip lines," *IEEE Trans. Microwave Theory Tech.*, vol. 44, pp. 1167–1170, July 1996.

16. H. M. Chen and K. L. Wong, "Characterization of cylindrical microstrip gap discontinuities," *Microwave Opt. Technol. Lett.*, vol. 9, pp. 260–263, Aug. 5, 1995.

17. J. H. Lu and K. L. Wong, "Equivalent circuit of an inside cylindrical microstrip gap discontinuity," *Microwave Opt. Technol. Lett.*, vol. 10, pp. 115–118, Oct. 5, 1995.

18. H. C. Su and K. L. Wong, "Dispersion characteristics of cylindrical coplanar waveguides," *IEEE Trans. Microwave Theory Tech.*, vol. 44, pp. 2120–2122, Nov. 1996.

19. H. C. Su and K. L. Wong, "Full-wave analysis of the effective permittivity of a coplanar waveguide printed inside a cylindrical substrate," *Microwave Opt. Technol. Lett.*, vol. 12, pp. 94–97, June 5, 1996.

20. T. Kitamura, T. Koshimae, M. Hira, and S. Kurazono, "Analysis of cylindrical microstrip lines utilizing the finite-difference time-domain method," *IEEE Trans. Microwave Theory Tech.*, vol. 42, pp. 1279–1282, July 1994.

21. C. J. Reddy and M. D. Deshpande, "Analysis of coupled cylindrical striplines filled with multilayered dielectrics," *IEEE Trans. Microwave Theory Tech.*, vol. 1301–1310, Sept. 1988.

22. B. N. Das, A. Chakrabarty, and K. K. Joshi, "Characteristic impedance of elliptic cylindrical strip and microstriplines filled with layered substrate," *IEE Proc.*, pt. H, vol. 130, pp. 245–250, June 1983.

23. R. E. Collin, *Field Theory of Guided Wave*, 2nd ed., IEEE Press, New York, 1991, pp. 273–279.

24. J. R. Brews, "Characteristic impedance of microstrip lines," *IEEE Trans. Microwave Theory Tech.*, vol. 35, pp. 30–34, Jan. 1987.

25. E. H. Fooks and R. A. Zakarevicius, *Microwave Engineering Using Microstrip Circuits*, Prentice Hall, Upper Saddle River, N.J., 1989, pp. 285–287.

26. R. K. Hoffmann, *Handbook of Microwave Integrated Circuits*, Artech House, Norwood, Mass., 1991, Chap. 9.

27. K. L. Wong and Y. C. Chen, "Resonant frequency of a slot-coupled cylindrical-rectangular microstrip structure," *Microwave Opt. Technol. Lett.*, vol. 7, pp. 566–570, Aug. 20, 1994.

28. N. Herscovici and D. M. Pozar, "Full-wave analysis of aperture-coupled microstrip lines," *IEEE Trans. Microwave Theory Tech.*, vol. 39, pp. 1108–1114, July 1991.

29. J. H. Lu and K. L. Wong, "Characteristics of slot-coupled double-sided microstrip lines with various coupling slots," *Microwave Opt. Technol. Lett.*, vol. 13, pp. 227–229, Nov. 1996.

30. N. G. Alexopoulos and S. C. Wu, "Frequency-independent equivalent circuit model for microstrip open-end and gap discontinuities," *IEEE Trans. Microwave Theory Tech.*, vol. 42, pp. 1268–1272, July 1994.

31. M. Maeda, "Analysis of gap in microstrip transmission lines," *IEEE Trans. Microwave Theory Tech.*, vol. 20, pp. 390–396, June 1972.

32. M. I. Aksun, S. L. Chuang, and Y. T. Lo, "Coplanar waveguide-fed microstrip antennas," *Microwave Opt. Technol. Lett.*, vol. 4, pp. 292–295, July 1991.

33. W. Menzel and W. Grabherr, "A microstrip patch antenna with coplanar feed line," *IEEE Microwave Guided Wave Lett.*, vol. 1, pp. 340–342, Nov. 1991.

34. L. Giauffret, J.-M. Laheurte, and A. Papiernik, "Study of various shapes of the coupling slot in CPW-fed microstrip antennas," *IEEE Trans. Antennas Propagat.*, vol. 45, pp. 642–647, Apr. 1997.

35. G. Ghione and C. Naldi, "Coplanar waveguides for MMIC applications: effect of upper shielding, conductor backing, finite-extent ground planes, and line-to-line coupling," *IEEE Trans. Microwave Theory Tech.*, vol. 35, pp. 260–267, Mar. 1987.

36. R. W. Jackson, "Considerations in the use of coplanar waveguide for millimeter-wave integrated circuits," *IEEE Trans. Microwave Theory Tech.*, vol. 34, pp. 1450–1456, Dec. 1986.

APPENDIX A

Curve-Fitting Formulas for Complex Resonant Frequencies of a Rectangular Microstrip Patch with a Superstrate

The curve-fitting formulas presented here can determine with good accuracy the complex resonant frequencies of the superstrate-loaded rectangular microstrip structure shown by Figure A.1. The rectangular patch has dimensions of $2L \times 2W$. The substrate has a thickness of h and a relative permittivity of ϵ_1; the superstrate has a thickness of t and a relative permittivity of ϵ_2. These curve-fitting formulas have the form of a multivariable polynomial and are developed using the database generated by a full-wave approach incorporating Galerkin's moment-method

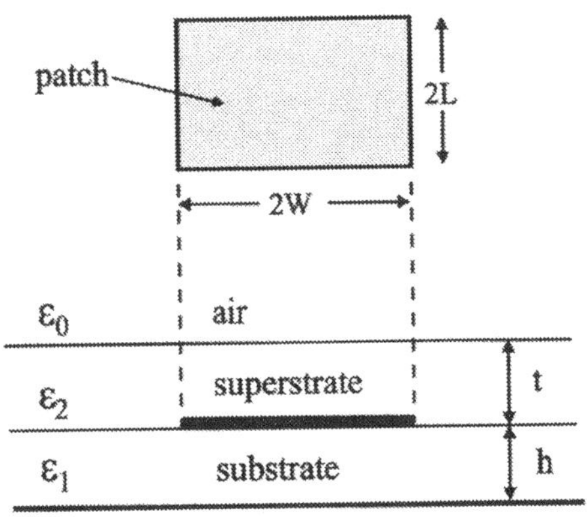

FIGURE A.1 Geometry of a superstrate-loaded rectangular microstrip structure.

calculation [1]. In the range of $1 < \epsilon_1$, $\epsilon_2 < 10$, $0.9 < W/L < 2.0$, $0 < h/2L < 0.2$, and $0 < t/h < 10$ for the ordinary design parameters of rectangular microstrip antennas, these formulas can rapidly reproduce the complex resonant frequency of the TM_{01} mode with an error of less than 1% compared with full-wave solutions [2].

A.1 REAL PARTS OF COMPLEX RESONANT FREQUENCY

The formula for the real parts of complex resonant frequencies is written as

$$\mathrm{Re}\left(\frac{f}{f_{01}}\right) = 1 + \sum_{i=0}^{1}\sum_{j=1}^{2}\sum_{k=0}^{2} A(i, j, k)\left(\frac{h}{2L}\right)^i\left(\frac{W}{L}\right)^j(\sqrt{\epsilon_1})^{-k}$$

$$+\sum_{n=1}^{4}\sum_{p=0}^{2}\sum_{q=0}^{3} B(n, p, q)\sqrt{\frac{h}{2L}}\left(\frac{t}{2L}\right)^n\left(\frac{\epsilon_2}{\epsilon_1}\right)^p(\sqrt{\epsilon_1})^q. \quad \text{(A.1)}$$

In (A.1), f_{01} is the cavity-model resonant frequency in the TM_{01} mode and is given as

$$f_{01} = \frac{7.5}{L\sqrt{\epsilon_1}} \quad \text{GHz} \quad \text{(A.2)}$$

where L is in centimeters. There are 12 coefficients for $A(i, j, k)$ and 48 coefficients for $B(n, p, q)$. When no superstrate is present ($t = 0$), the last term of (A.1) vanishes and the results obtained from (A.1) represent the complex resonant frequencies of a rectangular microstrip patch without a superstrate layer. The coefficients are given as follows: For $A(i, j, k)$,

$A(0, 1, 0) = 0.67537070692642$
$A(0, 1, 1) = -0.64058184912009$
$A(0, 1, 2) = -1.5496014282907$
$A(0, 2, 0) = -0.45155895111021$
$A(0, 2, 1) = 2.5612665743221\text{D-}02$
$A(0, 2, 2) = 0.30806585566060$
$A(1, 1, 0) = -2.8705647839616$
$A(1, 1, 1) = 0.74417125168338$
$A(1, 1, 2) = 5.3355177249185$
$A(1, 2, 0) = 1.0251452411942$
$A(1, 2, 1) = -7.1061408858346\text{D-}02$
$A(1, 2, 2) = -1.3536061323130$

and for $B(n, p, q)$,

$B(1, 0, 0) = -2.3324689443901$
$B(1, 0, 1) = 4.0045472041425$
$B(1, 0, 2) = -1.3183143615585$
$B(1, 0, 3) = 8.3061022322665\text{D-}02$
$B(1, 1, 0) = 6.4828893994388$
$B(1, 1, 1) = -7.8417750208709$
$B(1, 1, 2) = 0.62387888464163$
$B(1, 1, 3) = 0.36152721610731$
$B(1, 2, 0) = 1.5676029219638$
$B(1, 2, 1) = -4.4586669449210$

$B(1, 2, 2) = 3.6025996060135$
$B(1, 2, 3) = -0.79994833701351$
$B(2, 0, 0) = 38.417139637949$
$B(2, 0, 1) = -53.243750016582$
$B(2, 0, 2) = 21.312586733028$
$B(2, 0, 3) = -2.5772758786906$
$B(2, 1, 0) = -56.415890909913$
$B(2, 1, 1) = 69.242495734527$
$B(2, 1, 2) = -19.680739411304$
$B(2, 1, 3) = 0.83623253751615$
$B(2, 2, 0) = -4.3241920831186$
$B(2, 2, 1) = 16.570425073018$
$B(2, 2, 2) = -14.750641285815$
$B(2, 2, 3) = -3.5320721737926$
$B(3, 0, 0) = -53.958749122022$
$B(3, 0, 1) = 75.233670234451$
$B(3, 0, 2) = -31.395907411410$
$B(3, 0, 3) = 4.0511333575234$
$B(3, 1, 0) = 77.160332917202$
$B(3, 1, 1) = -98.939087318064$
$B(3, 1, 2) = 33.643739470696$
$B(3, 1, 3) = -2.8099858201492$
$B(3, 2, 0) = 2.0732444026241$
$B(3, 2, 1) = -13.595291745499$
$B(3, 2, 2) = 13.706318717575$
$B(3, 2, 3) = -3.5131745120997$
$B(4, 0, 0) = 18.931906950472$
$B(4, 0, 1) = -26.632625310224$
$B(4, 0, 2) = 11.363995809880$
$B(4, 0, 3) = -1.5109344249222$
$B(4, 1, 0) = -26.846135739375$
$B(4, 1, 1) = 35.368344376586$
$B(4, 1, 2) = -12.955944241853$
$B(4, 1, 3) = 1.2787440088034$
$B(4, 2, 0) = -0.32519494476210$
$B(4, 2, 1) = 3.5317689367863$
$B(4, 2, 2) = -3.8184718571032$
$B(4, 2, 3) = 1.0184635896788$

A.2 IMAGINARY PARTS OF COMPLEX RESONANT FREQUENCY

The formula for the imaginary parts of complex resonant frequencies is written as

$$\mathrm{Im}\left(\frac{f}{f_{01}}\right) = \sum_{i=0}^{2}\sum_{j=1}^{3}\sum_{k=0}^{3} C(i, j, k)\left(\frac{h}{2L}\right)^{i}\left(\frac{W}{L}\right)^{j}(\sqrt{\epsilon_1})^{-k}$$
$$+\sum_{n=1}^{4}\sum_{p=0}^{2}\sum_{q=0}^{3} D(n, p, q)\sqrt{\frac{h}{2L}}\left(\frac{t}{2L}\right)^{n}\left(\frac{\epsilon_2}{\epsilon_1}\right)^{p}(\sqrt{\epsilon_1})^{q}. \tag{A.3}$$

Equation (A.3) consists of 36 coefficients for $C(i, j, k)$ and 48 coefficients for $D(n, p, q)$. The first term determines the imaginary parts of complex resonant frequency for the case without a superstrate presence ($t = 0$), and the effect of superstrate loading on the imaginary resonant frequency is included in the second term. The coefficients of (A.3) are given as follows: For $C(i, j, k)$,

$C(0, 1, 0) = -0.29578897839040$
$C(0, 1, 1) = 1.6717157351621$
$C(0, 1, 2) = -3.0050757728590$
$C(0, 1, 3) = 1.6179611921551$

$C(0, 2, 0) = 0.39839211700415$
$C(0, 2, 1) = -2.0100317068747$
$C(0, 2, 2) = 2.5849326276806$
$C(0, 2, 3) = -1.2954248932276$
$C(0, 3, 0) = -1.0454441398277\text{D-}01$
$C(0, 3, 1) = 0.50509149824534$
$C(0, 3, 2) = -0.72569464605479$
$C(0, 3, 3) = 0.46005868419902$
$C(1, 1, 0) = 4.4761331867605$
$C(1, 1, 1) = -24.383753567307$
$C(1, 1, 2) = 39.385422714451$
$C(1, 1, 3) = -20.653011293177$
$C(1, 2, 0) = -5.7408446954338$
$C(1, 2, 1) = 30.926374646758$
$C(1, 2, 2) = -50.887927131029$
$C(1, 2, 3) = 28.247564514590$
$C(1, 3, 0) = 2.0477706859574$
$C(1, 3, 1) = -11.143877341285$
$C(1, 3, 2) = 19.214943282870$
$C(1, 3, 3) = -11.196248363425$
$C(2, 1, 0) = -12.885054203290$
$C(2, 1, 1) = 69.756002297855$
$C(2, 1, 2) = -118.68632227509$
$C(2, 1, 3) = 65.207685026849$
$C(2, 2, 0) = 19.038347133116$
$C(2, 2, 1) = -104.512652804801$
$C(2, 2, 2) = 182.13726114813$
$C(2, 2, 3) = -104.104118078570$
$C(2, 3, 0) = -7.2237408019063$
$C(2, 3, 1) = 39.992139901407$
$C(2, 3, 2) = -70.960227092976$
$C(2, 3, 3) = 41.052300439945$

and for $D(n, p, q)$,

$D(1, 0, 0) = -1.4085727141218$
$D(1, 0, 1) = 3.0328305018095$
$D(1, 0, 2) = -1.5182048123634$
$D(1, 0, 3) = 0.20153784397280$
$D(1, 1, 0) = 5.0878063575406$
$D(1, 1, 1) = -9.2365604366427$
$D(1, 1, 2) = 4.3783281257715$
$D(1, 1, 3) = -0.53262958968389$
$D(1, 2, 0) = -2.0788171782632$
$D(1, 2, 1) = 3.5350702871662$
$D(1, 2, 2) = -1.5882838762484$
$D(1, 2, 3) = 0.14416708373454$
$D(2, 0, 0) = -0.94255721135689$
$D(2, 0, 1) = -5.0948524304769$
$D(2, 0, 2) = 4.0231854421087$
$D(2, 0, 3) = -0.65010496223383$
$D(2, 1, 0) = -16.973295235916$
$D(2, 1, 1) = 36.262990723749$
$D(2, 1, 2) = -19.963775181377$
$D(2, 1, 3) = 2.8539918086838$
$D(2, 2, 0) = 9.2680582497742$
$D(2, 2, 1) = -17.072840829316$
$D(2, 2, 2) = 8.6921564335655$
$D(2, 2, 3) = -1.0787728697262$
$D(3, 0, 0) = 8.2121120565851$
$D(3, 0, 1) = -3.4290410499335$
$D(3, 0, 2) = -1.1113407029483$
$D(3, 0, 3) = 0.36079462081839$
$D(3, 1, 0) = 10.984368801967$
$D(3, 1, 1) = -31.851391249397$
$D(3, 1, 2) = 20.550940904614$
$D(3, 1, 3) = -3.3518488166087$
$D(3, 2, 0) = -10.891475410727$
$D(3, 2, 1) = 21.542948025956$
$D(3, 2, 2) = -12.099770193235$
$D(3, 2, 3) = 1.8253083212532$

$D(4, 0, 0) = -4.8789744574113$

$D(4, 0, 1) = 4.0710100522776$

$D(4, 0, 2) = -0.81993256259685$

$D(4, 0, 3) = 2.4425615679777\text{D-}02$

$D(4, 1, 0) = -0.74693726883219$

$D(4, 1, 1) = 6.9448192411127$

$D(4, 1, 2) = -5.6541613790039$

$D(4, 1, 3) = 1.0500298215804$

$D(4, 2, 0) = 3.5545519087800$

$D(4, 2, 1) = -7.4833519608238$

$D(4, 2, 2) = 4.4989987429771$

$D(4, 2, 3) = -0.75303625431323$

REFERENCES

1. J. S. Row and K. L. Wong, "Resonance in a superstrate-loaded rectangular microstrip structure," *IEEE Trans. Microwave Theory Tech.*, vol. 41, pp. 1349–1355, Aug. 1993.
2. H. J. Lin, Curve-Fitting Formulas for Fast Determination of Accurate Resonant Frequency of a Rectangular Microstrip Patch Antenna, M.S. thesis, National Sun Yat-Sen University, Kaohsiung, Taiwan, 1993.

APPENDIX B

Modified Spherical Bessel Function

The equation to be solved here is

$$\left[\frac{d}{dr}\left(r^2\frac{d}{dr}\right)+(kr)^2-n(n+1)\epsilon_{AR}\right]G_n(r,r')=-\delta(r-r')\,, \tag{B.1}$$

where the parameters in (B.1) are as defined in Section 3.2.2. Note that (B.1) is closely related to a Bessel equation with a source term on the right-hand side. We start by solving the homogeneous solutions to (B.1). Given a Bessel equation of standard form as follows:

$$t\frac{d}{dt}\left(t\frac{dZ}{dt}\right)+(\lambda^2t^2-\nu^2)Z=0\,, \tag{B.2}$$

whose solution is written as $B_\nu(\lambda t)$, a Bessel function, let $Z=yt^{1/2}$. Then (B.2) becomes

$$\frac{d}{dt}\left(t^2\frac{dy}{dt}\right)+(\lambda^2t^2-\nu^2+\tfrac{1}{4})y=0\,. \tag{B.3}$$

By comparing (B.1) and (B.3), a homogeneous solution to (B.1) can be obtained as

$$G_n\sim\frac{1}{\sqrt{r}}B_\nu(kr)\,, \tag{B.4}$$

with

$$\nu=\sqrt{n(n+1)\epsilon_{AR}+\tfrac{1}{4}}\,. \tag{B.5}$$

Next, the particular solution of (B.1) can be expressed in terms of the homogeneous solution. By considering the following general second-order differential equation,

$$\left\{\frac{1}{f(z)}\frac{d}{dz}\left[f(z)\frac{d}{dz}\right]+q(z)\right\}G(z,z')=\frac{-\delta(z-z')}{f(z)}, \tag{B.6}$$

with its particular solution given by

$$G(z,z')=\begin{cases}\dfrac{g_1(z')g_2(z)}{f(z')\Delta(z')}, & z<z',\\[2ex] \dfrac{g_1(z)g_2(z')}{f(z')\Delta(z')}, & z>z',\end{cases} \tag{B.7}$$

where $g_1(z)$ and $g_2(z)$ are homogeneous solutions for the regions $z>z'$ and $z<z'$, respectively, and $\Delta(z')$ is the Wronskian of $g_1(z')$ and $g_2(z')$, given by

$$\Delta(z')=\begin{vmatrix} g_1(z') & g_2(z') \\ g_1'(z') & g_2'(z') \end{vmatrix}. \tag{B.8}$$

By comparing (B.6) and (B.1), we can obtain a solution for (B.1), expressed as

$$G_n(r,r')=\begin{cases}\dfrac{1}{jkrr'}\tilde{\tilde{J}}_n(kr)\tilde{\tilde{H}}_n^{(2)}(kr'), & r<r',\\[2ex] \dfrac{1}{jkrr'}\tilde{\tilde{J}}_n(kr')\tilde{\tilde{H}}_n^{(2)}(kr), & r>r'.\end{cases} \tag{B.9}$$

To obtain (B.9), we select

$$f=r^2, \tag{B.10}$$

$$g_1=\frac{1}{\sqrt{r}}H_\nu^{(2)}(kr), \tag{B.11}$$

$$g_2=\frac{1}{\sqrt{r}}J_\nu(kr), \tag{B.12}$$

which gives

$$\Delta=\frac{2j}{\pi r^2}. \tag{B.13}$$

By substituting (B.10)–(B.13) into (B.7), the solution given by (B.9) is obtained, and $\tilde{\tilde{J}}_n(kr)$ and $\tilde{\tilde{H}}_n^{(2)}(kr)$ are defined by (3.47).

APPENDIX C

Curve-Fitting Formulas for Complex Resonant Frequencies of a Circular Microstrip Patch with a Superstrate

Similar to the case presented in Appendix A, the complex resonant frequencies of a superstrate-loaded circular microstrip structure, shown in Figure C.1, can be reproduced with good accuracy by a multivariable polynomial. The circular patch has a radius of r_d, and the substrate again has a thickness of h and a relative permittivity of ϵ_1. The superstrate has a thickness of t and a relative permittivity of ϵ_2. The curve-fitting formulas shown here are developed using a database generated by a full-wave approach [1]. In the range of $1 < \epsilon_1$, $\epsilon_2 < 10$, $0 < h/r_d < 0.24$, and $0 < t/h < 8$ for the ordinary design parameters of circular microstrip

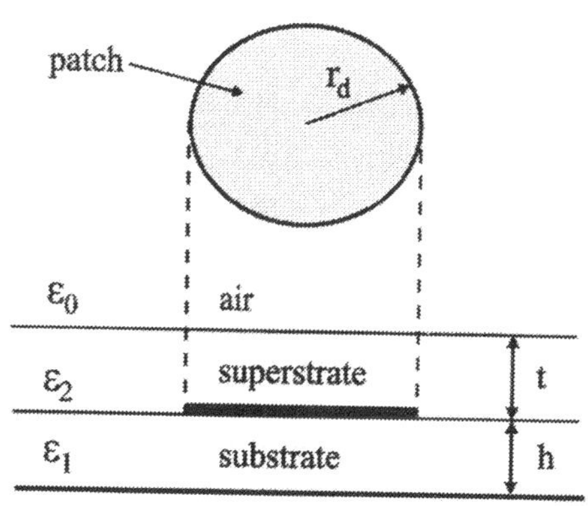

FIGURE C.1 Geometry of a superstrate-loaded circular microstrip structure.

antennas, the formulas can rapidly reproduce the complex resonant frequency of the TM_{11} mode with an error of less than 1% compared with full-wave solutions.

C.1 REAL PARTS OF COMPLEX RESONANT FREQUENCY

The formula for the real parts of complex resonant frequencies is written as

$$\mathrm{Re}\left(\frac{f}{f_{11}}\right) = \sum_{i=1}^{2}\sum_{j=0}^{3} A(i, j)\left(\frac{h}{r_d}\right)^i (\sqrt{\epsilon_1})^j + \sum_{m=1}^{4}\sum_{n=0}^{2}\sum_{p=0}^{1}\sum_{q=0}^{1}$$

$$\times B(m, n, p, q)\left(\frac{t}{h}\right)^m \left(\frac{h}{r_d}\right)^n \left(\sqrt{\frac{\epsilon_2}{\epsilon_1}}\right)^p (\sqrt{\epsilon_1})^q . \tag{C.1}$$

In (C.1), f_{11} is the cavity-model resonant frequency in the TM_{11} mode and is given as

$$f_{11} = \frac{8.7906}{r_d\sqrt{\epsilon_1}} \quad \text{GHz} \tag{C.2}$$

where r_d is in centimeters. There are 10 coefficients for $A(i, j)$ and 48 coefficients for $B(m, n, p, q)$. The first term is for the case without a superstrate layer; superstrate effects on resonant frequency are included in the second term. The coefficients are given as follows: For $A(i, j)$,

$A(1, 0) = -3.5237244299618$
$A(1, 1) = 3.3555135153225$
$A(1, 2) = -1.2004637573267$
$A(1, 3) = 0.14979869493340$
$A(2, 0) = 8.1528899939829$
$A(2, 1) = -8.7556426849333$
$A(2, 2) = 3.1716774408703$
$A(2, 3) = -0.39708525777831$

and for $B(m, n, p, q)$:

1. For the range of $1.5 < \epsilon_1 < 3.5$ and $1.5 < \epsilon_2 < 9.5$,

$B(1, 0, 0, 0) = 7.1032652028448\text{D-}03$
$B(1, 0, 0, 1) = -3.6177859328158\text{D-}03$
$B(1, 0, 1, 0) = -1.1083191152488\text{D-}02$
$B(1, 0, 1, 1) = 5.7712679826427\text{D-}03$
$B(1, 1, 0, 0) = 1.1804327478524$
$B(1, 1, 0, 1) = -0.16356561264695$
$B(1, 1, 1, 0) = -1.4926173273062$
$B(1, 1, 1, 1) = 6.9857583308978\text{D-}02$
$B(1, 2, 0, 0) = -2.4065835597519$
$B(1, 2, 0, 1) = 0.41187501248120$
$B(1, 2, 1, 0) = 2.9462560520606$
$B(1, 2, 1, 1) = -0.44801218677743$
$B(2, 0, 0, 0) = -2.4502944861996\text{D-}03$
$B(2, 0, 0, 1) = 1.6504491093187\text{D-}03$
$B(2, 0, 1, 0) = 8.8991013195096\text{D-}03$
$B(2, 0, 1, 1) = -4.9918437060999\text{D-}03$
$B(2, 1, 0, 0) = -0.32589664221675$
$B(2, 1, 0, 1) = 2.9307150440226\text{D-}02$

$B(2, 1, 1, 0) = -0.30016565608823$
$B(2, 1, 1, 1) = 5.2249163569479D\text{-}02$
$B(2, 2, 0, 0) = 0.53633923103178$
$B(2, 2, 0, 1) = -2.3250923005674D\text{-}02$
$B(2, 2, 1, 0) = -0.295068010664662$
$B(2, 2, 1, 1) = -0.14963396028026$
$B(3, 0, 0, 0) = 7.4631915634424D\text{-}04$
$B(3, 0, 0, 1) = -4.8087280363100D\text{-}04$
$B(3, 0, 1, 0) = -2.1734272358963D\text{-}03$
$B(3, 0, 1, 1) = 1.2132272673048D\text{-}03$
$B(3, 1, 0, 0) = 3.6674428253801D\text{-}02$
$B(3, 1, 0, 1) = 1.3635828856792D\text{-}03$
$B(3, 1, 1, 0) = -2.1031086739640D\text{-}02$
$B(3, 1, 1, 1) = -1.8397981869036D\text{-}02$
$B(3, 2, 0, 0) = -3.5429957437290D\text{-}02$
$B(3, 2, 0, 1) = -1.9017175575762D\text{-}02$
$B(3, 2, 1, 0) = -2.9299032398651D\text{-}02$
$B(3, 2, 1, 1) = 5.8749707606961D\text{-}02$
$B(4, 0, 0, 0) = -5.4355775179770D\text{-}05$
$B(4, 0, 0, 1) = 3.4296987437461D\text{-}05$
$B(4, 0, 1, 0) = 1.4585334263248D\text{-}04$
$B(4, 0, 1, 1) = -8.1403986057212D\text{-}05$
$B(4, 1, 0, 0) = -1.4798327338163D\text{-}03$
$B(4, 1, 0, 1) = -2.8667023504328D\text{-}04$
$B(4, 1, 1, 0) = 2.8325212532910D\text{-}04$
$B(4, 1, 1, 1) = 1.3485564348947D\text{-}03$
$B(4, 2, 0, 0) = 1.6101497663339D\text{-}04$
$B(4, 2, 0, 1) = 1.8499238673112D\text{-}03$
$B(4, 2, 1, 0) = 3.9572767759295D\text{-}03$
$B(4, 2, 1, 1) = -4.3299584564840D\text{-}03$

2. For the range of $3.5 < \epsilon_1 < 9.5$ and $1.5 < \epsilon_2 < 9.5$,

$B(1, 0, 0, 0) = 6.3724853325918D\text{-}03$
$B(1, 0, 0, 1) = -1.1249269119723D\text{-}03$
$B(1, 0, 1, 0) = -1.1016114599924D\text{-}02$
$B(1, 0, 1, 1) = 3.8858196234994D\text{-}03$
$B(1, 1, 0, 0) = 1.1961999519834$
$B(1, 1, 0, 1) = -0.26042443016644$
$B(1, 1, 1, 0) = -1.5229799354185$
$B(1, 1, 1, 1) = 0.147275589001113$
$B(1, 2, 0, 0) = -2.4927275063566$
$B(1, 2, 0, 1) = 0.72910444187676$
$B(1, 2, 1, 0) = 3.0339483236365$
$B(1, 2, 1, 1) = -0.67310900324478$
$B(2, 0, 0, 0) = -3.8854432606586D\text{-}03$
$B(2, 0, 0, 1) = 1.0311249482807D\text{-}03$
$B(2, 0, 1, 0) = 6.7423877735092D\text{-}03$
$B(2, 0, 1, 1) = -2.611462381525D\text{-}03$
$B(2, 1, 0, 0) = -0.28368002951650$
$B(2, 1, 0, 1) = 4.49488088626823D\text{-}02$
$B(2, 1, 1, 0) = 0.33025609456787$
$B(2, 1, 1, 1) = 4.8050219674103D\text{-}03$
$B(2, 2, 0, 0) = 0.42731106880135$
$B(2, 2, 0, 1) = -8.6011682171836D\text{-}02$
$B(2, 2, 1, 0) = -0.37134190728269$
$B(2, 2, 1, 1) = -2.6013358849602D\text{-}02$
$B(3, 0, 0, 0) = 8.7900780355080D\text{-}04$
$B(3, 0, 0, 1) = -2.8386601426407D\text{-}04$
$B(3, 0, 1, 0) = -1.4873008225781D\text{-}03$
$B(3, 0, 1, 1) = 6.0549915791522D\text{-}04$
$B(3, 1, 0, 0) = 2.9821174806113D\text{-}02$
$B(3, 1, 0, 1) = -2.5485383232291D\text{-}03$
$B(3, 1, 1, 0) = -3.0093665629033D\text{-}02$
$B(3, 1, 1, 1) = -7.8787104370029D\text{-}03$
$B(3, 2, 0, 0) = -1.9447704928935D\text{-}02$
$B(3, 2, 0, 1) = -6.3922480880827D\text{-}03$
$B(3, 2, 1, 0) = -4.6995417791164D\text{-}03$
$B(3, 2, 1, 1) = 3.1070915863402D\text{-}02$
$B(4, 0, 0, 0) = -5.8834841104296D\text{-}05$
$B(4, 0, 0, 1) = 2.0592605408135D\text{-}05$

$B(4,0,1,0) = 9.6317604997956\text{D-}05$

$B(4,0,1,1) = -4.0267484568319\text{D-}05$

$B(4,1,0,0) = -1.0914283231774\text{D-}03$

$B(4,1,0,1) = -5.8195326852407\text{D-}05$

$B(4,1,1,0) = 9.3480591356312\text{D-}04$

$B(4,1,1,1) = 6.7049606104917\text{D-}04$

$B(4,2,0,0) = -6.9738212271085\text{D-}04$

$B(4,2,0,1) = 1.0938267608707\text{D-}03$

$B(4,2,1,0) = 2.1576360584190\text{D-}03$

$B(4,2,1,1) = -2.5402018070469\text{D-}03$

C.2 IMAGINARY PARTS OF COMPLEX RESONANT FREQUENCY

The formula for the imaginary parts of complex resonant frequencies is written as

$$\mathrm{Im}\left(\frac{f}{f_{11}}\right) = \sum_{i=1}^{2}\sum_{j=0}^{4} C(i,j)\left(\frac{h}{r_d}\right)^i (\sqrt{\epsilon_1})^j + \sum_{m=1}^{4}\sum_{n=0}^{2}\sum_{p=0}^{1}\sum_{q=0}^{1}$$
$$\times D(m,n,p,q)\left(\frac{h}{r_d}\right)^m \left(\frac{t}{h}\right)^n \left(\sqrt{\frac{\epsilon_2}{\epsilon_1}}\right)^p (\sqrt{\epsilon_1})^q . \qquad \text{(C.3)}$$

There are 10 coefficients for $C(i, j)$ and 48 coefficients for $D(m, n, p, q)$. The first term again determines the imaginary resonant frequency of the microstrip patch without a superstrate, and superstrate effects on the imaginary resonant frequency are considered in the second term. The coefficients of (C.3) are given as follows: For $C(i, j)$,

$C(1,0) = 0.657712602497$

$C(1,1) = -0.40619272609907$

$C(1,2) = 8.9618092215648\text{D-}02$

$C(1,3) = 2.1838065873158\text{D-}04$

$C(1,4) = -1.4832820053691\text{D-}03$

$C(2,0) = -1.4716735001156$

$C(2,1) = 2.1506012080339$

$C(2,2) = -1.1717384408388$

$C(2,3) = 0.28694870651761$

$C(2,4) = -2.6751195741620\text{D-}02$

and for $D(n, p, q)$:

1. For the range of $1.5 < \epsilon_1 < 3.5$ and $1.5 < \epsilon_2 < 9.5$,

$D(1,0,0,0) = -0.52293899334699$

$D(1,0,0,1) = 0.17879443252257$

$D(1,0,1,0) = 0.91007567507384$

$D(1,0,1,1) = -0.31571089061107$

$D(1,1,0,0) = 0.72516210494699$

$D(1,1,0,1) = -0.23023999549361$

$D(1,1,1,0) = -1.0067567023183$

$D(1,1,1,1) = 0.35795021761490$

$D(1,2,0,0) = -0.10782707285721$

$D(1,2,0,1) = 3.7372512165575\text{D-}02$

$D(1,2,1,0) = 0.14994846464840$

$D(1,2,1,1) = -5.7751603167127\text{D-}02$

$D(2,0,0,0) = 11.426318047856$

$D(2,0,0,1) = -3.9817686361971$

$D(2,0,1,0) = -20.160473614117$

$D(2,0,1,1) = 7.2486351343574$

$D(2, 1, 0, 0) = -16.886526512535$
$D(2, 1, 0, 1) = 5.6920284066995$
$D(2, 1, 1, 0) = 24.07220569933$
$D(2, 1, 1, 1) = -9.0344046096448$
$D(2, 2, 0, 0) = 2.0241137934979$
$D(2, 2, 0, 1) = -0.74164379713251$
$D(2, 2, 1, 0) = -2.7773190139657$
$D(2, 2, 1, 1) = 1.1552678779889$
$D(3, 0, 0, 0) = -75.312691573735$
$D(3, 0, 0, 1) = 26.715543000276$
$D(3, 0, 1, 0) = 133.78239571742$
$D(3, 0, 1, 1) = -49.554657736410$
$D(3, 1, 0, 0) = 114.76475976330$
$D(3, 1, 0, 1) = -39.847003346305$
$D(3, 1, 1, 0) = -163.24270752822$
$D(3, 1, 1, 1) = 63.296202781529$
$D(3, 2, 0, 0) = -11.956730607976$
$D(3, 2, 0, 1) = 4.4267544290541$
$D(3, 2, 1, 0) = 15.742142843201$
$D(3, 2, 1, 1) = -6.6863209739983$
$D(4, 0, 0, 0) = 153.63999400224$
$D(4, 0, 0, 1) = -55.310098249439$
$D(4, 0, 1, 0) = -273.19591356775$
$D(4, 0, 1, 1) = 103.465551195511$
$D(4, 1, 0, 0) = -239.91371770260$
$D(4, 1, 0, 1) = 84.627645433449$
$D(4, 1, 1, 0) = 340.05623041169$
$D(4, 1, 1, 1) = -133.73927982527$
$D(4, 2, 0, 0) = 22.562349551243$
$D(4, 2, 0, 1) = -8.3698664201829$
$D(4, 2, 1, 0) = -28.493850927545$
$D(4, 2, 1, 1) = 12.210739620791$

2. For the range of $3.5 < \epsilon_1 < 9.5$ and $1.5 < \epsilon_2 < 9.5$,

$D(1, 0, 0, 0) = -0.54524310376090$
$D(1, 0, 0, 1) = 0.19257573519456$
$D(1, 0, 1, 0) = 0.92064986373897$
$D(1, 0, 1, 1) = -0.32176704192421$
$D(1, 1, 0, 0) = 0.89005709670965$
$D(1, 1, 0, 1) = -0.32238162171598$
$D(1, 1, 1, 0) = -1.1260386105934$
$D(1, 1, 1, 1) = 0.42475019127480$
$D(1, 2, 0, 0) = -0.12995940494659$
$D(1, 2, 0, 1) = 4.9657123829008D\text{-}02$
$D(1, 2, 1, 0) = 0.16611373921484$
$D(1, 2, 1, 1) = -6.6767096190994D\text{-}02$
$D(2, 0, 0, 0) = 11.963112126686$
$D(2, 0, 0, 1) = -4.3145695941922$
$D(2, 0, 1, 0) = -20.398737844184$
$D(2, 0, 1, 1) = 7.3845232503262$
$D(2, 1, 0, 0) = -20.148327438759$
$D(2, 1, 0, 1) = 7.5133568985201$
$D(2, 1, 1, 0) = 26.357631358465$
$D(2, 1, 1, 1) = -10.3485479205586$
$D(2, 2, 0, 0) = 2.4588508720906$
$D(2, 2, 0, 1) = -0.98257001318592$
$D(2, 2, 1, 0) = -3.0952405855456$
$D(2, 2, 1, 1) = 1.3323803366311$
$D(3, 0, 0, 0) = -79.212803279525$
$D(3, 0, 0, 1) = 29.141743107926$
$D(3, 0, 1, 0) = 135.43266486078$
$D(3, 0, 1, 1) = -50.494466395650$
$D(3, 1, 0, 0) = 134.88563219183$
$D(3, 1, 0, 1) = -51.099026086928$
$D(3, 1, 1, 0) = -177.64505191124$
$D(3, 1, 1, 1) = 71.347686400618$
$D(3, 2, 0, 0) = -14.618621038715$
$D(3, 2, 0, 1) = 5.9026188510615$
$D(3, 2, 1, 0) = 17.684804278400$
$D(3, 2, 1, 1) = -7.7685577420595$

$D(4, 0, 0, 0) = 162.04377461993$

$D(4, 0, 0, 1) = -60.526221263993$

$D(4, 0, 1, 0) = -276.62534342712$

$D(4, 0, 1, 1) = 105.40007524252$

$D(4, 1, 0, 0) = -279.00550906928$

$D(4, 1, 0, 1) = 106.51058327415$

$D(4, 1, 1, 0) = 367.87273081449$

$D(4, 1, 1, 1) = -149.28385502054$

$D(4, 2, 0, 0) = 27.702000592993$

$D(4, 2, 0, 1) = -11.219910909803$

$D(4, 2, 1, 0) = -32.236565617822$

$D(4, 2, 1, 1) = 14.295074933299$

REFERENCE

1. Y. S. Chang, Curve-Fitting Formulas for Fast Determination of Accurate Resonant Frequency of Circular Microstrip Patches with Superstrate, M.S. thesis, National Sun Yat-Sen University, Kaohsiung, Taiwan, 1993.

Index

Printed and bound by CPI Group (UK) Ltd, Croydon, CR0 4YY
16/09/2026
14963441-0003